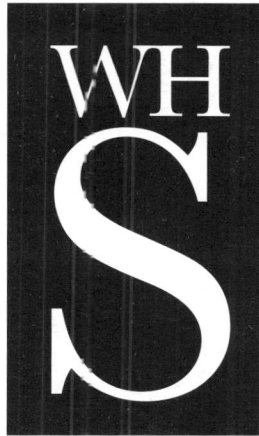

WH
S

Revision

AS and A Level

geography

Key to symbol

As you read through this revision guide, you will
notice the signpost symbol which occurs frequently
throughout the book. This is to help you identify
cross references to other parts of this book that are
relevant to the topic you are studying.

First published 2000
exclusively for WHSmith by

Hodder and Stoughton Educational
338 Euston Road
LONDON NW1 3BH

Text © David Jones and Laurence Kimpton 2000

A CIP record for this book is available from the British Library

Text: David Jones and Laurence Kimpton, with Tony Buzan
Mind Maps: The Buzan Centres
Illustrations: David Hancock

ISBN 0-340-74693-9

   10 9 8 7 6 5 4 3 2
Year   2005   2004   2003   2002   2001   2000

Typeset by Wearset, Boldon, Tyne and Wear

Printed and bound in Great Britain for Hodder & Stoughton Educational by
Redwood Books, Trowbridge, Wilts.

CONTENTS

Each section begins with a summary of the **specification content** for that subject. Use this to help you plan your revision.

You are now in the most important educational stage of your life and are soon to take exams that may have a major impact on your future career and goals. As one A Level student put it: 'It's crunch time!'

At this crucial stage of your life, the thing you need even more than subject knowledge is the knowledge of **how** to remember, **how** to read faster, **how** to comprehend, **how** to study, **how** to take notes and **how** to organise your thoughts. You need to know how to **think**; you need a basic introduction on how to use that super computer inside your head – your brain.

The next few pages contain a goldmine of information on how you can achieve success, both at school and in your A Level exams, as well as in your professional or university career. These pages will give you information on memory, thinking skills, speed reading and study that will enable you to be successful in all your academic pursuits. You will learn:

1   How to remember more *while* you are learning.

2   How to remember more *after* you have finished a class or a study period.

3   How to use special techniques to improve your memory.

4   How to use a revolutionary note-taking technique called Mind Maps that will double your memory and help you to write essays and answer exam questions.

5   How to read everything faster, while at the same time improving comprehension and concentration.

6   How to zap your revision.

How to understand, improve and master your memory

Your memory really is like a muscle. Don't exercise it and it will grow weaker; do exercise it and it will grow incredibly more powerful. There are really only four main things you need to understand about your memory in order to increase its power dramatically:

1 Recall during learning – you must take breaks!

When you are studying, your memory can concentrate, understand and remember well for between 20 and 45 minutes at a time. Then it needs a break. If you carry on for longer than this without one, your memory starts to break down. If you study for hours non-stop, you will remember only a fraction of what you have been trying to learn and you will have wasted valuable revision time.

So, ideally, *study for less than an hour*, then take a five- to ten-minute break. During this break, listen to music, go for a walk, do some exercise or just daydream. (Daydreaming is a necessary brainpower booster – geniuses do it regularly.)

During the break your brain will be sorting out what it has been learning and you will go back to your study with the new information safely stored and organised in your memory banks.

Make sure you take breaks at regular intervals as you work through your *Revise AS and A Level* book.

2 Recall after learning – surfing the waves of your memory

What do you think begins to happen to your memory straight after you have finished learning something? Does it immediately start forgetting? No! Your brain actually *increases* its power and carries on remembering. For a short time after your study session, your brain integrates the information making a more complete picture of everything it has just learnt. Only then does the rapid decline in memory begin, and as much as 80% of what you have learnt can be forgotten in a day.

However, if you catch the top of the wave of your memory, and briefly review back what you have been revising at the correct time, the memory is stamped in far more strongly and stays at the crest of the wave for much longer. To maximise your brain's power to remember, take a few minutes and use a Mind Map to review what you have learnt at the end of a day. Then review it at the end of a week, again at the end of a month and, finally, a week before the exams. That way you'll surf-ride your memory wave all the way to your exam, success, and beyond!

3 The memory principle of association

The muscle of your memory becomes stronger when it can **associate** – when it can link things together.

Think about your best friend and all the things your mind automatically links with that person. Think about your favourite hobby and all the associations your mind has when you think about (remember) that hobby.

When you are studying, use this memory principle to make associations between the elements in your subjects and to thus improve both your memory and your chances of success.

4 The memory principle of imagination

The muscle of your memory will improve significantly if you can produce big **images** in your mind. Rather than just memorising the name of an historical character, **imagine** that character as if you were a video producer filming that person's life.

In *all* your subjects, use the **imagination** memory principle.

Your new success formula: Mind Maps®

You have noticed that when people go on holidays or travels they take maps. Why? To give them a general picture of where they are going, to help them locate places of special interest and importance, to help them find things more easily and to help them remember distances, locations and so on.

It is exactly the same with your mind and with study.

If you have a 'map of the territory' of what you have to learn, then everything is easier. In learning and study, the Mind Map is that special tool.

As well as helping you with all areas of study, the Mind Map actually *mirrors the way your brain works*. Your Mind Maps can be used for taking notes from your study books, taking notes in class, preparing your homework, presenting your homework, reviewing your tests, checking your and your friends' knowledge in any subject, and for *helping you understand anything you learn.*

As you will see, Mind Maps use, throughout, Imagination and Association. As such, they automatically strengthen your memory muscle every time you use them. Throughout this *Revise AS and A Level* book you will find Mind Maps that summarise the most important areas of the subject you are studying. Study them, add some colour, personalise them, and then have a go at drawing your own – you will remember them far better! Put them on your walls and in your files for a quick and easy review of the topic.

Using Mind Maps

Mind Maps are a versatile tool – use them for taking notes in class or from books, for solving problems, for brainstorming with friends, and for reviewing and revising for exams – their uses are infinite! You will find them invaluable for planning essays for coursework and exams. Number your main branches in the order in which you want to use them and off you go – the main headings for your essay are done *and* all your ideas are logically organised.

Super speed reading and study

What happens to your comprehension as your reading speed rises? 'It goes down.' Wrong! It seems incredible, but it has been proved that the faster you read, the more you comprehend and remember.

So here are some tips to help you to practise reading faster – you'll cover the ground much more quickly, remember more *and* have more time for revision and leisure activities.

How to make study easy for your brain

When you are going somewhere, is it easier to know beforehand where you are going, or not? Obviously it is easier if you do know. It is the same for your brain and a book. When you get a new book, there are seven things you can do to help your brain get to 'know the territory' faster.

1   Scan through the whole book in less than 20 minutes, as you would do if you were in a shop thinking whether or not to buy it. This gives your brain control.

2   Think about what you already know about the subject. You'll often find out it's a lot more than you thought. A good way of doing this is to draw a quick Mind Map on everything you know after you have skimmed through it.

How to draw a Mind Map

1   Start in the middle of the page with the paper turned sideways. This gives your brain more radiant freedom for its thoughts.

2   Always start by drawing a picture or symbol. Why? Because **a picture is worth a thousand words to your brain**. Try to use at least three colours, as colour helps your memory even more.

3   Let your thoughts flow, and write or draw your ideas on coloured branching lines connected to your central image. These key symbols and words are the headings for your topic.

4   Next, add facts and ideas by drawing more, smaller, branches on to the appropriate main branches, just like a tree.

5   Always print each word clearly on its line. Use only one word per line.

6   To link ideas and thoughts on different branches, use arrows, colours, underlining and boxes.

How to read a Mind Map

1   Begin in the centre, the focus of your topic.

2   The words/images attached to the centre are like chapter headings; read them next.

3   Always read out from the centre, in every direction (even on the left-hand side, where you will have to read from right to left; instead of the usual left to right).

3 Ask who, what, why, where, when and how questions about what is in the book. Questions help your brain 'fish' the knowledge out.

4 Ask your friends what they know about the subject. This helps them review the knowledge in their own brains and helps your brain get new knowledge about what you are studying.

5 Have another quick speed through the book, this time looking for any diagrams, pictures and illustrations, and also at the beginnings and ends of chapters. Most information is contained in the beginnings and ends.

6 Build up a Mind Map as you study the book. This helps your brain organise and hold (remember) information as you study.

7 If you come across any difficult parts in your book, mark them and move on. Your brain *will* be able to solve the problems when you come back to them a little bit later, much like saving the difficult bits of a jigsaw puzzle for later. When you have finished the book, quickly review it one more time and then discuss it with friends. This will lodge it permanently in your memory banks.

Super speed reading

1 First read the whole text (whether it's a lengthy book or an exam paper) very quickly, to give your brain an overall idea of what's ahead and get it working. (It's like sending out a scout to look at the territory you have to cover – it's much easier when you know what to expect.) Then read the text again for more detailed information.

2 Have the text a reasonable distance away from your eyes. In this way your eye/brain system will be able to see more at a glance and will naturally begin to read faster.

3 Take in groups of words at a time. Rather than reading 'slowly and carefully', read faster, more enthusiastically. Your comprehension will rocket!

4 Take in phrases rather than single words while you read.

5 Use a guide. Your eyes are designed to follow movement, so a thin pencil underneath the lines you are reading, moved smoothly along, will 'pull' your eyes to faster speeds.

Helpful hints for exam revision

To avoid exam panic, cram at the start of your course, not the end. It takes the same amount of time, so you may as well use it where it is best placed!

Use Mind Maps throughout your course and build a Master Mind Map for each subject – a giant Mind Map that summarises everything you know about the subject.

Use memory techniques, such as mnemonics (verses or systems for remembering things like dates and events or lists).

Get together with one or two friends to revise, compare Mind Maps and discuss topics.

And finally . . .

- *Have fun while you learn* – studies show that those people who enjoy what they are doing understand and remember it more and generally do better.

- *Use your teachers* as resource centres. Ask them for help with specific topics and with more general advice on how you can improve your all-round performance.

- *Personalise your **Revise AS and A Level** book* by underlining and highlighting, by adding notes and pictures. Allow your brain to have a conversation with it!

Your amazing brain and its amazing cells

Your brain is like a super computer. The world's best computers have only a few thousand or hundred thousand computer chips. Your brain has 'computer chips' too; they are called brain cells. Unlike the computer, you do not have only a few thousand computer chips – the number of brain cells in your head is a *million million*! This means you are a genius just waiting to discover yourself! All you have to do is learn how to get those brain cells working together, and you'll not only become more smart, you'll have more free time to pursue your other fun activities.

The more you understand your amazing brain, the more it will repay and amaze you!

	OCR A	OCR B	AQA A	AQA B	EDEXCEL A	EDEXCEL B	WJEC
Ch.1 LITHOSPHERE	Lithosphere	●	Geomorphological processes and hazards		Earth systems	●	●
Ch.2 HYDROLOGICAL PROCESSES AND SYSTEMS	Hydrological systems	Landform systems and people	Water on land	Physical geography: shorter term and local change	Fluvial environments	River environments	Drainage basins: hydrology and landforms
Ch.3 RIVER LANDFORMS	Fluvial environments						
Ch.4 ATMOSPHERIC SYSTEMS	Atmospheric systems	Atmospheric systems and people	●	Phys. geog. (atmosphere)	Atmospheric systems	The natural environment	●
Ch.5 ECOSYSTEMS	Ecosystems		Energy and life	Phys. geog. (soils, vegetation)	Ecosystems		Small-scale ecosystems
Ch.6 COASTAL ENVIRONMENTS	Coastal environments	Coastal systems and people	Coasts–processes and problems	Coastal environments	Coastal environments	Coastal environments	Landforms: process and management (coastal)
Ch.7 GLACIAL ENVIRONMENTS	Glacial and periglacial environments	Cold environments and human responses	Cold environments and human activity	Glacial environments	Glacial systems		Landforms: process and management (glacial)
Ch.8 HAZARDS	Lithosphere / Hazardous environments	Natural hazards and human responses	Geomorphological processes and hazards / Climatic hazards and change	Phys. geog. (plate tectonics) / People & Env. (hazards)	Earth systems / Atmospheric systems	Living with hazardous environments	Global tectonic processes and hazards / Climatic hazards
Ch.9 POPULATION PATTERNS	Populations: pattern, process & change	Population and development	Population dynamics	Human geog. (migration) / People & Env. (population & resources)	Population characteristics / Population movements	Population and the economy	Dynamic populations
Ch.10 POPULATION AND RESOURCES	Resources and population	Sustainable development	Population pressure & resource management	People & Env. (resource conflict)	●	Environments and resources	Sustainable development
Ch.11 SETTLEMENTS: PATTERNS AND PROCESSES	Rural and urban settlement	Settlement dynamics	Settlement processes and patterns	Human geog. (urbanisation)	Settlement patterns / Rural–urban inter-relationships	Rural environments / Urban environments	Aspects of rural change / Changing urban environments
Ch.12 SETTLEMENTS: URBAN AND RURAL ISSUES	Managing urban environments	Changing urban places	Managing cities, challenges and issues	Urban change			
Ch.13 INDUSTRY	Manufacturing industry	Economic activity and change / Globalisation of economic activity	Economic activity	Human geog. (TNCs & NICs) / Urban change	Economic systems	Population and the economy	Changing geographies of economic activities
Ch.14 SERVICES	Service activities						
Ch.15 TOURISM AND RECREATION	Tourism, recreation	Leisure and tourism	Recreation and tourism				
Ch.16 DEVELOPMENT		Population & development / Sustainable development		Human geog. (development)	Development processes	Development and disparity	Inequalities in development / Sustainable development

(AQA B column Ch.11–Ch.16 also carries: Human geog. – changes in the UK)

Legend:

In each column the examination specification modules or units which relate to this book's chapters are listed.

(shaded)	AS module / unit
(white)	A2 module / unit
(shaded)	Option module / unit
●	Chapter includes relevant background material

Using the book

Skim through the book to get a good idea of the overall layout. Check the chapter headings and sub-sections against your course (see below). Don't expect the book to follow your course exactly; some topics will come under different headings and there'll be some that aren't in your course. All through the book we mention case studies; they are very important. As you do them during the course, build up a list and for each one make a note of the topic, issue, problem or place. Just doing that will help to reinforce it in your memory and, to be really sure, carry out regular reviews of your case studies (see Tony Buzan's section and below). Remember that you can usually use one case study to answer many different questions.

Your specification

Check that you know your examination board and the specification you are following. Each board will have at least two options, so make sure you know which option you are doing.

Check the details of your specification option, making sure which parts of the overall specification are Advanced Subsidiary (AS) and which parts are Advanced (A2). How will you be assessed? How many examination papers will there be? What format will the examination papers follow? Is your coursework an investigative project or a major essay?

An overview of your specification

It's easier to fit new work into your existing knowledge if you have an idea of the overall course. Make a large chart (see Figure 0.1 on the following page) showing the content of your course and do it in the order of the course rather than the order in the specification documents – they are not likely to be the same. Use this as the basis for your review timetable.

Make a **Mind Map** of your course; in a visual form it will be easier to make connections between different parts of the course and easier to keep track of where you are. Use a large (A3) piece of paper and you will have room to add extra lines and notes as the course progresses.

During the course

The message in Tony Buzan's introductory section of the book is the key to success – follow his advice and there will be no panic in the run-up to the examination becase you will be confident about your knowledge.

Start preparations for the examination early in the course.

Use this book as a means of linking to key ideas and as a way of reviewing your work. Remember that this book must be used with your course notes. In particular, your **case studies** are necessary for answering examination questions.

Review tasks in each section of the book specifically ask you to use your own course materials as well as the material from the book itself.

Use **Mind Maps** for summarising and reviewing work. They give you an 'at-a-glance' view of a topic.

Make the most of your time by following a **review plan**.

Review plan

How much time is there to the next examination? The end of the year comes very quickly, so make the most of your time by having a review plan.

1 Review Tony Buzan's section on how your memory works. This explains why you need a review plan if you are going to do your best.
2 Use your chart of the course to record your reviews.
3 As you complete a section of the course, review it thoroughly, using this book and your course notes, working through review tasks and producing a summary. A Mind Map is the ideal summary.
4 Review it again after ten minutes – quickly! Five minutes will be enough.
5 Review it again the next day, then the next week and then the next month – again spending only five to ten minutes. If there's a long time to the final examination, a short review every two months would be a good idea, too.
6 By now, you really know the work but another review a week before the examination will give you even more confidence.

Specification topic	First review	Second review	Third review	Fourth review	Fifth review (one week before examination)
Population patterns	(20 Oct) ✓	(21 Oct) ✓	(28 Nov)	(date)	(date)
Population, resources and environment	(7 Nov) ✓				
Urban environments					

Figure 0.1 Review record

Review record

This is a starting point; obviously your chart will be much bigger and you will need to sub-divide the specification topics.

Brainstorming is a great way to test yourself. Take any topic you have studied and work through the following points.

1 Jot down anything you think of to do with the topic.
2 Sort the items into groups.
3 Add more points.
4 Discard the useless ones – there are bound to be some but they serve their purpose by getting your brain remembering!
5 By now you have a pretty good summary of the topic.

6 Remember to include case studies as well as general principles.
7 Present the results of your brainstorming session as a Mind Map and you have the perfect revision tool!

Brainstorming works really well with a group because people spark off ideas in the others. This makes the whole exercise quicker, more interesting and more complete.

Examinations

The style of question varies for most courses between the AS and A2 papers. Make sure you are familiar with the papers for **your** course. Your teachers will have sample question papers.

PREVIEW

What you need to know:

- **Processes of physical, chemical and biological weathering**
- **Factors controlling weathering**
- **Processes of mass movement**
- **Slope systems**
- **Slope form and development**
- **Tectonic processes and how plate movements influence volcanic activity, earthquakes and landforms**

Also, you will need to be able to:

- **apply your knowledge of weathering processes to landforms in limestone and granite areas;**
- **transfer your understanding of weathering processes, mass movement and slope development to other geomorphological topics such as glaciation, river landforms and coastal landforms;**
- **analyse geomorphological systems diagrams;**
- **apply your knowledge and understanding to examples.**

The study of the lithosphere at AS level aims to be an introduction to the processes that form landscapes, in particular weathering, mass movement and slopes together with the study of the structure of the earth.

Weathering

Weathering is the disintegration and decomposition of rocks *in situ* (that is, in their original position). The key difference from erosion is that weathering does not usually involve the movement of rock, other than fragments falling because of gravity. **Physical (or mechanical) weathering** involves rock disintegrating into fragments and **chemical weathering** involves the decomposition of rock through chemical change. In addition, there is **biological weathering** which has both physical and chemical aspects. Weathering is important because the breakdown of rocks facilitates erosion and transport by rivers, ice, the sea and the wind.

Physical (or mechanical) weathering

- **Frost shattering** or **freeze-thaw** action occurs when water in joints and cracks freezes, expands and so

exerts pressure. The process is repeated as temperature fluctuates and so cracks are wedged further open and rock fragments are broken off. The process is particularly important in sub-arctic and alpine environments. In upland glacial landscapes the broken rock fragments form **scree** or **talus** slopes.

- **Salt crystallisation** involves slightly saline water entering pore spaces and cracks. As the water evaporates, salt crystals form, expand and exert stresses which cause the rock to disintegrate. The process is important in deserts (where evaporation is considerable) and on coasts.

- **Pressure release** affects rocks such as granite which have been formed under conditions of extreme pressure. When overlying rocks are removed by erosion, the release of their weight causes the rock to expand and fracture parallel to the surface. Other processes will then assist in the peeling away or **exfoliation** of rock.

- **Insolation weathering** involves rocks being heated and cooled. In theory, this may have two effects:

  1 as outer layers of rock expand and contract, stresses are set up parallel to the surface and exfoliation occurs;

  2 stresses are set up and disintegration occurs because different minerals expand and contract at different rates.

  However, experiments have shown that heating and cooling alone is ineffective in breaking up rocks. Rock disintegration in deserts once attributed to insolation weathering is now considered to result from chemical weathering processes.

Two effects rather than processes of mechanical weathering are **block disintegration** (rock breaks up into large fragments, as caused by frost shattering) and **granular disintegration** (rock disintegration on the scale of individual grains).

Chemical weathering

- **Hydration** is the absorption of water by rock minerals causing them to swell. The swelling of clay minerals in shales can cause up to 60 per cent expansion. Frequent wetting and drying will therefore cause considerable stresses in some rocks. Although water is absorbed into the crystal lattice of minerals, there is no fundamental chemical change to them when they absorb water. Therefore, this process is sometimes put into the category of mechanical weathering.

- **Hydrolysis** involves water chemically combining with minerals. H^+ and OH^- ions combine with mineral ions. The rate of hydrolysis depends on the amount of H^+ present. In the case of granite, the mineral orthoclase feldspar forms kaolin (china clay), soluble silica and soluble potassium carbonate. The soluble substances are removed in solution and the kaolin is left behind. The other minerals in granite (quartz and mica) remain untouched but the granite will have been significantly weakened.

- **Oxidation** occurs when oxygen in air and water chemically combines with metallic ions. Thus **ferrous** iron changes into a **ferric** state – in effect iron rusting takes place. The result is a weakening of rock structure.

- **Carbonation** is the action of rainwater, a weak carbonic acid, on rocks which contain calcium carbonate ($CaCO_3$), in particular limestone. Calcium bicarbonate ($Ca(HCO_3)_2$) is produced and carried off in solution. Carbonation is also known, less precisely, as limestone solution.

- **Solution** is the dissolving in water of soluble minerals such as rock salt.

The various processes of chemical weathering act together to weaken rock. The processes are accelerated in wetter and warmer environments (see below). The **sheeting** of outer layers of masses of rock such as granite to produce **exfoliation domes** represents a combination of pressure release, heating and cooling and chemical weathering. On a smaller scale, **spheroidal weathering** of blocks of rock is usually a result of a combination of physical and chemical processes.

Biological weathering

The action of tree roots penetrating joints and bedding planes and so widening them is a physical or mechanical means of biological weathering. **Chelation** is a biochemical process important in soil formation; decaying organic matter in the soil releases organic (humic) acids which remove iron and aluminium from clay and rock particles.

Page 46

Review task

Construct a Mind Map to show and classify the different types of weathering. Include key words which will remind you of the key points of each type of weathering.

Factors controlling weathering

Chemical composition of rocks

As rocks vary in their chemical composition, so they vary in their susceptibility to different types of chemical weathering. Limestone, being composed of calcium carbonate, is particularly susceptible to carbonation. As a result, distinctive **karst** landforms are produced (Figure 1.1). With a sedimentary rock such as sandstone, the chemical composition of the cement that binds grains together is important. For example, an iron oxide based cement is quickly weathered by oxidation and a calcareous cement by carbonation; thus the rock structure is quickly broken down. Igneous rocks are composed of a variety of minerals, some of which are more easily weathered than others. Thus granite, mica and feldspar weather more easily than quartz. The weathering of just one mineral type in a rock seriously weakens its structure.

Rock structure

The existence of pores in a rock such as sandstone allows water to penetrate the rock and allows weathering to take place. Joints and bedding planes allow water to enter the rock and accelerate weathering. The well-jointed nature of granite and the pronounced joints and bedding planes of many limestones contribute to these rocks having distinctive landforms resulting from weathering (Figures 1.1 and 1.2).

EXAM TIP

It is important that you are able to apply your knowledge and understanding of limestone and granite weathering and resultant landforms to case studies.

Climate

As temperature increases, the rates of chemical reactions and therefore chemical weathering also increase. If high temperatures are combined with the presence of much water, weathering is particularly intense. Besides influencing the rate of weathering, different climatic conditions influence the type of weathering processes that operate. Figure 1.3 shows how rates and types of weathering vary in different climates.

Human activities

Where human activities release gases into the atmosphere (for example, through power station and

Features of limestone scenery (karst) result from weathering (especially carbonation), solution by running water (the same chemical process as carbonation) and other erosion processes.

A zone of percolation, solution and collapse

B zone of stream water solution and corrasion

C zone of solution by phreatic water (water under pressure) below the saturation level

surface drainage
swallow hole
IMPERMEABLE ROCK
solution dolines
pothole
closer joints: solution more effective
surface drainage
collapse doline
limestone pavement
A
B
C
IMPERMEABLE ROCK
cave
grike
clint
limestone scar
resurgence

TROPICAL KARST: CONE KARST (e.g. Jamaica)

large hollows (cockpits)   cone-shaped hills   thin vegetation on hills

TROPICAL KARST: TOWER KARST (e.g. Southern China)

tower   active caves   river

In tropical areas, high precipitation accelerates both carbonation and erosion by surface water.
High temperatures also accelerate weathering.

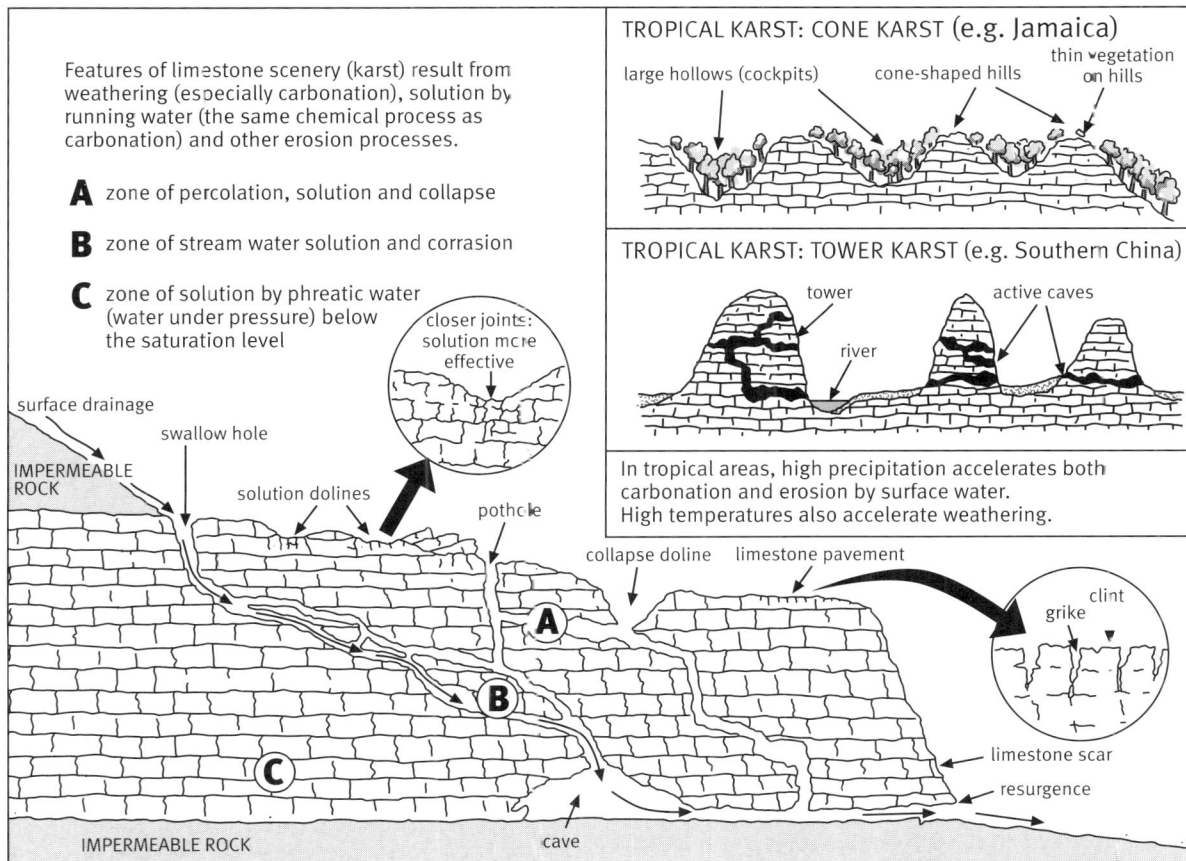

Figure 1.1 Karst landforms

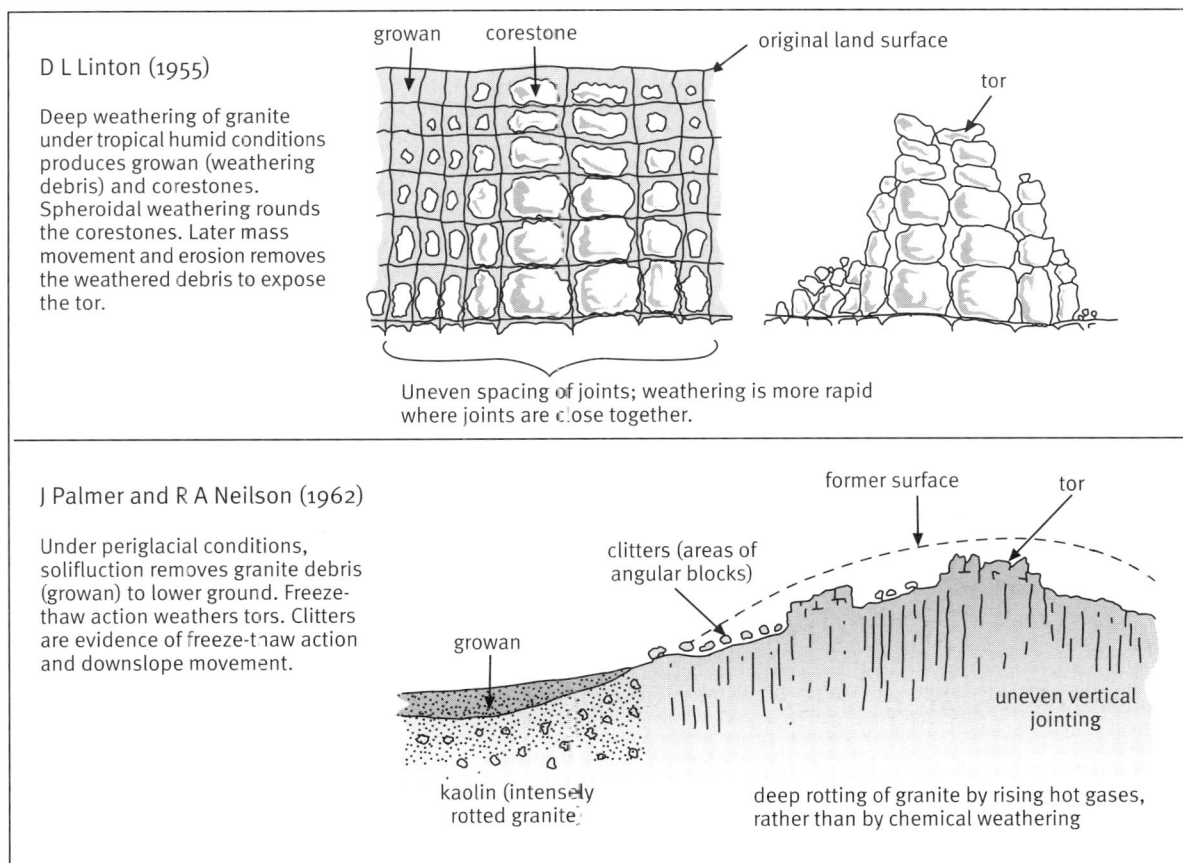

D L Linton (1955)

Deep weathering of granite under tropical humid conditions produces growan (weathering debris) and corestones. Spheroidal weathering rounds the corestones. Later mass movement and erosion removes the weathered debris to expose the tor.

growan   corestone   original land surface   tor

Uneven spacing of joints; weathering is more rapid where joints are close together.

J Palmer and R A Neilson (1962)

Under periglacial conditions, solifluction removes granite debris (growan) to lower ground. Freeze-thaw action weathers tors. Clitters are evidence of freeze-thaw action and downslope movement.

former surface   tor
clitters (areas of angular blocks)
growan
uneven vertical jointing
kaolin (intensely rotted granite)
deep rotting of granite by rising hot gases, rather than by chemical weathering

Figure 1.2 Weathering of granite: two theories of tor formation

ENVIRONMENT	PERIGLACIAL	TEMPERATE HUMID	SEMI-ARID AND ARID	TROPICAL WET/DRY	EQUATORIAL
ANNUAL PRECIPITATION	100–180 mm	180–1000 mm	50–250 mm	200–1800 mm	1800–2600 mm
MEAN ANNUAL TEMPERATURE	2°C	3–10°C	8–20°C	15–20°C	18–25°C
DEPTH OF WEATHERING	up to 2 metres	up to 7 metres	up to 2 metres	up to 16 metres	up to 28 metres
WEATHERING PROCESSES	Frost action dominant	Moderate chemical weathering and frost action	Slight physical and chemical weathering	Moderate chemical weathering	Intense chemical weathering

Figure 1.3 Variations in weathering in different environments

vehicle emissions releasing sulphur dioxide, carbon dioxide and nitrogen oxides) the levels of acidity in rainwater are increased; the processes of chemical weathering are thus accelerated. (Consider why the weathering of limestone will be particularly accelerated.)

Other factors

Other factors influencing rates of weathering include relief, depth of regolith and vegetation cover. In a humid tropical environment the **regolith** (layer of weathered rock) is deep (see Figure 1.3). As it becomes deeper, water movement down to the bedrock is restricted and the rate of weathering is reduced. On mountain slopes in the same environment, mass movement processes and erosion remove weathered material as it forms, bedrock is continually exposed and rates of weathering are therefore high. Vegetation cover contributes to soil development and the protection of rock against weathering. On the other hand, forest vegetation in a tropical environment produces large amounts of humic acids which accelerate weathering.

Review task

1 **Add to your Mind Map the rock types, rock structures and climatic conditions where each type of weathering will be particularly effective.**
2 **Check carefully your case studies of limestone and granite scenery. Make sure that you are able to describe the relevant processes of weathering operating on these rocks, besides the actual landforms.**

Processes of mass movement

Mass movement is the downslope movement of rock and the regolith in response to gravity. It does *not* include movements when the material is carried by water, ice or the wind. Mass movement can be broadly classified into **slides**, **flows** and very slow movements, sometimes described as **heave**. Figure 1.4 shows how the various types of mass movement may be fitted into this classification. Notice that the movements, besides varying in speed, vary in their water content; thus there are dry movements, such as rockfalls, and fluid movements, such as mudflows. Figure 1.5 shows the main types of mass movement.

Very rapid movements and slides

The most rapid movements include **rockfalls** and **debris avalanches**. As Figure 1.5 shows, a rockfall may cause a debris avalanche. **Slides** of rock or debris involve the sliding material maintaining its shape. The slide may be planar, leaving behind a flat slip plane, or involve a rotational movement along a curved slip plane. Such rotational slides are also called **slumps**.

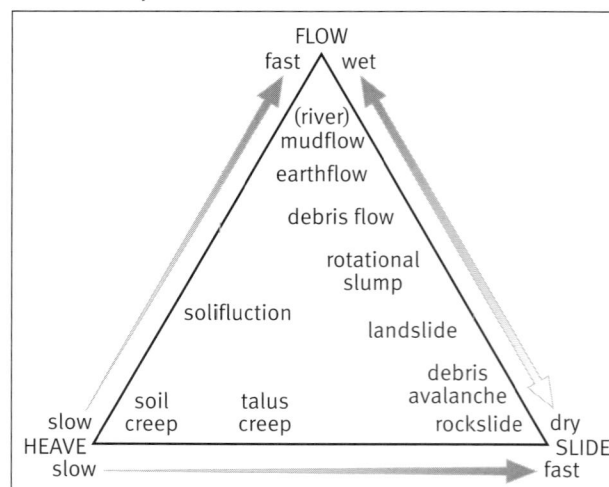

Figure 1.4 Classification of mass movement processes

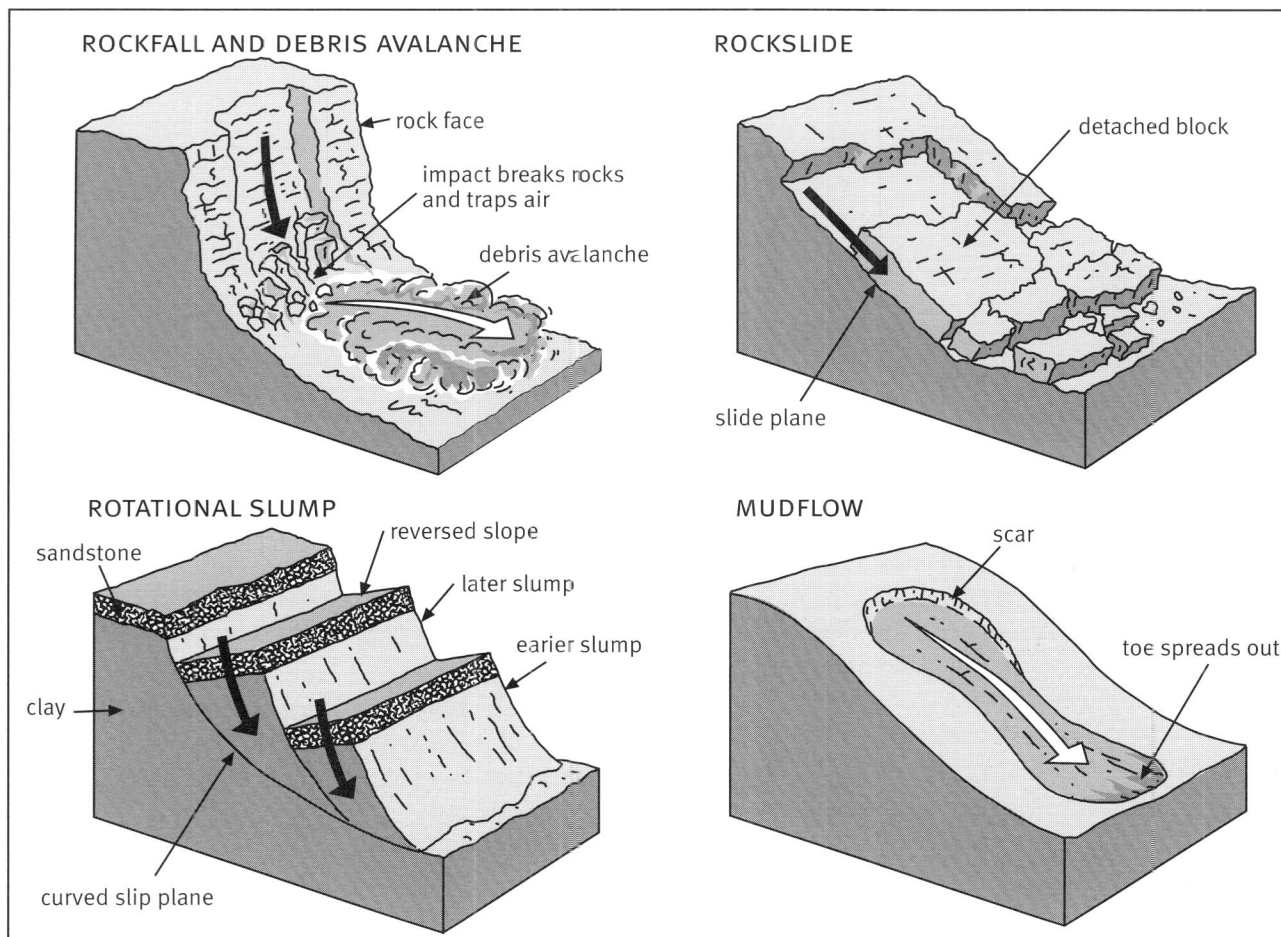

Figure 1.5 Types of mass movement

Flow movements

In contrast to slides, flows suffer from internal deformation. They have a higher water content than slides and vary in speed from relatively slow **earthflows** to rapid **mudflows** and **debris flows**.

Slow movements

The slowest movement is **soil creep**, at a rate of under 1 cm a year. Although slow, creep is very important as the movement is almost continuous and acts on all weathered slopes. Individual particles are pushed to the surface by wetting, heating or freezing of soil moisture. They move at right angles to the surface (Figure 1.6). The particles fall vertically, under the influence of gravity, as they dry out or cool or as soil moisture thaws. The result of millions of such movements over time is a net movement downslope. As the processes of creep involve a lifting of the surface, the term **heave** is sometimes used to describe soil creep.

Pages 74–5

Solifluction (soil flow) is a faster movement (5 cm to 1 m a year), particularly associated with periglacial conditions (see Chapter 7). In summer the

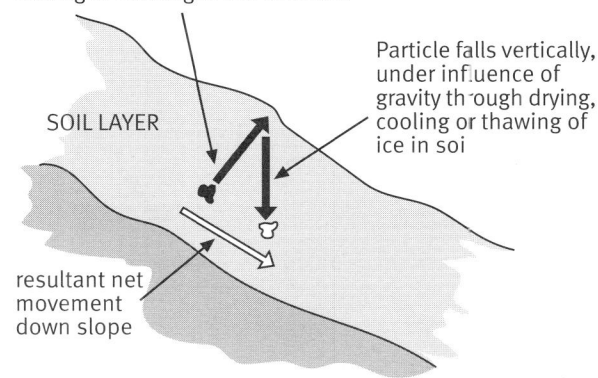

Figure 1.6 Soil creep: the movement of a soil particle

subsoil remains frozen, as the permafrost layer, but the surface layer thaws. Water cannot penetrate the underlying frozen layer, so the surface layer becomes saturated and flows as an **active layer**.

Although the various types of mass movement may be classified neatly, as in Figures 1.4 and 1.5, in reality movements are often more complex. Many flows exhibit sliding at their base. The water which saturates

rock to trigger a rotational slump may help to produce an earthflow at the toe of the slumped mass.

Which type of mass movement is dominant in a particular location depends upon:

- rock type and structure
- slope angle
- depth of regolith
- vegetation cover
- amount of water present
- processes at work at a slope's base (as with a sea cliff)
- the occurrence of tectonic activity (such as an earthquake triggering a rapid movement)

Mass movement on slopes is augmented by water eroding and transporting weathered debris. The impact of raindrops (**rain-splash erosion**), **sheetflow** (water moving downslope as a thin film) and water flowing in **rills** and **gullies** are all of importance. However, remember that these are *not* processes of mass movement.

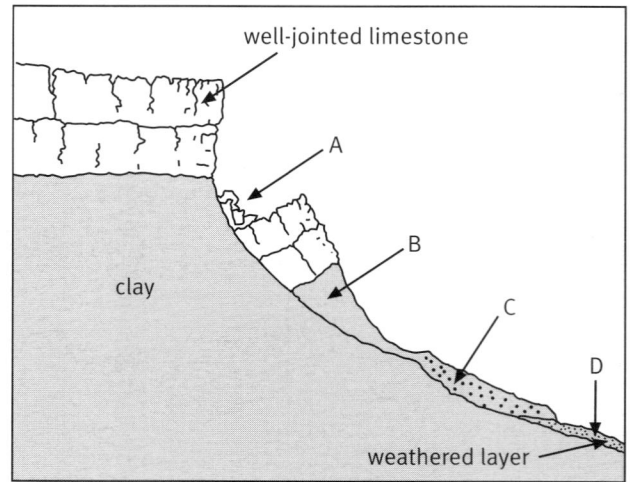

Figure 1.7 A slope with several types of mass movement

Review task

Figure 1.7 shows an example of a slope affected by various types of mass movement. Identify the types of movement shown (A to D) and, using the evidence on the diagram, explain why they have occurred.

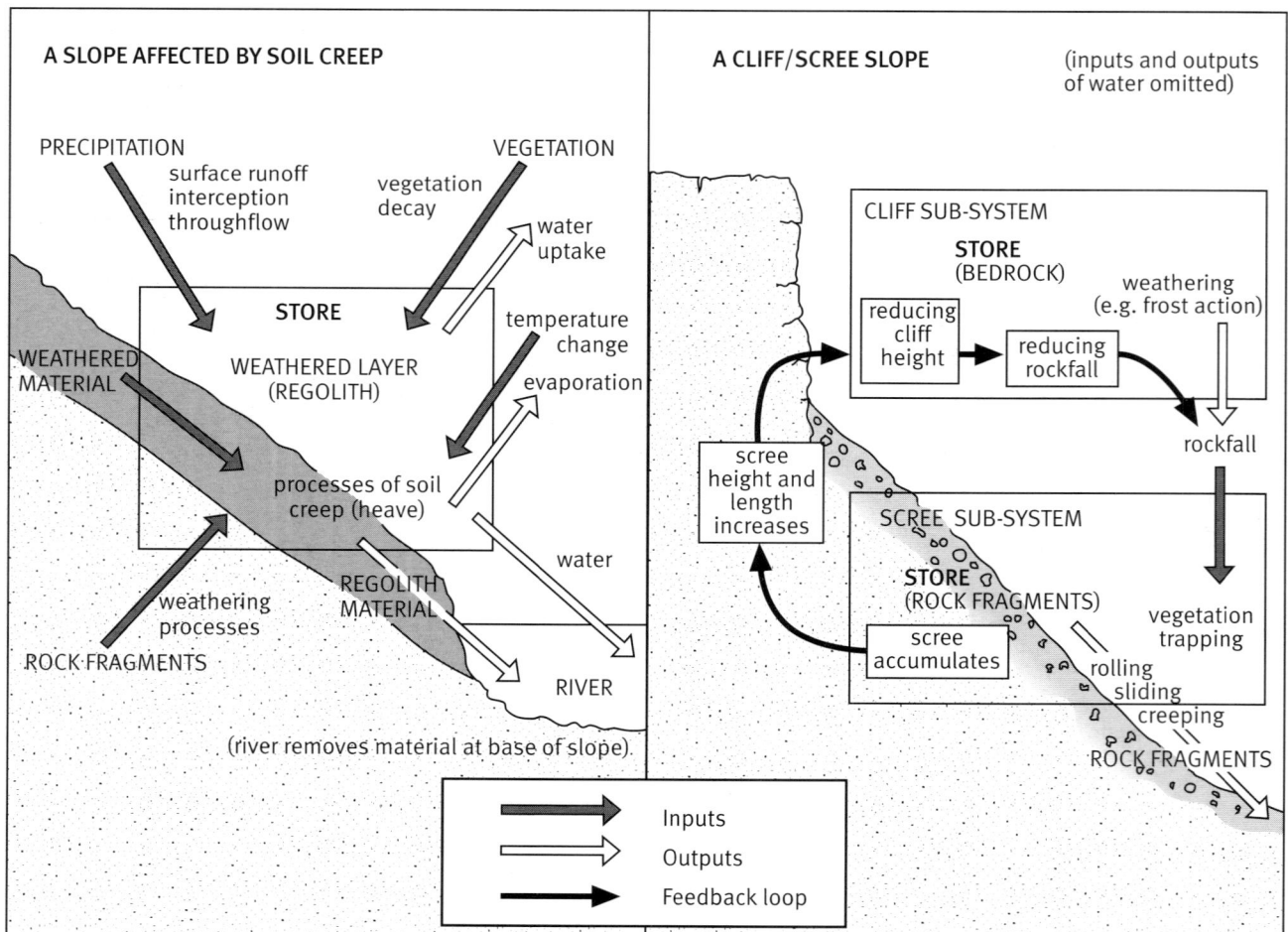

Figure 1.8 Slope systems

Slopes

Slope systems

All landforms can be viewed as open systems with inputs of energy and materials, various stores, throughputs and processes and, finally, outputs. Figure 1.8 applies this idea to a slope where the dominant process is soil creep and to a slope made up of a cliff with a scree slope below. The second example brings in the idea of a **feedback loop**; as the scree builds up, the cliff is reduced in height and so rockfall is reduced.

Slope form and development

Slopes are the result of processes operating within the earth (**endogenetic processes**: volcanism, faulting, folding) and processes operating at or near the surface (**exogenetic processes**: weathering, mass movement, processes of erosion and deposition by water on land, the sea, ice and wind). Slopes are usually made up of a number of elements as in the three models in Figure 1.9. The differences between the slopes in Figure 1.9 reflect the interplay of a variety of factors which control slope form and development. The main factors are: geology (rock type and structure), climate, nature of regolith, vegetation cover and human activity. In addition the time over which slopes have developed under these various influences is an important consideration. Three main theories have been proposed as to how slopes develop over time: **slope decline, slope replacement** and **slope retreat**. In Figure 1.9 they are linked to the three slope element models and to different environments.

CONVEX-CONCAVE SLOPE	RECTILINEAR SLOPES	
thick regolith present; convex (shedding slope); transfer slope; concave (receiving slope)	free face; debris slope (scree)	resistant cap rock; free face; pediment
Slope development: SLOPE DECLINE (W M DAVIS) Slopes initially steep because of vertical erosion by rivers. Erosion and mass movement reduces slope angles over time. Eventually an almost flat *peneplain* is formed.	**Slope development:** SLOPE REPLACEMENT (W PENCK) Free face retreats through weathering. Lower slopes extend upwards at a constant angle and so gradually replace upper slopes.	**Slope development:** PARALLEL RETREAT (L C KING) Free face and debris slope retreat through weathering, at a constant angle. Their retreat leaves behind a gently concave pediment slope (which may be eroded by sheet floods).
Environments: Humid temperate climates	**Environments:** Mountain areas of rapid tectonic uplift	**Environments:** Semi-arid and arid areas

Figure 1.9 Slope forms and slope development

Review task

1   Refer to Figure 1.8. First, check carefully your understanding of both systems diagrams. Then draw a similar systems diagram for the slope shown in Figure 1.7. Include just one store in your diagram: the slumped mass B.

2   Consider the three slopes in Figure 1.9. For each one, note the probable influences of geology, climate, nature of regolith and vegetation cover on their form. (You will need to bear in mind likely weathering processes and types of mass movement operating on them.)

3   Consider a steep slope (at an angle of about 20°) with a thick regolith in a humid temperate environment (such as in the UK).
    a) What may be the possible effects on such a slope of:
       i) building a road that follows the contour of the slope and that will require excavation into the slope?
       ii) constructing buildings on the slope?
    b) How might engineers attempt to avoid possible problems when constructing roads or buildings on such a slope?

Tectonic processes

If your examination specification has a distinct AS unit on the lithosphere or earth systems, you will require an understanding of the earth's structure, in particular plate tectonics. You will need to know how the movement of the plates of the earth's crust influences volcanic activity, earthquakes and landforms. In other specifications, plate tectonics forms part of the content of AS and A2 units on hazardous environments. In Chapter 8 you will find that tectonic processes are considered with particular reference to earthquake and volcanic hazards.

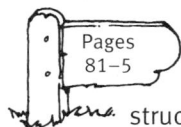

Pages 81–5

Review task

You will need to refer to your course notes or textbooks, together with Figures 8.7, 8.9 and 8.10 in Chapter 8, as you work through this task. For each of the descriptions below draw a diagram with explanatory labels. For the descriptions of different plate boundaries (descriptions 3 to 7) give named examples of locations and major landforms such as fold mountains.

1   The earth's crust and the upper part of the mantle is rigid and is known as the *lithosphere*. The lithosphere is made up of plates which float on the underlying semi-molten mantle known as the *asthenosphere*.

2   Plates are made of two types of crust: *continental crust* (sial) which consists of less dense, granitic rock and *oceanic crust* (sima) which consists of denser basaltic rock. *Convection currents* generated by heat from within the earth cause plates to move.

3   *Destructive margin*. Where a plate of oceanic crust moves towards a plate of continental crust, the denser oceanic crust sinks below the lighter continental crust and is destroyed in the *subduction zone*. In some locations fold mountains and volcanoes are formed on the edge of the continent; in other locations, volcanic island arcs and deep ocean trenches are formed.

4   *Constructive margin*. Where two plates move away from each other, new oceanic crust is formed. Sea floor spreading occurs. Ocean ridges with volcanic islands develop.

5   In some locations continental plates are splitting apart to form *rift valleys* (sometimes occupied by lakes and seas) with associated volcanoes.

6   *Collision zone*. Where two plates of continental crust move together, neither plate can sink and crust is compressed and forced up into fold mountains.

7   *Conservative margin*. Where two plates move sideways past each other, crust is neither formed nor destroyed, but there will be considerable earthquake activity.

What you need to know:

- **The drainage basin system (inputs, stores, flows and processes, outputs)**
- **The concept of water balance**
- **River discharge and river regimes**
- **Storm hydrographs**
- **Physical and human influences on hydrological processes**
- **Causes and effects of flooding and low flow conditions**
- **How drainage basins are managed**

Also, you will need to be able to:

- **analyse water budget graphs;**
- **interpret river regime graphs;**
- **analyse storm hydrographs;**
- **use quantitative techniques to describe drainage basins;**
- **apply your understanding and knowledge to case studies.**

The drainage basin system

An understanding of how water works in the global environment can be developed through the framework of the hydrological cycle. The **hydrological cycle** is a **closed system** through which water circulates (in the forms of liquid, solid or gas) through the major stores of the atmosphere, lithosphere, hydrosphere and biosphere (Figure 2.1).

A **drainage basin** is part of the hydrological cycle and is a manageable unit within which to understand many of the processes of the hydrological cycle. A drainage basin is the catchment area for water which drains into a river and its tributaries. The boundary

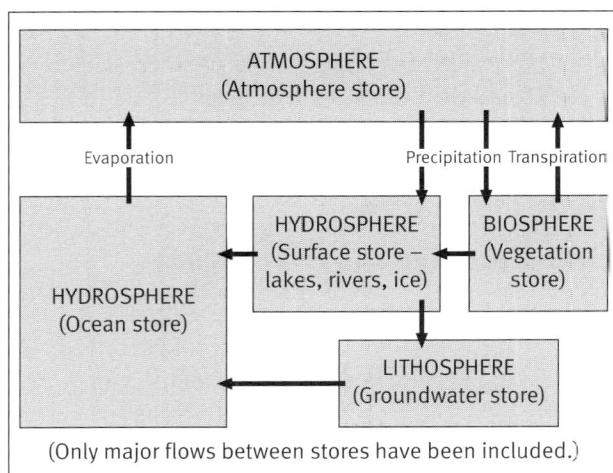

Figure 2.1 The hydrological cycle

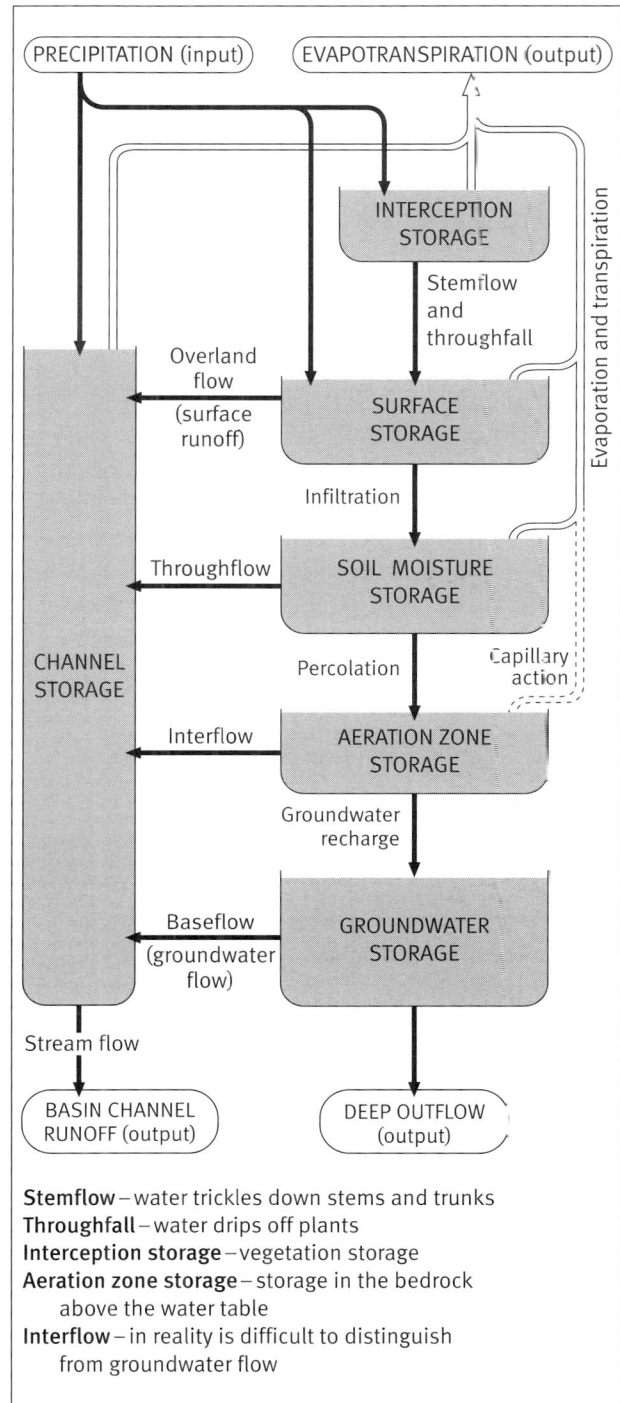

Stemflow – water trickles down stems and trunks
Throughfall – water drips off plants
Interception storage – vegetation storage
Aeration zone storage – storage in the bedrock above the water table
Interflow – in reality is difficult to distinguish from groundwater flow

Figure 2.2 The drainage basin system

between one drainage basin and another is called a **watershed**. Figure 2.2 shows the hydrological inputs, outputs, stores and flows of the **drainage basin system**. This system is an **open system** as the input of precipitation comes from outside the drainage basin and the outputs of river runoff and **evapotranspiration** (evaporation from surface water plus transpiration from plants) leave the system.

Figure 2.2 shows the drainage basin system as a 'cascading system'. Textbooks and examination papers may depict the system in different ways; however, all the components and flows will be the same. Remember that hydrological processes do not operate in isolation. They are linked with geomorphological processes such as weathering, mass movement and erosion. Thus sediment and flows of energy will also move through the drainage basin system.

In the drainage basin system, as Figure 2.2 shows, there are a number of **stores**. Each one is capable of absorbing and holding water as it flows through the system. As a store fills with water, it may release some at the same time. Eventually it releases as much water as it receives (for example, in the case of soil when it is saturated). A system of **flows** or **transfers** links the stores together; many of these flows can also be described as **hydrological processes**. Figure 2.3 shows these flows or processes in diagrammatic form. Notice in Figures 2.2 and 2.3 that although channels act as stores, they are also a major means in which water moves through the system.

The water balance

The relationship between precipitation, evapotranspiration and storage (in the form of soil moisture and groundwater) can be expressed as the **water balance** equation:

precipitation (P) = streamflow (Q) + evapotranspiration (E) + change in storage (S)

A **water budget graph** is a useful way to look at the water balance of a location over a year. Precipitation and evapotranspiration rates are plotted on to a single graph (Figure 2.4). To understand a water budget graph you need to know the following terms:

- **Actual evapotranspiration** (AET). The loss of moisture to the atmosphere by the processes of evaporation and transpiration which actually takes place.

- **Potential evapotranspiration** (PET). The maximum amount of evapotranspiration which would occur if an adequate supply of water were continuously available. Thus in deserts PET is very much higher than AET because the amount of water that is available is very limited.

- **Soil moisture surplus.** Occurs when the soil water store is full and thus there is surplus water for plants, runoff and groundwater recharge. (Precipitation exceeds PET.)

- **Soil moisture utilisation.** Plants (and people) use moisture stored in the soil, leaving it depleted. (AET exceeds precipitation.)

- **Soil moisture deficiency.** Equivalent to the extra water which would be needed to maintain maximum plant growth. There is little or no water available for plant growth (irrigation could make good this deficit). (PET exceeds AET.)

- **Soil moisture recharge.** The soil water store starts to fill again after a period of deficiency. (Precipitation exceeds PET.)

- **Field capacity.** The moisture a freely drained soil can hold after all free or gravity water has drained away. Such moisture is held by tension around soil particles, mainly as **capillary water**.

River discharge

River discharge and river regimes

Discharge or **streamflow** is the major output of a drainage basin system. Discharge is calculated by multiplying the velocity of a river (in metres per second) by its cross-sectional area (measured in square metres). It is expressed in cumecs (cubic metres per second). The annual variation in the discharge of a river is called its **regime**. The river

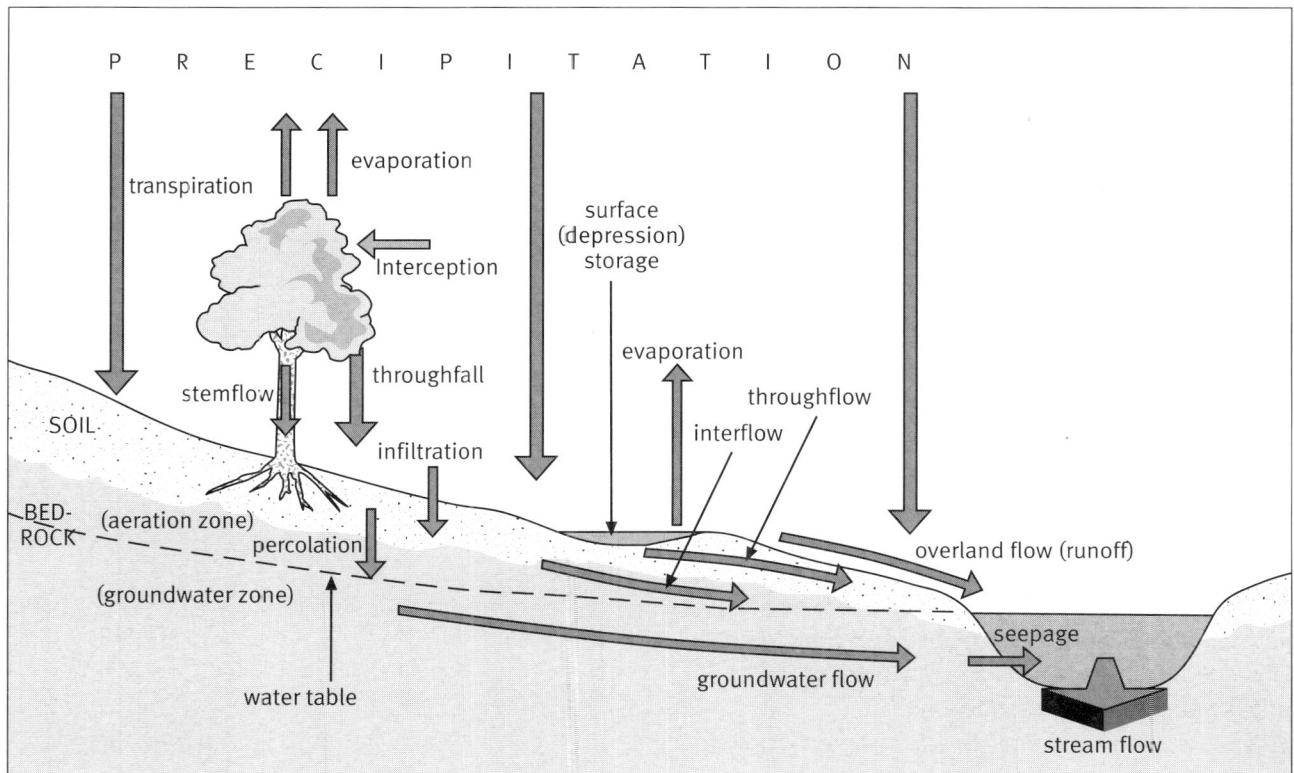

Figure 2.3 Hydrological processes in a drainage basin

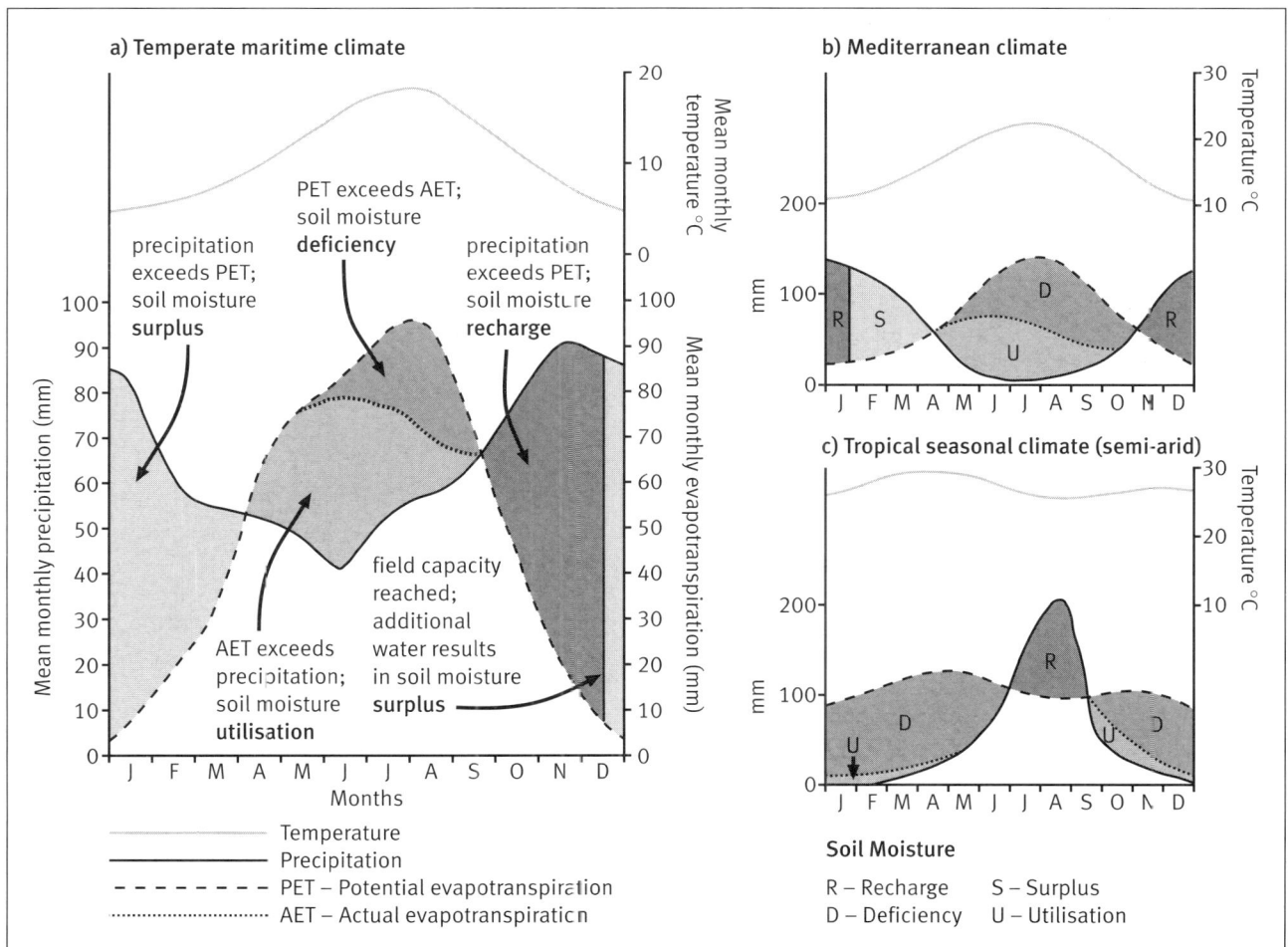

Figure 2.4 Water budget graphs

regime is controlled by changes in the water balance equation (as shown in Figure 2.4) and changes in the various flows (see Figure 2.3) in the drainage basin system. Snowmelt and human activity are additional important influences on river regimes. Rivers with **simple regimes** (Figure 2.5) have distinct periods of high and low flows; their headwater catchments have a uniform environment. Large rivers such as the Mississippi often have **complex regimes** as their basins include headwater catchments with different environments and runoff patterns.

The rivers in Figure 2.5 are all **perennial rivers,** flowing throughout the year. Even when a period of low precipitation occurs and overland flow or throughflow into the channel ceases, slow flow from the groundwater store is enough to maintain **base flow.** In arid and semi-arid environments **ephemeral streams** occur. Occasional sudden downpours give rise to considerable overland flow into stream channels. Channel flow is short lived and downstream water quickly infiltrates the channel bed. **Intermittent streams** occur where climates have marked wet and dry periods and strong seasonal contrasts in the water balance.

> ## Review task
>
> 1 **Refer to the water budget graphs in Figure 2.4.**
>   a) **Check your understanding of the terms used on the graphs. Describe and explain each graph. (For each start at a key month, e.g. April for the Mediterranean climate.)**
>   b) **Why does graph C show no soil moisture surplus?**
> 2 **Suggest the likely pattern of river regimes for the locations of the water budget graphs.**
> 3 **Construct a Mind Map to show the influences (including human) on river regimes.**

Precipitation and river discharge; the storm hydrograph

River regimes reflect relationships between river discharge and precipitation (together with other influences) over a long time period. It is useful to compare precipitation with discharge over the short term (i.e. the response of a river to an individual rainfall event) for practical reasons such as predicting the timing and severity of floods. This comparison is usually done by means of a **storm hydrograph** (Figure 2.6). The **base flow** on the hydrograph is sustained by a continual supply of groundwater. The **storm flow** is

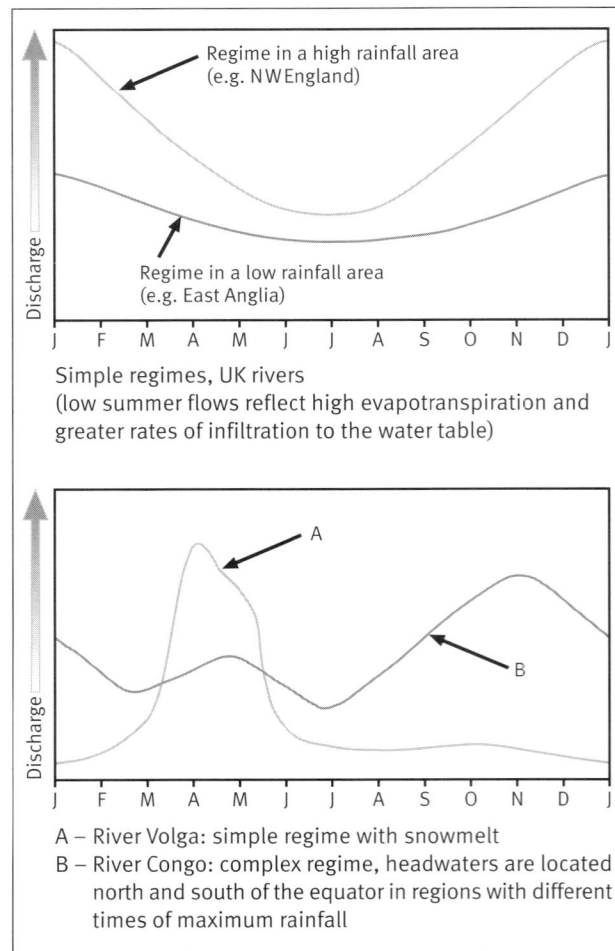

Simple regimes, UK rivers
(low summer flows reflect high evapotranspiration and greater rates of infiltration to the water table)

A – River Volga: simple regime with snowmelt
B – River Congo: complex regime, headwaters are located north and south of the equator in regions with different times of maximum rainfall

Figure 2.5 River regimes

the discharge resulting from a single rainfall event and is supplied by overland flow and quick throughflow. A certain level of discharge on a hydrograph will be associated with **bankfull discharge** (when a river's water level reaches the top of the channel banks) and therefore higher levels of discharge will cause flooding.

Factors influencing hydrological processes

Many physical and human factors influence hydrological processes. The shapes of storm hydrographs reflect these factors. The Mind Map in Figure 2.7 summarises the influences and also shows that many of the factors are interrelated.

Physical influences

- **Basin size.** Discharge increases with the size of drainage basin. Lag time also increases as the water has further to travel to reach the main channel.

- **Basin shape.** A circular basin will have a higher peak flow and a shorter lag time than an elongated basin (Figure 2.8).

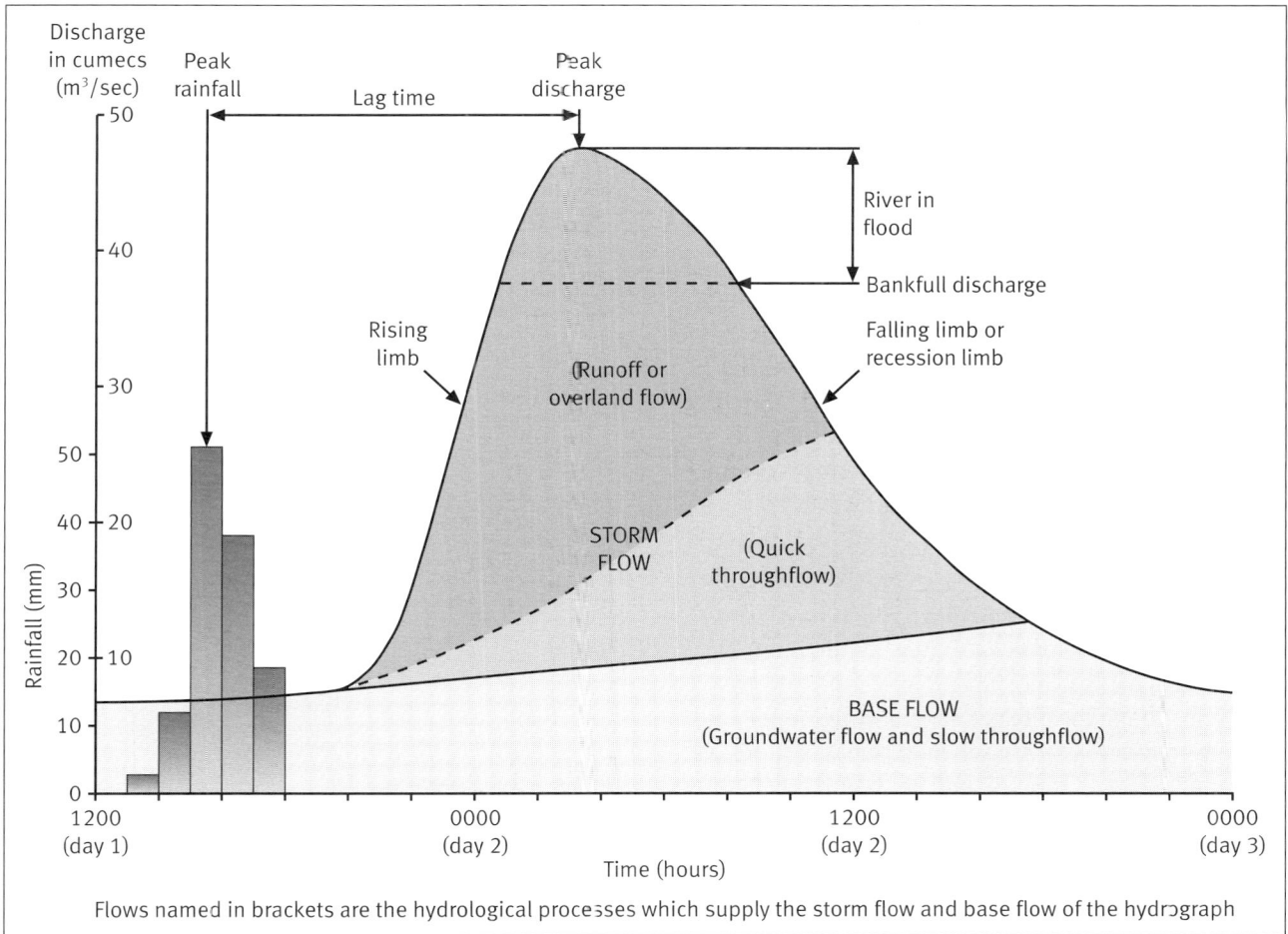

Figure 2.6 The storm hydrograph

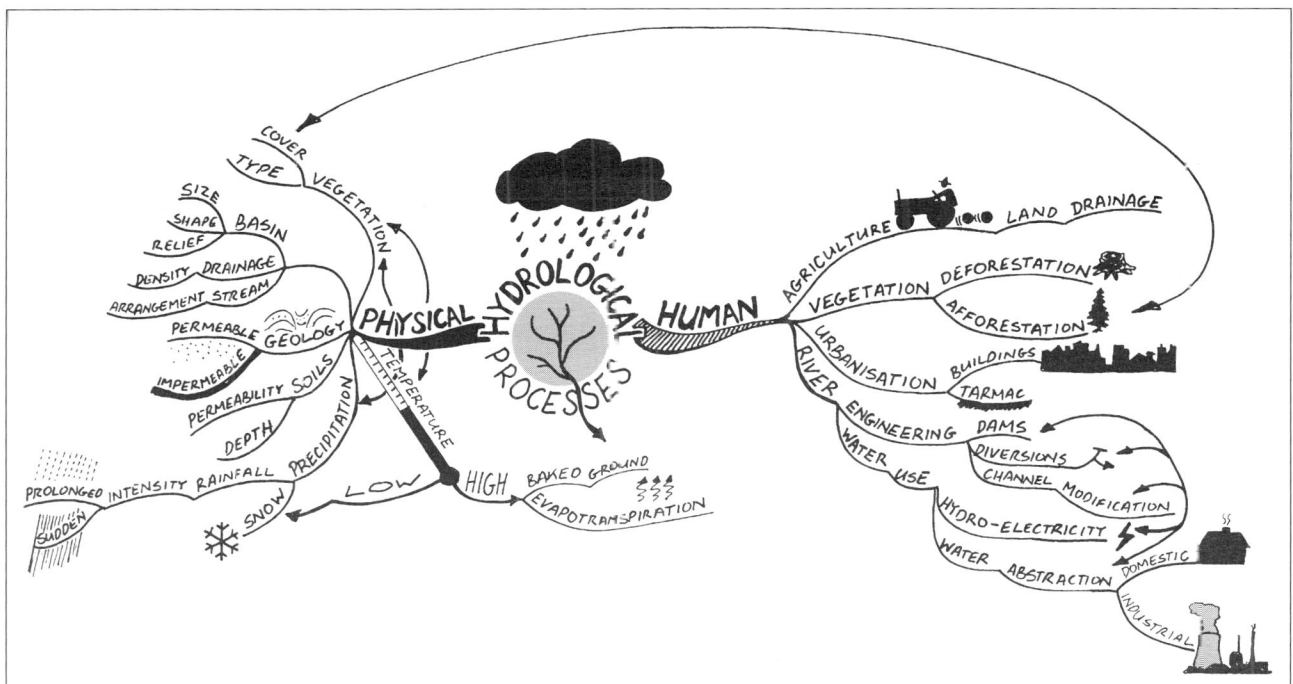

Figure 2.7 Influences on hydrological processes

- **Basin relief.** Water will reach a river more quickly where there are steep-sided valleys in a basin.

- **Drainage density and stream arrangement.** Drainage density is the sum of the lengths of all streams in a drainage basin divided by the area of the basin. A high drainage density (typically developed on impermeable rocks) causes a more rapid response in river discharge to a rainfall event. A stream arrangement with a complex branching pattern, as with basin B in Figure 2.8, is very efficient at moving water through and produces a high peak discharge. A basin with a less branching pattern, as with basin A in Figure 2.8, is less efficient and so produces a lower peak discharge, but an extended falling limb on a hydrograph.

- **Geology. Permeable** rocks allow rapid infiltration and thus there is little surface runoff. Permeable rocks include: 1) **porous** rocks which have pores that are able to store water (e.g. sandstone); 2) **pervious** rocks which have joints and bedding planes that allow the passage of water (e.g. Carboniferous limestone). Impermeable rocks do not allow water through and so there is very little infiltration and greater surface runoff.

- **Soils.** Soils vary in permeability; thus a sandy soil with large pore spaces allows quicker infiltration and a greater volume of throughflow than a clay soil. Soil depth is also important – more water can be stored or flow through a deeper soil.

- **Tidal conditions.** High tides restrict the flow of a river into the sea. Thus flood water builds up in the river's lower course. High tides combined with conditions of low atmospheric pressure and onshore winds result in exceptionally high water levels known as **storm surges**.

- **Rainfall intensity.** Prolonged, steady rainfall allows water stores to fill up gradually and efficiently. However, once the ground has become saturated, further rainfall will result in rapid runoff and flooding. Intense rainfall can exceed soil infiltration and vegetation interception capacities and so causes rapid runoff.

- **Temperature.** If temperatures are low enough to cause precipitation to fall as snow, much water is held in storage on the surface. If the temperature then rises rapidly, rivers are quickly swollen by meltwater (especially if the ground remains frozen so restricting infiltration). Very high temperatures can result in baked ground and thus restricted infiltration. On the other hand, high temperatures increase evapotranspiration rates, so that their effect needs to be considered in combination with other influences on hydrological processes such as vegetation cover.

- **Vegetation cover.** Vegetation intercepts rainfall and stores moisture on its leaves. Some of this water is then evaporated and the rest is delivered to the soil more slowly than by direct rainfall (Figure 2.9). Plant roots take up water from the soil and so reduce throughflow. The net effect of these processes is to restrict runoff and reduce the likelihood of flooding in forested areas.

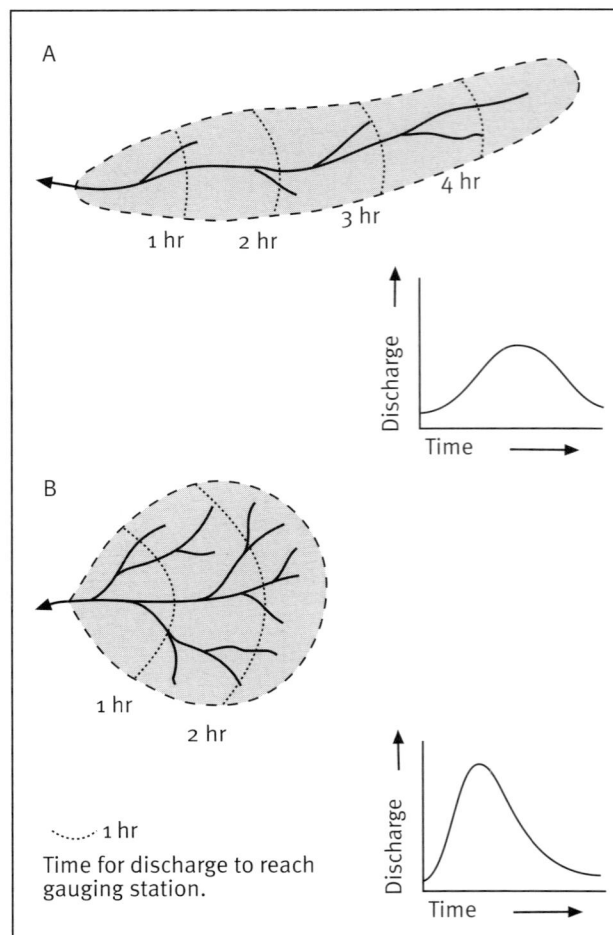

Figure 2.8 The effect of basin shape on the hydrograph

Human influences

- **Agriculture and land drainage.** Agriculture involves changes to the vegetation cover which affects interception, runoff, infiltration and percolation. Ploughing and other activities alter soil texture and thus infiltration rates and storage capacities. Soil conservation measures such as contour ploughing are specifically aimed at reducing runoff (and therefore soil erosion). Irrigation, especially if it involves considerable engineering works, leads to major impacts on hydrological processes. Land drainage aims to lower the water table and prevent waterlogging by the use of pipes laid below the surface and deepened ditches. In the case of most soils these measures result in more rapid throughflow, shorter lag times and higher hydrograph peak flows. (In clay soils surface runoff is reduced by land drainage.)

- **Deforestation and afforestation.** Deforestation accelerates runoff and increases the risk of flooding (Figure 2.9). Greater runoff flows accelerate soil erosion, and the resulting greater sediment loads increase flood levels further. Deforestation in the Himalayas where the Ganges and its tributaries have their headwaters has resulted in greater flood frequency and intensity in Bangladesh. Afforestation has become an important method used to reduce runoff (and erosion) and even out river discharges.

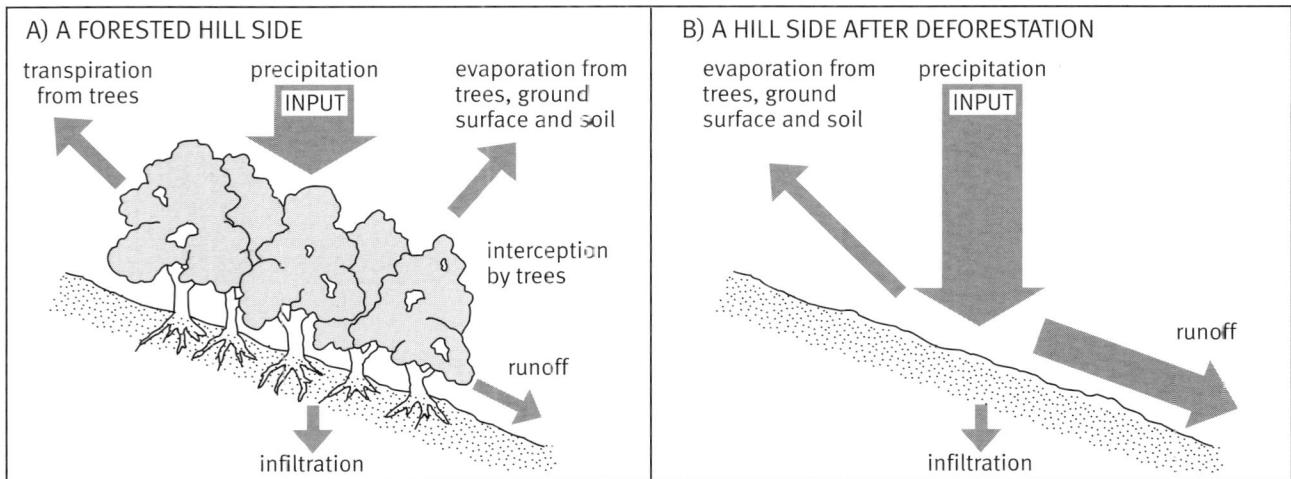

Figure 2.9 The effect of vegetation of hydrological processes

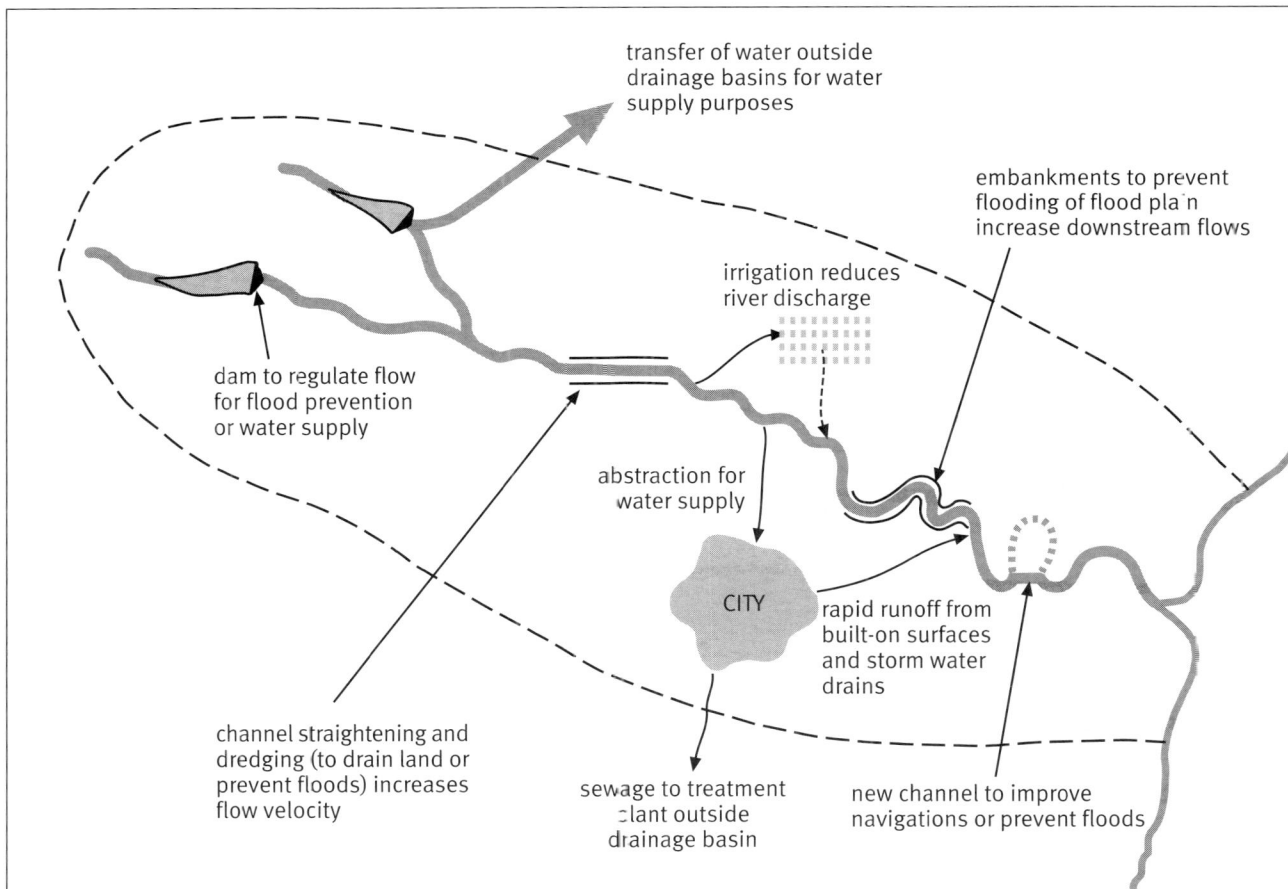

Figure 2.10 The impact of river engineering on river flows

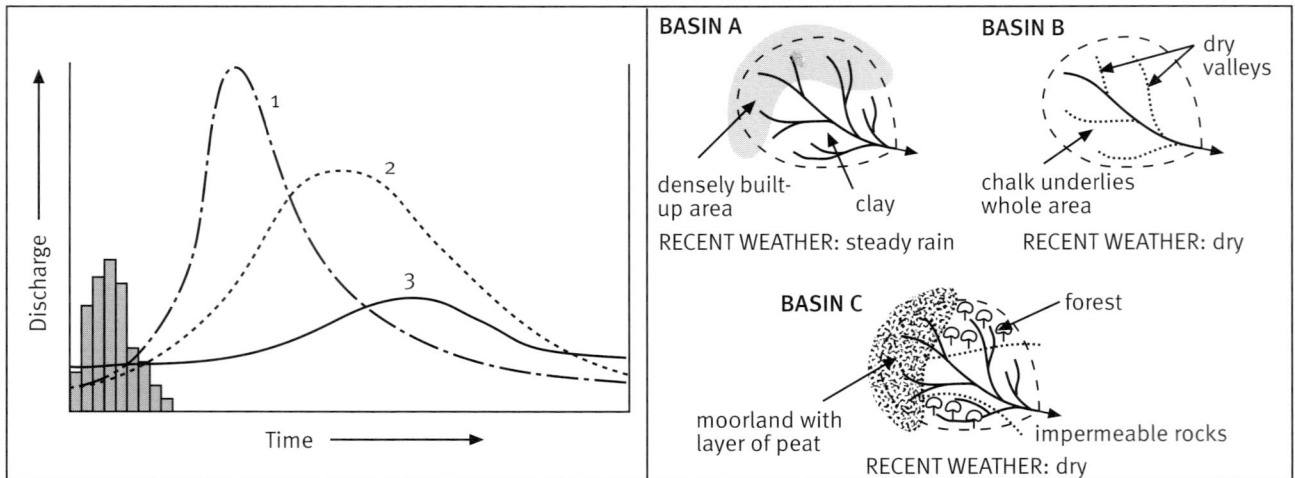

Figure 2.11 Hydrographs of contrasting drainage basins

- **Urbanisation.** Urban development replaces vegetation and permeable surfaces with impermeable tarmac and concrete. Thus runoff volumes grow and flood risk increases. The flood risk is aggravated by gutters and drains increasing flow velocities and reducing lag times. Less vegetation cover and reduced surface water storage lead to lower evapotranspiration rates and thus even more water available for runoff.

- **River engineering.** There is direct human impact on river flows through various engineering works to reduce flooding, generate hydro-electricity, provide irrigation water and supply water for industrial and domestic uses. Therefore dams are built, river channels altered and major diversions of water flows are made (Figure 2.10).

Review task

Refer to Figure 2.11. Match the drainage basins A, B and C with the three hydrographs. Check that you are able to give reasons for your decisions.

Floods and low flow conditions

Flooding

The differing physical characteristics of drainage basins means that some rivers are more likely to flood than others. Human action may increase or decrease the flood risk. River flooding in response to precipitation is a natural event which becomes a **hazard** when people live in areas liable to flooding. Figure 2.12 takes the various physical and human influences on hydrological processes and directly applies them to the matter of flooding.

EXAM TIP

You can use Figure 2.12 as a check list to analyse the causes of flooding in case studies you are familiar with or data on a flood presented to you in an examination question. It also includes check lists on the effects of flooding and human responses to floods.

Major floods usually result from a critical combination of factors as the case of Bangladesh in Figure 2.13 shows.

Low flow conditions

Conditions of low river flows are also natural occurrences, especially in environments with marked seasonal variations in precipitation. In Britain, low river flows reflect periods of drought, such as in south-east England between 1988 and 1992. Surface runoff is lessened and reduced infiltration results in falling water tables. Human activity, such as excessive abstraction of groundwater and abstraction from rivers, reduces flows further. Effects of low river flows include the drying up of headwaters, decreasing water quality (as pollutants are more concentrated) and problems of water supply.

Drainage basin management

In the UK, the priorities in managing drainage basins tend to be concerned with urban water supply, flood prevention and, more recently, pollution control and other environmental considerations. On some major European rivers (for example, the Rhône) navigation

PHYSICAL INFLUENCES

1. NATURE OF DRAINAGE BASIN (size, shape, relief, drainage density)

2. GEOLOGY (especially rock permeability)

3. SOILS (permeability, depth)

4. VEGETATION COVER

5. PRECIPITATION (over a long time period; also incidence of storms)

6. SNOWMELT

7. ALTITUDE of a place relative to other parts of a drainage basin and to sea level

8. HIGH TIDES AND STORM SURGES (affect lower courses of rivers)

SEDIMENT LOAD

e.g. steep slopes lead to more erosion

Unconsolidated sands and silts easily eroded

Intense rain → more sediment

Dams trap sediment

MORE SEDIMENT INCREASES RIVER DISCHARGE

HUMAN INFLUENCES

1. FARMING PRACTICES (e.g. arable land on slopes → much runoff)

2. DEFORESTATION (→ increased runoff on slopes)

3. URBANISATION (→ increased runoff, short lag times)

4. CHANNEL MANAGEMENT (reduces flooding locally but speeds water to downstream locations and so can increase flood risk

5. DAMS reduce flood risk, but DAM FAILURE results in flooding

INCREASING THE FLOOD RISK

1. Farming practices on flood plains which are not compatible with flooding

2. Building on flood plains (flood plain encroachment)

FLOODS

EFFECTS

Primary effects

- Loss of life
- Crop damage
- Loss of livestock
- Damage to roads and railways
- River transport disrupted
- Sewage disposal disrupted
- Water, gas and electricity services damaged
- Houses damaged
- Commercial and industrial buildings damaged

Secondary effects

- Farm income reduced
- Food supply problems
- General disruption of local economy
- Disease threat
- Reconstruction costs
- Homelessness
- Polluted water

RESPONSES

- Emergency action
- Bear losses
- Insurance against losses
- Improve flood forecasting
- Government and international aid
- Flood plain land use zoning
- Engineering solutions
  1. Headwater dams, relief channels, storage basins
  2. Channel engineering (widening, deepening, straightening)
  3. Floodbanks

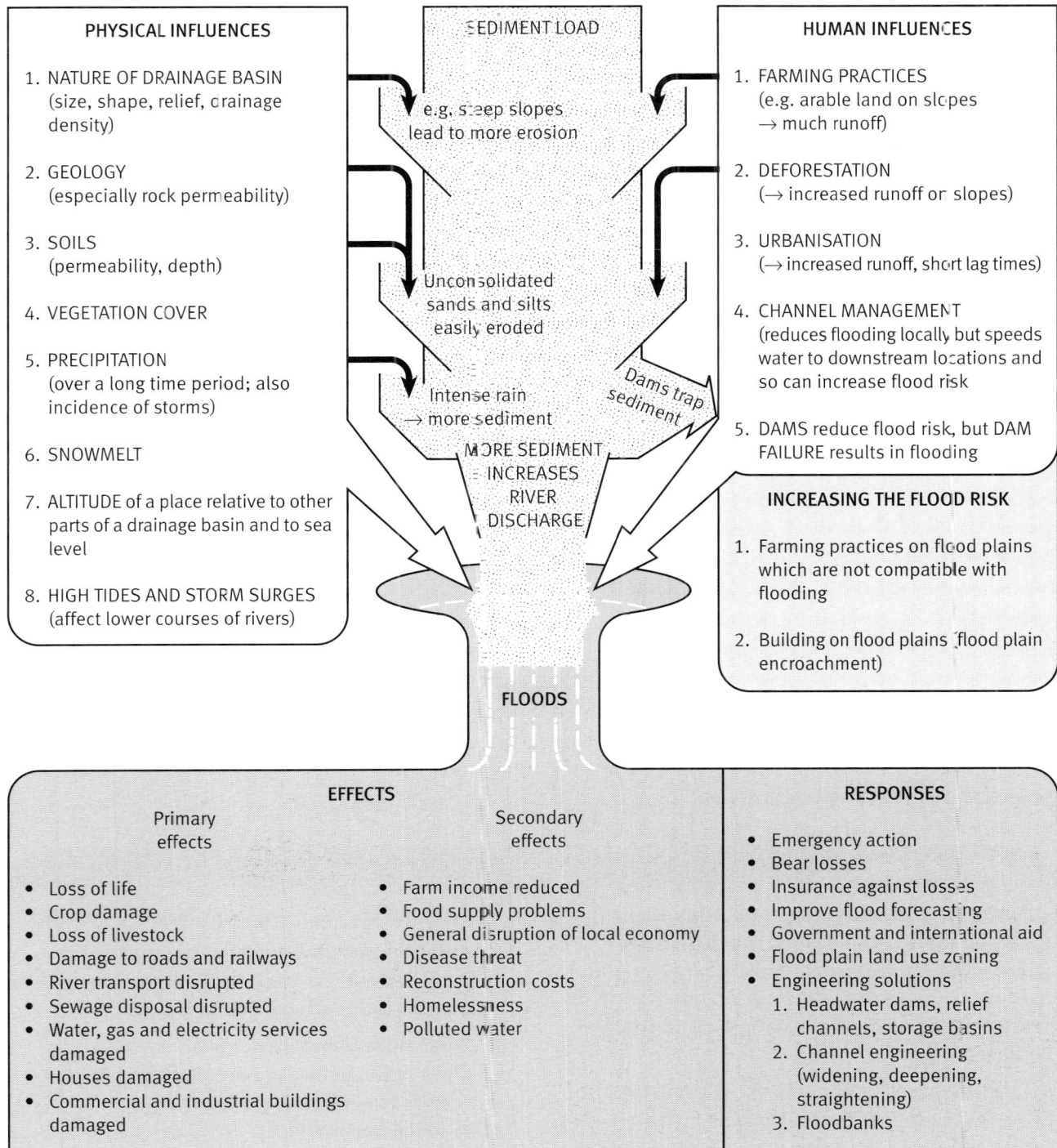

Figure 2.12 Floods: influences, effects and responses

and hydro-electric power development are of major importance. In many LEDCs (and in south-west USA) priority is given to irrigation and hydro-electricity. Flood prevention is the major priority in the lower courses of major rivers flowing through densely populated areas in both MEDCs and LEDCs (for example, the lower Mississippi and Ganges Delta respectively). Where international boundaries divide river basins, political factors may influence management (for example, the Colorado River flowing from USA to Mexico).

In recent years there has been a worldwide trend to integrated approaches to river basin management. In the UK, the Environment Agency (EA), formed in 1996, took on the responsibilities of the National Rivers Authority (NRA) to regulate and manage water resources, water quality, flood defence and other aspects of the use of rivers and water. The EA publishes Catchment Management Plans (CMPs) for all sizeable drainage basins in England and Wales. CMPs identify issues and conflicts and, after public consultation, produce action plans.

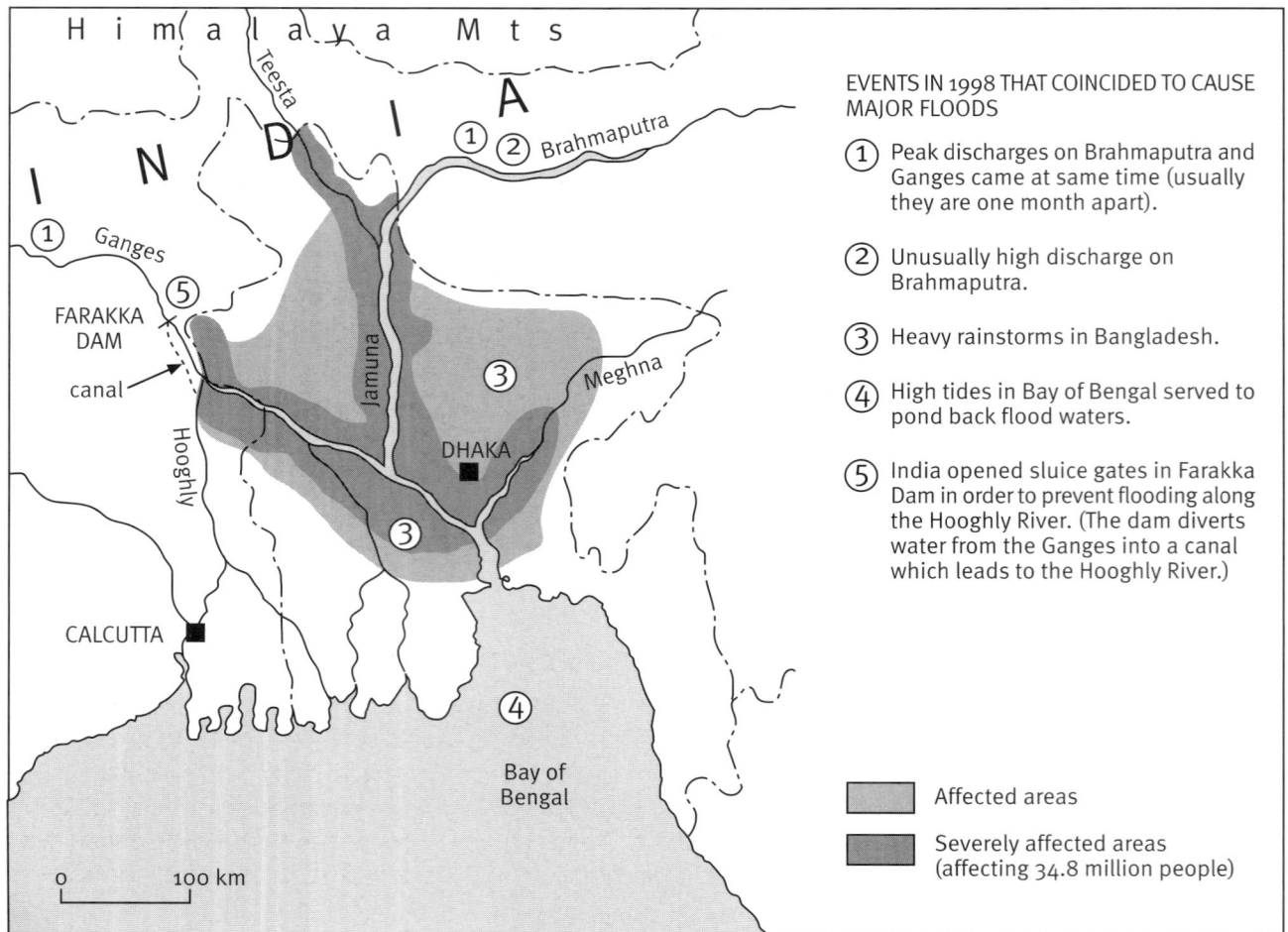

Figure 2.13 The 1998 Bangladesh floods

EVENTS IN 1998 THAT COINCIDED TO CAUSE MAJOR FLOODS

① Peak discharges on Brahmaputra and Ganges came at same time (usually they are one month apart).

② Unusually high discharge on Brahmaputra.

③ Heavy rainstorms in Bangladesh.

④ High tides in Bay of Bengal served to pond back flood waters.

⑤ India opened sluice gates in Farakka Dam in order to prevent flooding along the Hooghly River. (The dam diverts water from the Ganges into a canal which leads to the Hooghly River.)

Affected areas

Severely affected areas (affecting 34.8 million people)

Review task

Check your syllabus for the case studies of drainage basin management required: some demand one, usually in the UK, others two, the second being an example from outside Europe. To revise a case study of a drainage basin, approach it as follows:

1 Draw a basic sketch map of the basin to act as a focus for your thoughts. As you give answers to the questions below, add relevant labels to the map.

2 What are the main physical features of the basin? Does the main river of the basin flow through distinct sections?

3 What are the main physical influences on basin hydrology? (Use Figure 2.12 as a check list.)

4 What are the main human influences? (Use Figures 2.10 and 2.12 as check lists.)

5 Have there been major flooding incidents? If so, note details.

6 What are the main priorities for management?

7 Are there political considerations affecting management (for example, because the river crosses international boundaries)?

8 What management strategies have taken place in the following areas (not all will be relevant to every drainage basin):
a) flood prevention?
b) problems of low flow conditions?
c) water supply (for domestic use, industry, irrigation)?
d) hydro-electricity?
e) water quality?
f) navigation?
g) recreation?
h) fisheries?
i) conservation?

You may find that a Mind Map will be a convenient way of noting your key points, either for all of points 2 to 8 or just for point 8.

River channels

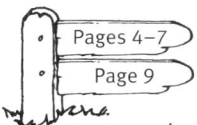

Pages 4–7
Page 9

River channels are open systems which interact with other open systems, particularly hill slope systems (see Chapter 1). They form part of drainage basin hydrological systems (see Chapter 2, Figure 2.2).

River energy and river flow

The stored or potential energy a river has available is proportional to its height above sea level. As it flows downslope the potential energy is converted into kinetic energy. In this conversion process about 95 per cent of the energy is lost to overcome friction: external friction between the water and banks, bed and air and internal between eddies of water within the river. The remaining energy is available to transport and erode debris. External friction results in variations of velocity through the cross-section of a river channel (Figure 3.1a). If water flows in a smooth, straight channel at slow velocity its flow is described as **laminar**. As river flow is usually in a rough, irregular channel, laminar flow seldom exists in reality; typical river flow is **turbulent** (Figure 3.1b), with chaotic water movement involving eddies.

Influences on river velocity and energy

Casual observation of rivers may suggest that an upland river with a steep gradient has greater velocity and energy and a lowland river flowing across a gently sloping flood plain. The picture is, in fact, much more complex. A river's velocity is influenced by channel cross section, channel roughness and channel angle of slope.

Channel cross section

In Figure 3.1c three of the river cross sections have the same cross-section area. The wide, shallow channel B has a longer **wetted perimeter** than the narrower, deeper channel A and therefore friction with bed and banks is greater and energy loss is greater. Channel A is therefore less efficient than channel B and its mean velocity will be less. **Hydraulic radius** (cross-section area divided by length of wetted perimeter) is a measurement of the efficiency of river channels. The semicircular channel C has the highest hydraulic radius and the greatest efficiency. When a river floods, the hydraulic radius increases and its efficiency is raised further, peaking at **bankfull discharge**. Substantially more energy is available for transport and erosion as Figure 3.1d shows. If, however, it overflows its banks, efficiency is decreased.

Channel roughness

The channel of a stream with projections of rock in the banks and a bed of angular rocks is clearly rougher, causing greater friction (and turbulence), than a channel with bed and banks of cohesive clays and silts. A very irregular channel course will add to loss of energy through friction. Although an upland stream may have one thread of rapidly flowing water, its mean velocity across its cross-section area may well

a) VARIATION OF FLOW IN A RIVER CHANNEL

air resistance reduces velocity at surface

maximum velocity

minimum velocity; resulting from friction with bed and banks

b) TYPES OF FLOW

LAMINAR FLOW

TURBULENT FLOW

eddies

maximum velocity

Smooth horizontal flow

Irregular flow with eddies

c) THE EFFICIENCY OF DIFFERENT CHANNELS

CHANNEL A

2 m · 2 m

5 m

wetted perimeter $= 2 + 5 + 2 = 9$ m

cross-section area $= 10$ m^2

CHANNEL B

1 m · 1 m

10 m

wetted perimeter $= 1 + 10 + 1 = 12$ m
cross-section area $= 10$ m^2

CHANNEL C

radius 2·52 m

wetted perimeter $= 7.91$ m

cross-section area $= \frac{\pi r^2}{2} = 10$ m^2

HYDRAULIC RADIUS

Channel A: $\frac{10}{9} = 1.11$ Channel B: $\frac{10}{12} = 0.83$

Channel C: $\frac{10}{7.91} = 1.26$

d) BANKFULL DISCHARGE INCREASES EFFICIENCY

bankfull discharge

normal discharge

1 m · 1 m
2 m · 2 m
5 m

hydraulic radius $= \frac{15}{11} = 1.36$

hydraulic radius $= \frac{10}{9} = 1.11$

With bankfull discharge, cross-section area increases from 10 m^2 to 15 m^2 and wetted perimeter from 9 m to 11 m

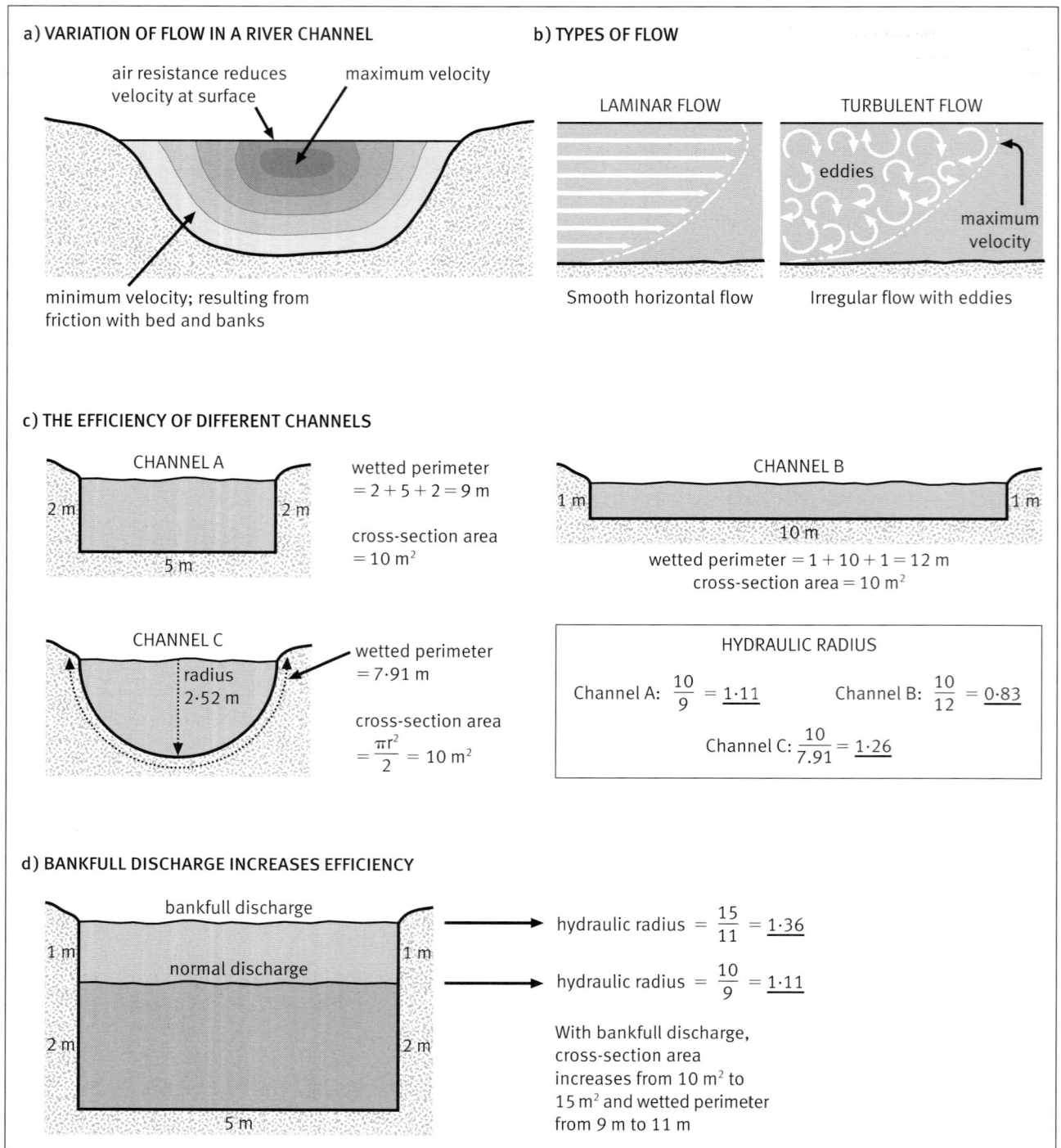

Figure 3.1 River flow and channel efficiency

be very low – because of considerable channel roughness. An index of channel roughness called **Manning's 'n'** can be calculated (Figure 3.2). It reflects the amount of friction that a particular cross-section and angle of slope will offer to the flow of water.

Channel angle of slope

River velocity is *not* simply a function of gradient because of the influences of channel shape and roughness. The characteristic concave long profile of a river (Figure 3.3) with a gradually decreasing gradient does *not* cause fast flow on the steep upper course and slow flow on the gentle lower course. A steep gradient is needed in the upper course for the river to overcome the great energy loss due to friction. In the lower course the river loses less energy due to friction because of a smoother channel, a more efficient channel shape and a larger cross-section area. Also, a greater discharge (see page 10) gives the river even more energy (kinetic energy is proportional both to velocity and to mass). Thus in the lower course a gentler gradient will maintain the same, or even a greater, velocity than upstream.

The relationship between channel roughness and velocity is expressed in the equation:

$$v = \frac{R^{\frac{2}{3}} S^{\frac{1}{2}}}{'n'}$$

v = mean velocity
R = hydraulic radius
S = channel slope
'n' = index of roughness

The higher the value of 'n', the rougher the channel:

e.g. uniform bed with sand and gravel: 'n' = 0.02
very irregular bed with boulders: 'n' = 0.10

Figure 3.2 Manning's 'n' formula

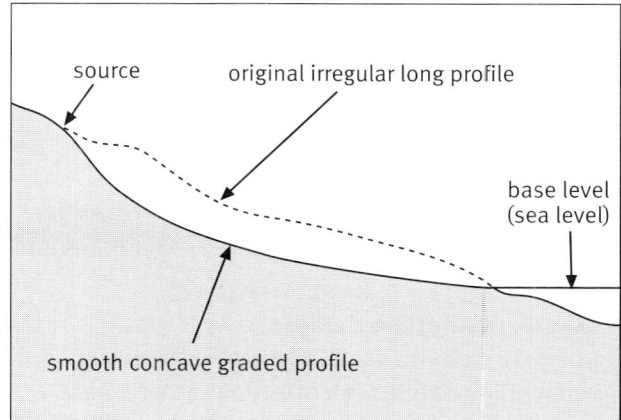

Figure 3.3 River long profiles

Hydraulic geometry

Hydraulic geometry is the study of the relationships which exist within river channels. Width, depth and velocity vary in their downstream relationship to discharge. Channel roughness, slope or gradient and the nature of the river's sediment load also change. Figure 3.4 summarises these variations and relationships. If one of the variables changes other

variables will adjust. The **river channel system** is therefore a complex interrelationship of variables, all linked to the central component of discharge. For example, a sudden increase in discharge in a section of river (perhaps after sudden rainfall) leads to an increase in velocity and load, and thus accelerated erosion, so altering the depth and width of the channel. When discharge returns to normal levels,

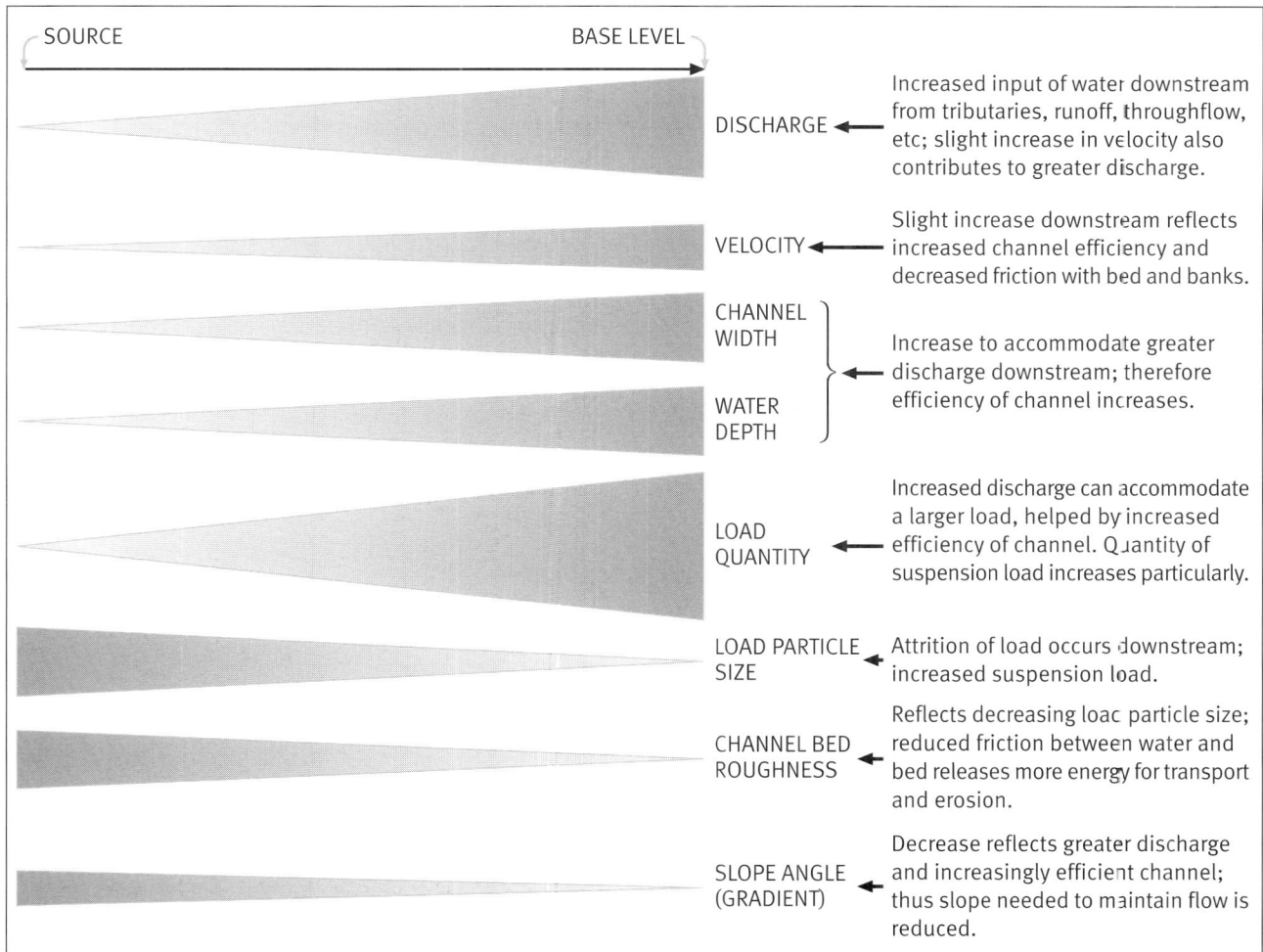

Figure 3.4 Hydraulic geometry: variations in river channel characteristics from source to mouth

deposition takes place. This type of self-repairing by the river is **negative feedback** within the river channel system; the river is being returned to its equilibrium with a minimum of disturbance.

> **Review task**
>
> Refer to Figure 3.4. Read through the explanations of the changes in hydraulic geometry; as you do so, add arrows to show where explanations are interlinked and where changes are linked to each other.

A river's long profile and the concept of grade

A river's state of equilibrium can also be seen in its long profile. We have already seen that the characteristic long profile of a river is a curve, steep in its upper course, gently sloping in its lower course (Figure 3.3). To achieve such a smooth profile, a river will have removed irregularities in its profile by erosion and deposition. A smooth curved profile with the river in a state of equilibrium is described as a **graded profile**. As originally defined by geomorphologists, the concept of grade emphasised the profile as being a slope delicately adjusted to provide just the velocity needed for the transportation of the river's load. In practice, the slope has to adjust constantly, through erosion and deposition, to changes in discharge and load.

Erosion, transportation and deposition

The processes that allow all the variables of hydraulic geometry to adjust to changes in discharge are **erosion**, **transportation** and **deposition**.

Erosion

The processes of river erosion

- **Corrasion** (or **abrasion**) involves the river using material of various sizes to scrape away and smooth its bed and banks. When turbulent eddies swirl pebbles around in hollows in the river bed, so deepening them, **potholes** are formed.
- **Hydraulic action** is the dislodging of fragments of unconsolidated materials (for example, silt in a river bank) by the force of the water itself. An extreme form of the process is **cavitation**: bubbles of air in turbulent water collapse and generate shock waves which dislodge material.

- **Solution** is the dissolving of soluble material by river water. The presence of carbonic acid in river water leads to the solution of limestone (chemically, calcium carbonate). Other chemical compounds in rocks, especially when in their weathered states, are also soluble.
- **Attrition** is erosion of the river's load itself. Fragments collide, and so fracture or become rounded. Thus the size of a river's load decreases downstream.

Besides eroding its bed and banks, a river can extend its channel upstream by **headwards erosion**. The main process involved is **spring sapping**, the collapse of rock (usually saturated) around an emerging spring.

Transportation

A river transports its load as **suspension load**, **bedload** or **solution load**. Figure 3.5 shows how these loads are moved along by a river. Suspension load is the most important of the three. The larger boulders of the bedload will be moved only in flood conditions. The maximum size of particle that a river can transport at a particular time is known as the river's **competence**. The total load actually carried is the river's **capacity**. Both competence and capacity vary with a river's velocity and discharge. **Hjülstrom's graph** (Figure 3.6) shows a critical velocity exists for the transport of particles of various size to be possible; it is shown by the fall velocity curve on the graph. The Hjülstrom graph also includes a critical erosion velocity curve showing the velocities necessary for erosion to take place. Notice that because clay particles stick together, the critical velocity is as high as that for pebbles. The importance of high discharges for transporting the load is

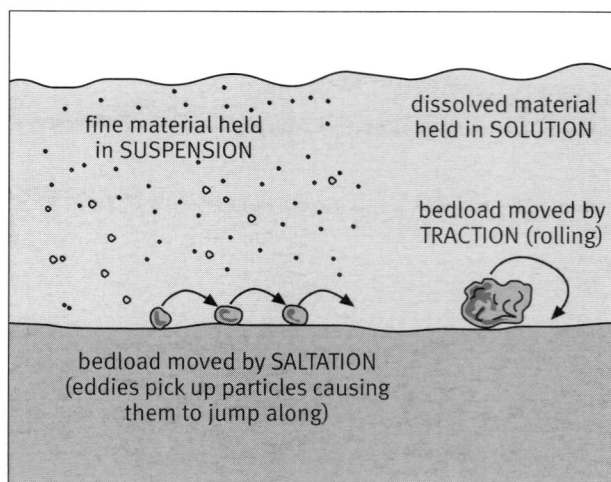

Figure 3.5 Transportation processes in a river

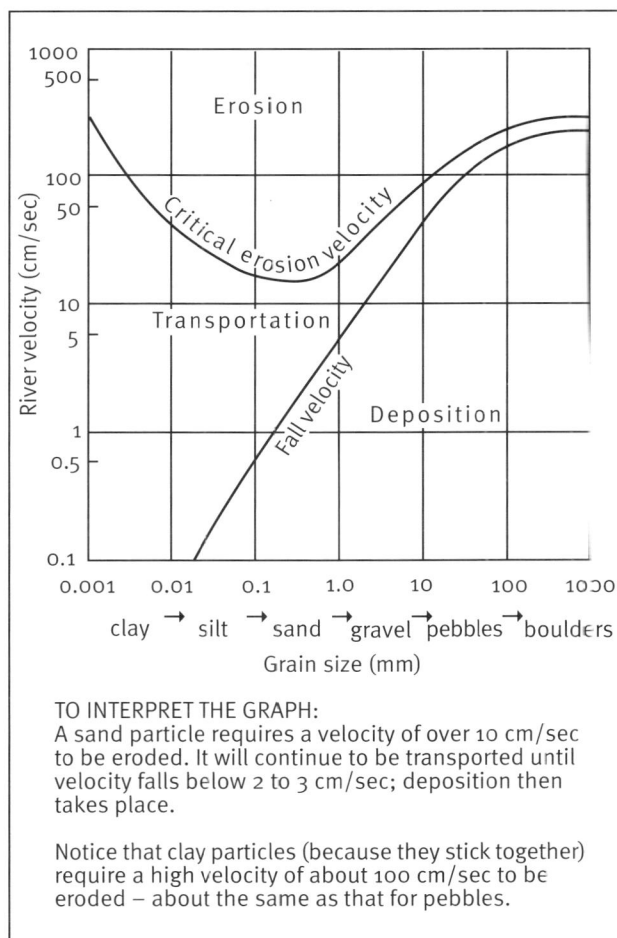

Figure 3.6 Hjülstrom's graph

TO INTERPRET THE GRAPH:
A sand particle requires a velocity of over 10 cm/sec to be eroded. It will continue to be transported until velocity falls below 2 to 3 cm/sec; deposition then takes place.

Notice that clay particles (because they stick together) require a high velocity of about 100 cm/sec to be eroded – about the same as that for pebbles.

demonstrated by **Hopkins' Sixth Power Law** which states that whatever factor discharge multiplies by in times of flood, then the mass of the maximum particle of load multiplies by that factor to the power six. Thus if discharge trebles, a river can carry particles 3^6 or 729 times as heavy.

Deposition

As Figure 3.6 shows, once river velocity drops below a critical fall value, particles will be deposited. As different sizes of particles will drop out of suspension at different velocities, **sorting** of deposits will occur.

Review task

Check your understanding of Hjülstrom's graph: what speed is necessary for a) a small clay particle, b) a pebble, to be eroded? Why are the speeds similar? At what velocity will a small pebble be deposited? Why will the clay particles be transported at velocities as low as 0.1 cm/sec?

As deposition results from decreasing velocity, it occurs at times of low flow, when a river enters the sea or a lake, on the inside of a meander and on a flood plain when a river overflows its banks.

Meanders, pools and riffles

Rivers meander rather than flow in straight channels. The exact cause of meanders is uncertain, but flowing water has a sinuous movement, even across a smooth plate in a laboratory tank. In an initially straight channel, the sinuous flow becomes emphasised by the development of shoals of deposition within the channel (Figure 3.7). **Pools** and **riffles** also develop. Pools are areas of deeper water where a reduction in friction increases the river's energy and so erosion takes place. Riffles are areas of shallow water, where energy is dissipated and so deposition of relatively coarse particles takes place. The deposition produces a localised steeper gradient immediately following the riffle. Pools and riffles are regularly spaced at five to seven times the channel width.

Considering the sinuous flow again, notice in Figure 3.7 how the line of deepest water (and highest velocity) – the **thalweg** – follows an even more sinuous course than the channel itself. Centrifugal force 'throws' water towards the outside of each bend. Where the thalweg comes close to bank, greater energy will cause erosion, thus developing a meander. Pools occur close to the outside bank of meanders, and riffles in the straighter sections at the points of inflexion. As the line of highest velocity swings from side to side of the river, water is piled up at the outside of bends; the resultant pressure causes water to move downwards and then across the bed towards the inside of the bend. When this cross movement of water is combined in three dimensions with the general downstream flow, the result is a corkscrew-like movement known as **helicoidal flow** (Figure 3.8). Reduced velocity on the inside of meanders results in deposition to form **point bars**; some geomorphologists believe that helicoidal flow assists this deposition process.

Studies of many rivers suggest that meander wavelength and the width of the meander belt is related to a river's bankfull discharge and indirectly to the channel width; thus a limit exists on the size of meanders. It is believed that a river develops meanders so as to balance its energy with the load it has to carry. A river's excess energy can be dissipated by reducing the slope and velocity between two points – this is achieved by the river flowing over a longer course, created by an enlarged meander.

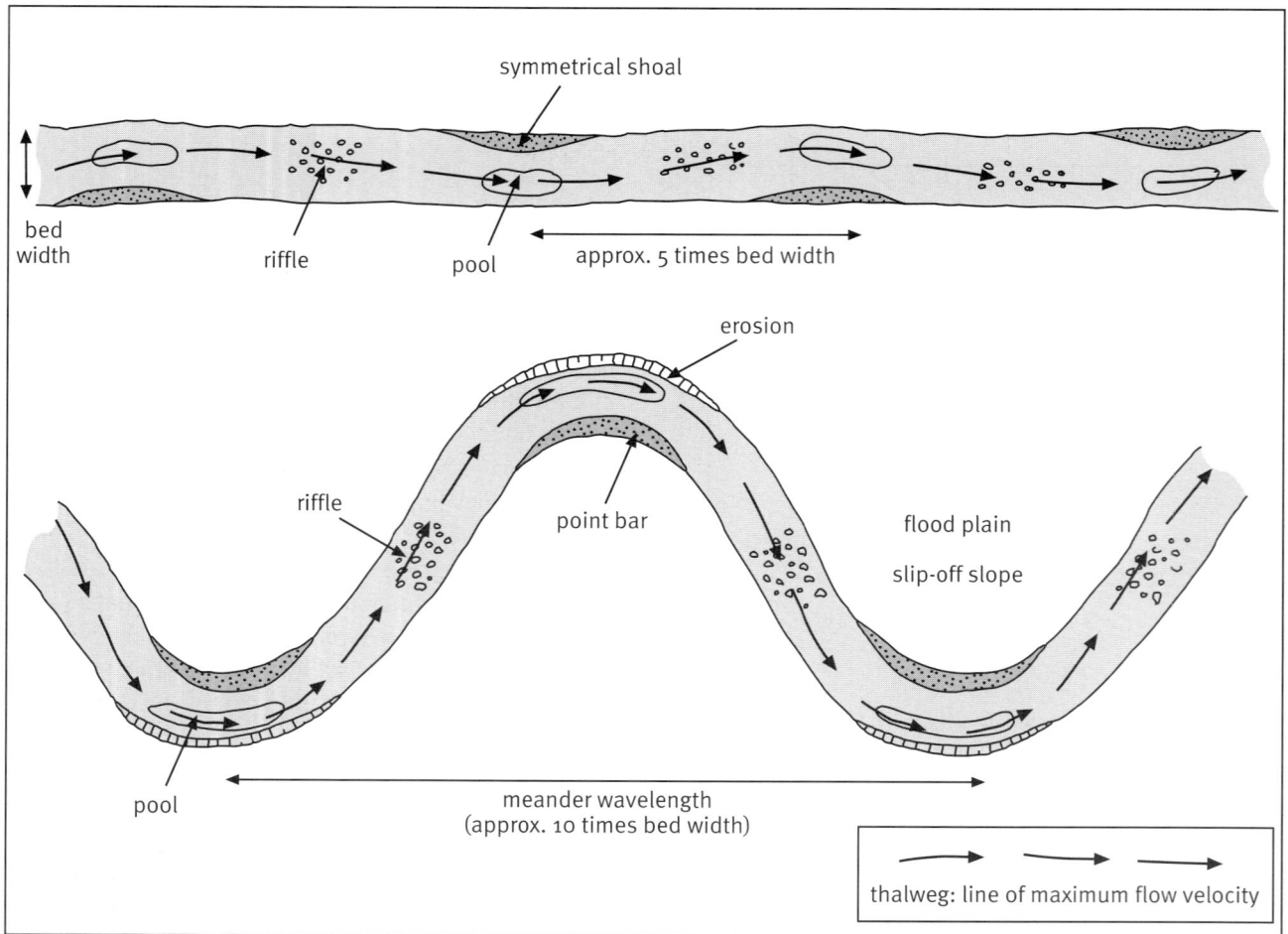

Figure 3.7 Features of river channels

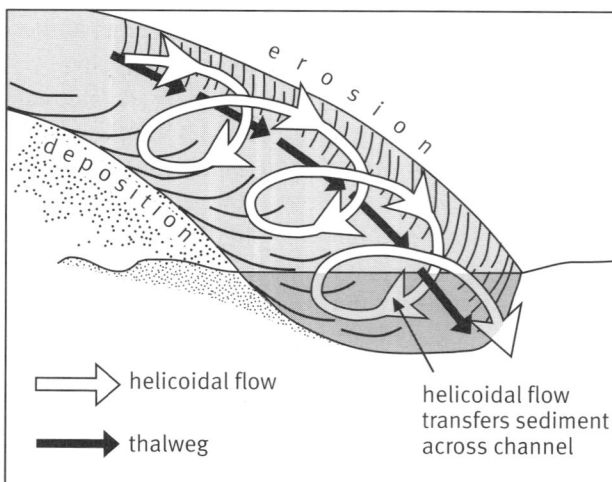

Figure 3.8 Helicoidal flow in a meander

Meanders and flood plains

Erosion on the outside banks of meanders tends to be at a maximum just downstream of their apex. This is even more marked during floods. Thus **meander migration** downstream takes place. Steeper land at the side of the valley is undercut to form **river cliffs** or **bluffs**. The flood plain is thus extended by this lateral erosion associated with meander migration.

On the inside banks of meanders, meander migration leads to the deposits of point bars being incorporated into the flood plain. Meander enlargement or different rates of migration between adjacent meanders can lead to cut-offs and the creation of **ox-bow lakes**.

Review task

1   In some meandering rivers, the meanders tend to migrate, but not to enlarge; in others, variable meander migration combined with meander enlargement takes place, sometimes leading to cut-offs. Why might rivers differ in such a way?

2   Artificial cut-offs are sometimes constructed to improve navigation or reduce flood hazards. What problems might this pose for the future?

Braided channels

When a river has a course divided into separate intertwining channels by alluvium or gravel it is described as being **braided**. Braided channels are associated with rivers with very variable discharge (for example, in monsoon areas or where high spring discharges are associated with snowmelt) combined with heavy loads. When peak discharge quickly reduces, rapid deposition follows, with deposits choking and splitting channels.

Fluvial landforms

The landforms or landscape features created by river action can be classified broadly into three groups:

1 Those formed by erosion.

2 Those formed by deposition.

3 Those formed by a combination of erosion and deposition.

Features of erosion

* **V-shaped valleys** are usually associated with the upper course of rivers where vertical erosion is the key process. As Figure 3.9 shows, slope processes and weathering remove most of the mass of material from the V-shape, delivering it to the river for removal. (Refer back to Chapter 1 for details on slope processes.) Where rivers follow a winding course, the V-shaped profile is associated with **interlocking spurs**.

Pages 4–8

* **Gorges** are also a result of vertical erosion. Their steep sides are due to rock being relatively resistant to the effects of slope processes and weathering.

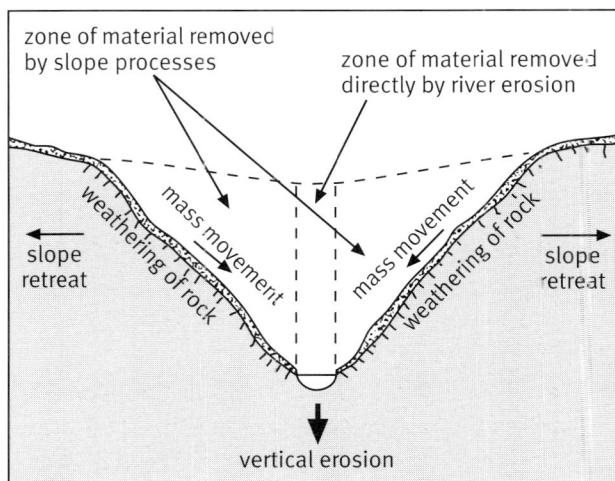

zone of material removed by slope processes

zone of material removed directly by river erosion

weathering of rock
mass movement
mass movement
weathering of rock

slope retreat

slope retreat

vertical erosion

Figure 3.9 The formation of a V-shaped valley

* **Waterfalls** form where a river flows over resistant rock and then reaches more easily eroded rock. Erosion of the softer rock undermines the resistant rock and the waterfall retreats upstream, leaving a **gorge of recession**.

* **River cliffs or bluffs** – see page 24.

* **Features resulting from rejuvenation** (such as terraces and incised meanders) – see page 26.

Features of deposition

* **Flood plains** may initially be developed by erosion of higher ground through meander migration (see page 24). However, flood plains are made up of layers of alluvium deposited when rivers overflow their banks. (When a river floods, the broad flood plain becomes, in effect, a very inefficient channel leading to an increase in friction, a decrease in velocity and energy and thus the deposition of suspended material.) Point bar deposits are also incorporated into the flood plain when meander migration occurs.

* **Levées** are raised river banks. As a river overflows its banks, the loss of energy leads to coarser material being deposited immediately, close to the channel. Levées are often artificially heightened and strengthened to prevent floods. At times of flood a weak point in a levée may be breached. Flood waters surge through the breach and spread a lobe of sediment, known as a **crevasse splay**, over the flood plain.

* **Deltas** are formed as a river loses energy as it enters the sea or a lake. In cross section, the form of the delta shows a distinct bedding of sediment according to grain size. In plan form, deltas may be classified as **arcuate** (a coastline curved by longshore drift), **cuspate** (pointed delta formed by one dominant channel) and **bird's foot** (where distributaries are edged by banks and so project as fingers into the sea). Delta formation is encouraged by large sediment load, small tidal range and relatively ineffective wave erosion or longshore drift.

* **Alluvial fans** are triangular-shaped masses of sediment found at abrupt changes of slope. For example, as a stream drops down the side of a glacial U-shaped valley and reaches the valley floor, it suddenly loses velocity and energy and so deposits much of its load.

Features of erosion and deposition

* **Meanders** – see page 23.

* **Ox-bow lakes** – see page 24.

Changes over time

Rejuvenation

The graded long profile of a river develops in relation to **base level**, usually sea level. Over time base level may change. The change may be worldwide (**eustatic change**) resulting from the growth and decay of ice sheets in recent geological history. Local changes, such as uplift resulting from the reduction in the weight of a melting ice sheet on the land, are **isostatic changes**. If sea level falls relative to the land (or land rises relative to the sea) the increased gradient of the river near the coast leads to an increase in energy and renewed erosion. This is **rejuvenation**. Erosion works its way upstream to form a new graded profile as shown in Figure 3.10. The point where the new profile meets the original graded profile is the **knick point**. The knick point may be marked by the presence of a waterfall, especially if its retreat upstream is held up at a band of resistant rock.

Downcutting by rejuvenated rivers combined with continued lateral erosion and meander migration leads to the formation of paired **river terraces** which are remnants of old flood plains. A further fall in base

level may lead to a further phase of rejuvenation and the creation of another flood plain and set of river terraces, as shown in Figure 3.10. Considerable lateral movement of meanders combined with continued downcutting following a steady, gradual fall in base level may give rise to **unpaired terraces**; a meander cuts into higher ground on one side of the valley, creating a terrace and then swings to the other side of the valley, creating a terrace at a slightly lower level.

Rapid and/or prolonged fall in sea level (or uplift of land) leads to the formation of **incised meanders**. There are two types of incised meander. An **entrenched meander** is caused by rapid downcutting and has a symmetrical cross-section. In the case of an **ingrown meander**, incision has been accompanied by lateral erosion and so the resultant valley has one side steeper than the other (Figure 3.10).

In reality the circumstances shown in Figure 3.10 are often more complex. Rejuvenation may be followed by a rise in base level. With the reduced gradient to the sea, the river's energy will decrease and resultant deposition may bury the new valley and flood plain created by rejuvenation. In the case of the river Thames, successive phases of rejuvenation led to a

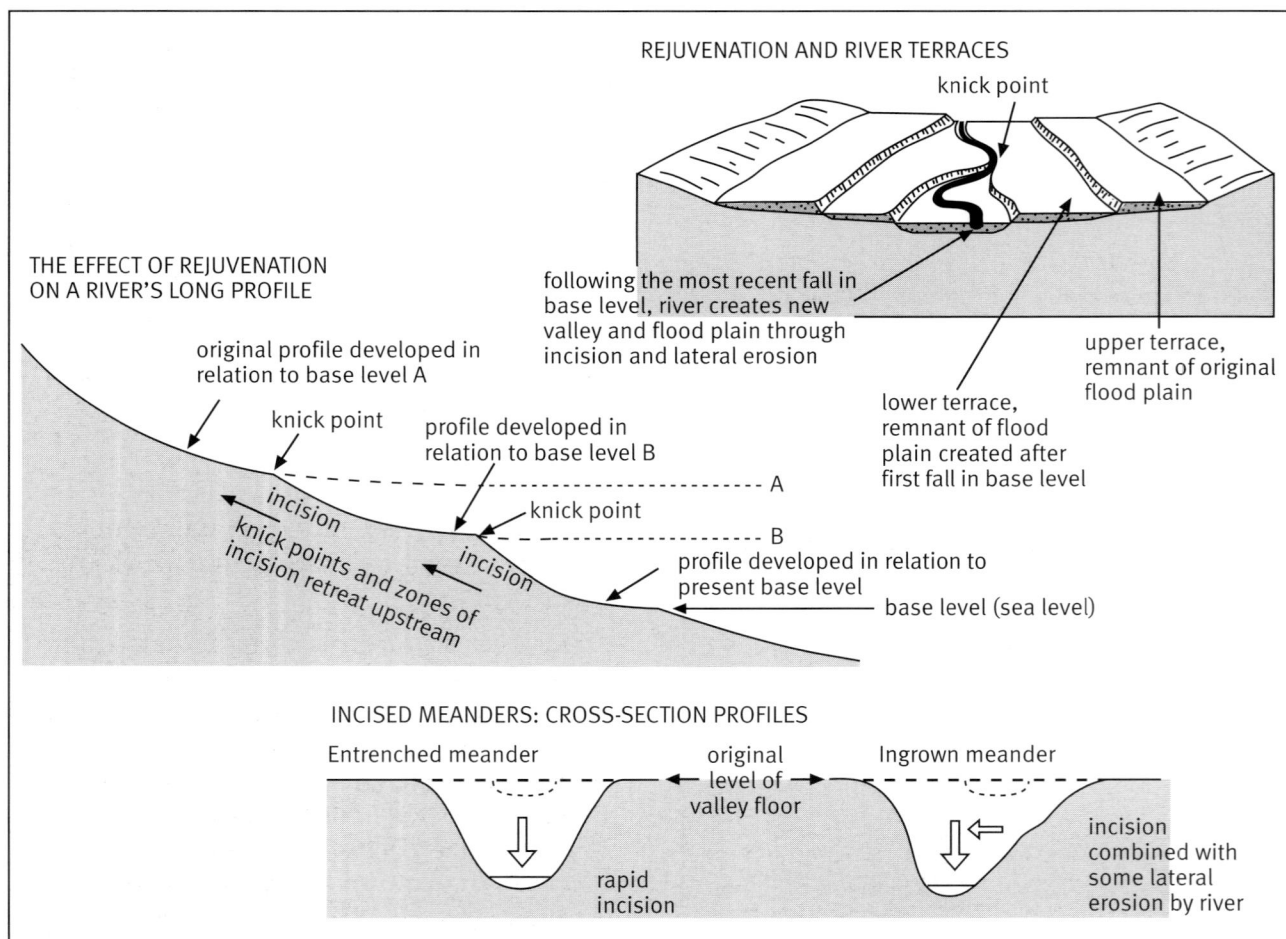

Figure 3.10 Rejuvenation

succession of terraces being formed and finally a new river valley with a bed 10 metres below the present day level. A later rise in sea level led to deposition filling up the valley formed by the last period of rejuvenation and the flooding of the lower course of the river to form the Thames estuary.

Changing drainage patterns

The arrangement of a river and its tributaries within their drainage basin is the **drainage pattern**. Various terms are used to describe especially distinctive drainage patterns:

- **Dendritic drainage** forms a branching, tree-like pattern; it usually develops in a drainage basin with little variation of rock type or structure.
- **Trellised drainage** forms a rectangular pattern where tributaries join the main river at right angles; it develops in areas of alternating bands of resistant and less resistant rock.
- **Radial drainage** is usually developed on a structural dome with streams radiating out from a central point.

Drainage patterns change over time. Figure 3.11 shows a trellised pattern where **river capture** has

taken place. Headward erosion, through spring sapping (see page 22) has led to the tributary of river A cutting back to the headwaters of river B and capturing its flow. There are many examples of river capture in the scarplands of south-east England. Drainage patterns may also be greatly altered by **glacial diversions** (see Chapter 7). Figure 7.10 explains how the eastern part of the Vale of Pickering was drained by a river flowing eastwards to the North Sea before the Ice Age, but is now drained by the river Derwent which flows south-westwards.

In some areas the drainage pattern itself may not have changed significantly, but the pattern's relationship with the underlying geological structure has. The radial drainage pattern of the Lake District involves rivers cutting across a wide variety of rock types and structures which appear to have had little effect on the courses of the rivers. This is because the original drainage pattern developed on a dome of relatively recent sedimentary rocks which overlaid the volcanic and ancient sedimentary rocks now at the surface. The more recent rocks on which the drainage pattern developed were stripped away by fluvial and glacial erosion to reveal the underlying complex pattern of ancient rocks. Such a drainage pattern unrelated to present rock types and structure is known as **superimposed drainage** because the drainage pattern has been superimposed upon the underlying rocks. Another drainage pattern unrelated to structure is **antecedent drainage**. Here the drainage pattern developed before structural movements such as the uplift of a plateau or the folding movements of mountain building. Vertical erosion by the river was able to keep pace with the uplift or folding across its course. Antecedence explains why the courses of rivers such as the Indus and the Brahmaputra lie across the Himalayas; they were able to cut down at the same rate as the Himalayas were being folded and uplifted.

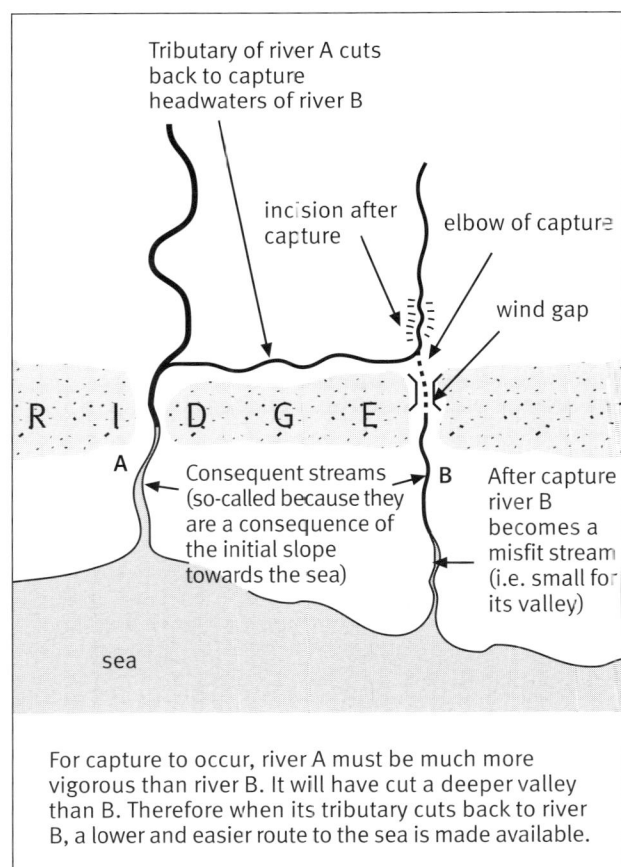

Tributary of river A cuts back to capture headwaters of river B

incision after capture

elbow of capture

wind gap

R I D G E

A

Consequent streams (so-called because they are a consequence of the initial slope towards the sea)

B

After capture river B becomes a misfit stream (i.e. small for its valley)

sea

For capture to occur, river A must be much more vigorous than river B. It will have cut a deeper valley than B. Therefore when its tributary cuts back to river B, a lower and easier route to the sea is made available.

Figure 3.11 River capture

Review task

Consider the various landforms associated with river erosion, river deposition and changes over time. Make sure that you are familiar with examples of the various landforms (many might be within the same river basin and so be interrelated). You can revise your knowledge of them by drawing simple (learnable) sketch maps and diagrams and then adding explanatory labels. Include also any human influences on the landforms, whether direct (such as building up levées) or indirect (such as reducing river flows).

What you need to know:

- **The structure of the atmosphere**
- **The energy budget and energy exchanges, including the general atmospheric circulation and local systems**
- **Air masses and weather systems**
- **Atmospheric water and the water cycle**
- **Temperature variations**
- **Local weather and climate including urban climate**

Also, you will need to be able to:

- **understand weather and climate as systems;**
- **distinguish between different scales of atmospheric activity;**
- **understand the effects of the different scales of atmospheric systems;**
- **analyse weather and climate data.**

Weather and climate – the atmospheric system

Weather is not constant and its study is concerned with the weather systems which bring about the varied conditions of the atmosphere from one moment to another.

- **Planetary-scale** systems are represented by waves in the general circulation of the atmosphere. They last for several weeks and are between 500 and 10 000 km in scale.

- **Synoptic-scale** systems last for about five days and are several thousand kilometres across, like mid-latitude depressions.

- **Mesoscale** systems involving features like thunderstorms, last for only a few hours and are about 10 km across.

- **Microscale** systems involve small areas like an urban area or a lake or even the area within the vegetation canopy.

Climate is usually described as the average of the weather conditions in any place but a more accurate definition is 'the long-term state of the atmosphere encompassing the aggregate effect of weather phenomena which includes extreme as well as mean values'.

Macroclimate examines regional and global climate, and **local** or **topo-climate** considers small areas like towns and cities, valleys and slopes.

Structure of the atmosphere

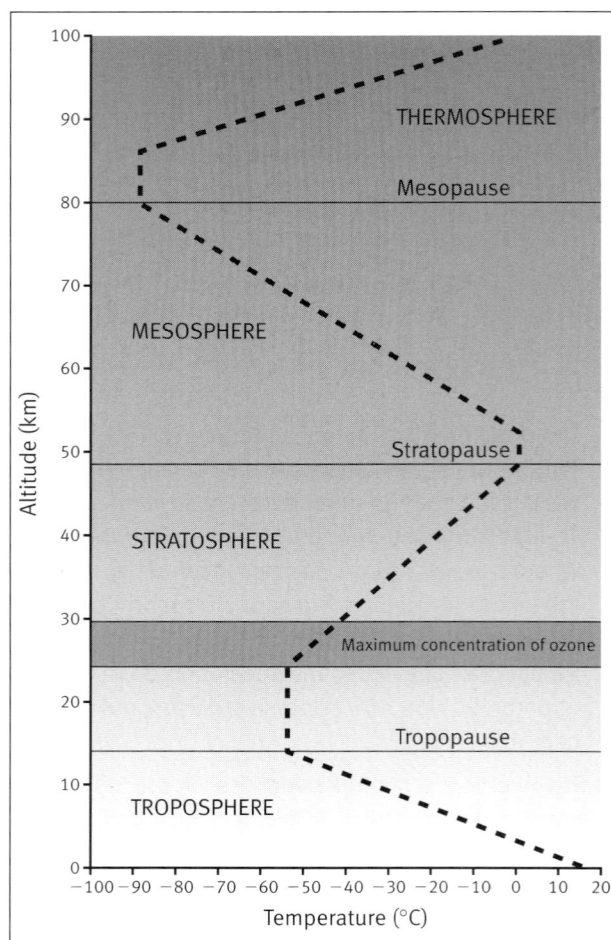

Figure 4.1 Vertical structure of the atmosphere

Review task

Answer these questions about the vertical structure of the atmosphere (Figure 4.1). Refer to your notes or a textbook if necessary.

1   How many layers make up the atmosphere?
2   Name them.
3   Name the boundaries between them.
4   Name five features of the troposphere.
5   What and where is the ozone layer?

Solar radiation

Solar radiation is the energy source that drives the atmosphere. Solar energy or **insolation** varies from place to place and with the seasons, but on the global scale, incoming radiation is balanced by outgoing radiation giving rise to the term **global energy budget** (or solar budget).

Most incoming radiation is short wave whereas outgoing radiation is long wave. The latter is partly absorbed by water vapour and carbon dioxide before being re-radiated and, as some of this goes back to earth, it keeps the surface warmer than if all radiation went directly into space. This is the **greenhouse effect** which keeps the global temperature at a habitable level. It is a natural process and is not a result of human-induced climate change (see Chapter 8).

Variations in solar energy received at the surface

Solar energy received from the sun varies for a number of reasons:

- **Atmosphere.** The atmosphere absorbs some of the incoming radiation, with water vapour and carbon dioxide being the most significant gases.

- **Albedo.** This is a measure of what percentage of incoming energy is reflected from different surfaces, like snow and ice compared with grass. Differences in albedo are particularly important in relation to the effects of water and land on total received solar energy (see below).

- **Altitude of the sun.** The angle between the sun's rays and the surface affects the intensity of heating at the surface. A greater concentration of heating intensity occurs in the tropics compared with nearer the poles, where the sun is at a much lower angle spreading the heat over a greater area, as well as passing through a greater thickness of atmosphere.

- **Seasons.** The effect of the seasons accentuates the effect of the altitude of the sun in addition to directly affecting the length of day, in the summer increasing it in the northern hemisphere and decreasing it in the southern hemisphere. As a result of the apparent movement of the sun with the seasons the maximum receipt of radiation is not at the equator but at the tropics.

- **Cloud cover.** Areas with greater cloud cover reflect more solar radiation than areas lacking cloud cover. Despite clouds having a high albedo they also retain heat that would be lost by radiation from the earth. This reduces the daily temperature range.

- **Land and sea.** Water stores heat it receives while land returns it quickly to the atmosphere, due partly to the different specific heats of land and sea since the specific heat of water is much greater than most other substances. Water absorbs and stores more heat than does land, but also gives off more heat on cooling, producing a seasonal effect on temperature differences between land and sea.

  The sea also reflects very little (except when the sun is at a low angle such as in poleward locations

where reflection increases). The sea generally absorbs 90 per cent of received radiation in its surface layers (down to about 9 m) and transfers that heat to greater depths by currents. The term **continentality** summarises the effect of the difference in heating and cooling of land and sea with greater seasonal extremes of temperature over land than over sea.

- **Altitude and aspect.** These are more local in effect. High-altitude locations receive a greater input of solar radiation because it passes through less of the atmosphere, although increased cloud cover could counteract this. Aspect produces sun-facing slopes and shadowed slopes. The former receive a more intense concentration of solar radiation intensity than does flat land.

The overall result of these factors is that the polar regions receive less insolation than the tropics, and equatorial areas receive less insolation than areas straddling the tropics.

Review task

Add examples from your course notes to illustrate each of the factors listed above. There's no need to go into more than the barest detail, but make a note of where to find the information.

Transfer of heat

The great differences in heat energy received at the surface means that there must be mechanisms for transferring heat, otherwise some areas would become warmer and warmer and others colder and colder.

Heat energy is transferred by:

- **Radiation** in the form of short-wave radiation and long-wave radiation. On entering the atmosphere some radiation is **absorbed** by gases. Some is **scattered** where the direction of radiation is altered by interaction with (or bouncing off) gas molecules or aerosols (small particles) and water droplets.

- **Conduction** where the heat passes through a substance. Air is a poor conductor so conduction is not significant in the atmosphere but it is important on the surface.

- **Convection** circulates and spreads heat throughout liquids and gases and is the main means of atmospheric heat transfer. Energy is transferred as:

* **sensible heat,** by the rising and mixing of warmed air;
* **latent heat,** when water is converted into water vapour by evaporation, using heat, and then releasing it when the water is condensed back into the atmosphere.

Review task

Draw simple diagrams to summarise the three heat energy transfer methods. Keep the labelling simple and clear.

Global energy budget

This is the balance between incoming solar energy and outgoing radiant energy. The amount received at the edge of the atmosphere is roughly constant, so is called the **solar constant.** Overall total values show that incoming and outgoing radiation is balanced. This is what we call the **solar** or **global energy budget.** Figure 4.2 sums it up.

Energy exchanges

The balancing energy exchange needed to stop polar regions becoming colder and tropical regions becoming hotter is shown in Figure 4.3. The actual energy transfers needed to balance out polar and tropical regions are:

* Horizontal, through winds (80 per cent) and ocean currents (20 per cent). Depressions and anticyclones in mid-latitudes are particularly important in this transfer.

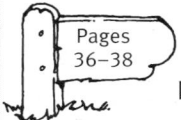

Pages 36–38

* **Vertical,** through convection, conduction and radiation as well as the transfer through latent heat after evaporation (and released as sensible heat by condensation).

Temperature patterns

Two groups of factors affect the spatial and seasonal patterns of temperatures and of temperatures ranges and anomalies:

1 Those to do with variations in solar energy received at the surface (see page 29):
   * latitude (including seasons and altitude of the sun)
   * altitude
   * land and sea distribution
2 Those to do with energy transfers (see page 29):
   * winds
   * ocean currents

Atmospheric circulation

The uneven distribution of energy or heat creates the movement of atmosphere and oceans and, in turn, redistributes the energy or heat from areas of surplus to areas of deficit. The movements create the familiar pattern of winds, pressure belts and ocean currents.

At the global scale the pattern of circulation of the atmosphere is described by **global circulation models.**

Review task

Terms to remember

Complete the definitions of the terms below. Refer back to the last few sections if necessary.

Insolation is the incoming short-wave radiant ....

Albedo measures the ....reflected from land, sea or cloud surface.

Scattering changes the direction ....

....

Reflection is measured by the ....

Radiation is the transfer of ....

....

Condensation is the process by which water is changed ....

Absorption is the process whereby, at particular wavelengths, energy is taken up ....

....

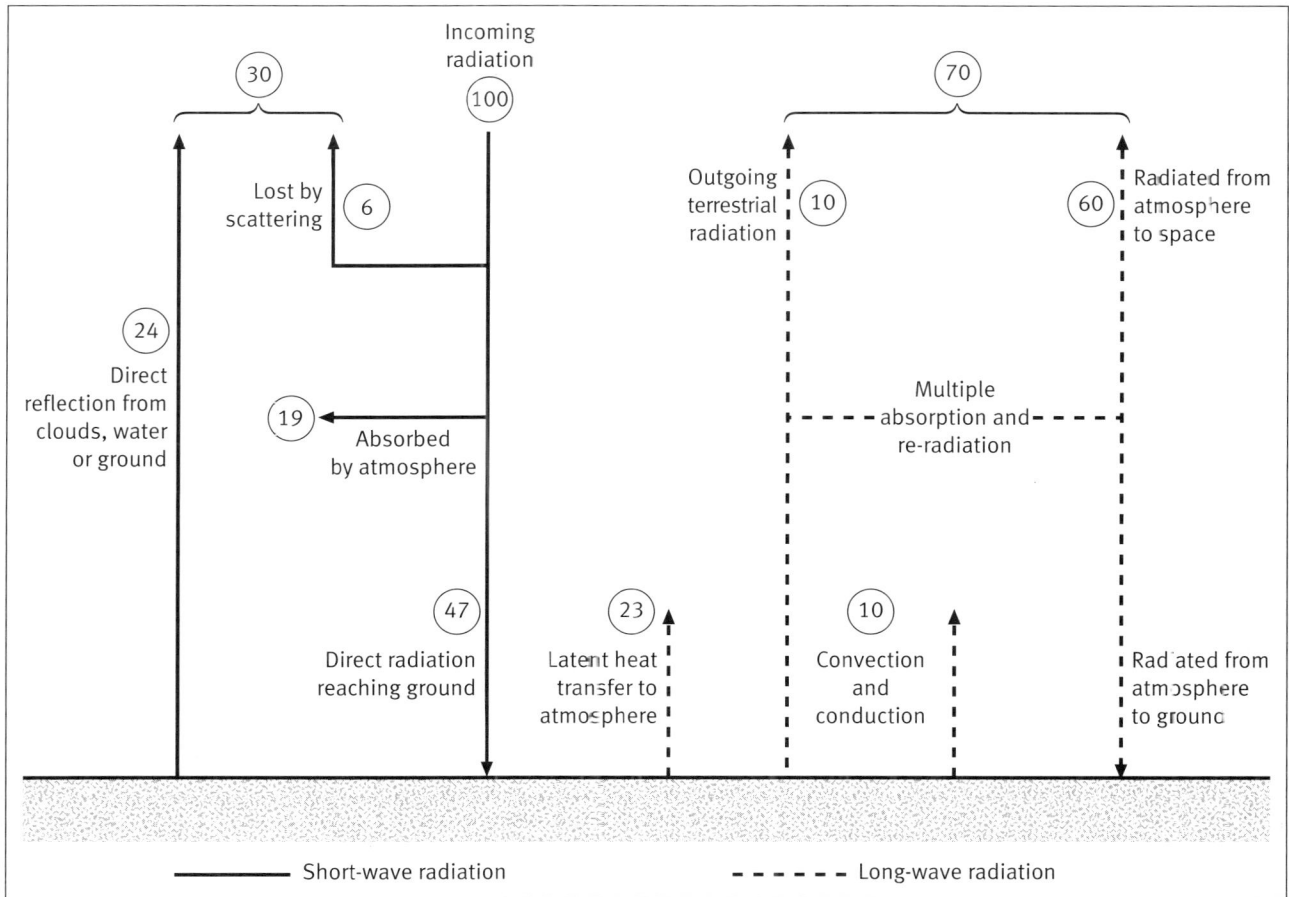

Figure 4.2 The balance of the solar or energy budget

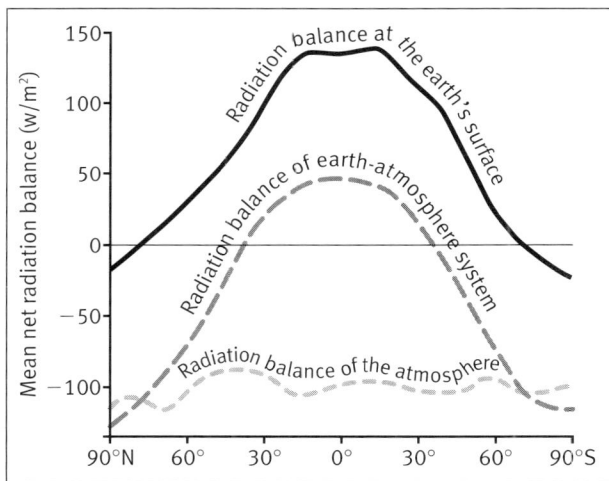

Figure 4.3 Average annual distribution of solar radiation by latitude

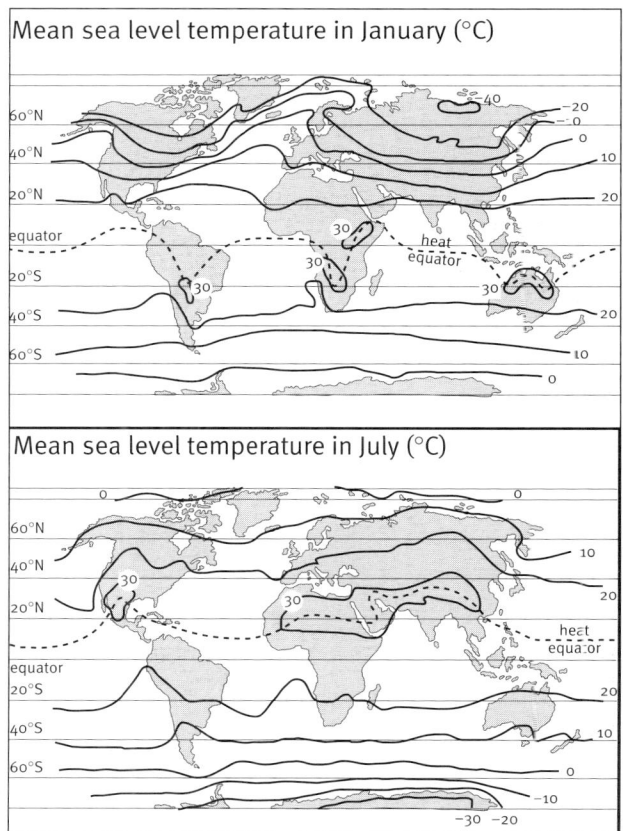

Figure 4.4 Temperature patterns

Review task

Use the maps of world mean temperature patterns in Figure 4.4 and refer back to the section on variations in solar energy received at the surface. Identify the main features of world-scale temperature patterns and their causes.

Models of global circulation

Figure 4.5 incorporates the new ideas of today's basic global circulation model.

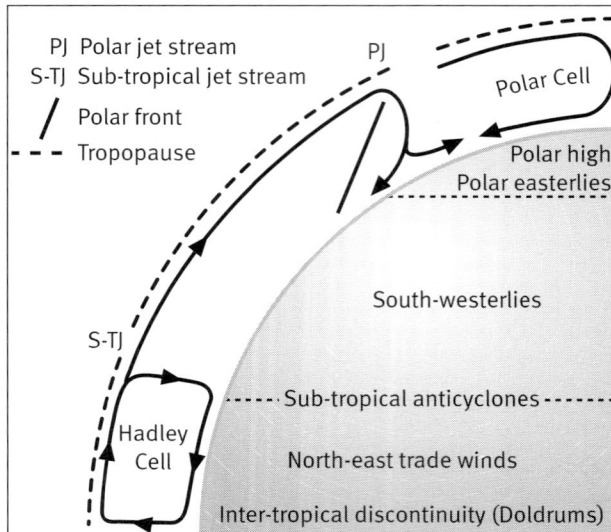

Figure 4.5 New tri-cellular global model

Global circulation model features

- **Inter-Tropical Convergence Zone (ITCZ)** or **Inter-Tropical Discontinuity (ITD)** which is a discontinuous belt of low pressure, marked by clouds and rain. Sometimes called the heat equator, it moves with the movement of the overhead sun.

- **Hadley cell,** whose surface characteristic is the trade wind belt. Its vertical structure is more complex than that of the simple convection cell, and in places it has an east–west movement in contrast to the north–south movement formerly thought to be common throughout the Hadley cell.

- **Sub-tropical jet,** marking a break in the tropopause, and forming the boundary between the Hadley cell and the Ferrel westerlies.

- **Ferrel westerlies,** also known as the circumpolar vortex, involve eastward moving depressions and anticyclones which have a major effect on the weather.

- **Polar jet,** marking another break in the tropopause.

- **Polar cell,** an area of persistent high pressure producing generally equatorward winds.

Mid-latitudes

The belt of meandering upper-air westerly winds in mid-latitudes are given the name **Rossby waves**. There are between two and five waves varying with the seasons. Their form is related to the thermal differences between tropical and polar regions, the rotation of the earth and possibly major relief barriers such as the Rockies. These upper air waves, together with jet streams, are the direct cause of surface depressions and anticyclones.

Jet streams are very high-speed narrow bands of wind which mark the boundaries between the major wind belts and are located where warm and cold air mix where the temperature and pressure gradients are greatest.

The Polar Front Jet Stream, which lies at about 35° to 55°, separates polar air from tropical air in waves of troughs and ridges. The troughs bring cold air which descends in a clockwise direction to give dry, stable conditions associated with areas of high pressure at the surface. The ridges take warm air up in an anticlockwise direction to give areas of unstable, low pressure at the surface (Figure 4.6). This shows that the jet streams are major guiding forces for the depressions and anticyclones which dominate the weather in mid-latitudes.

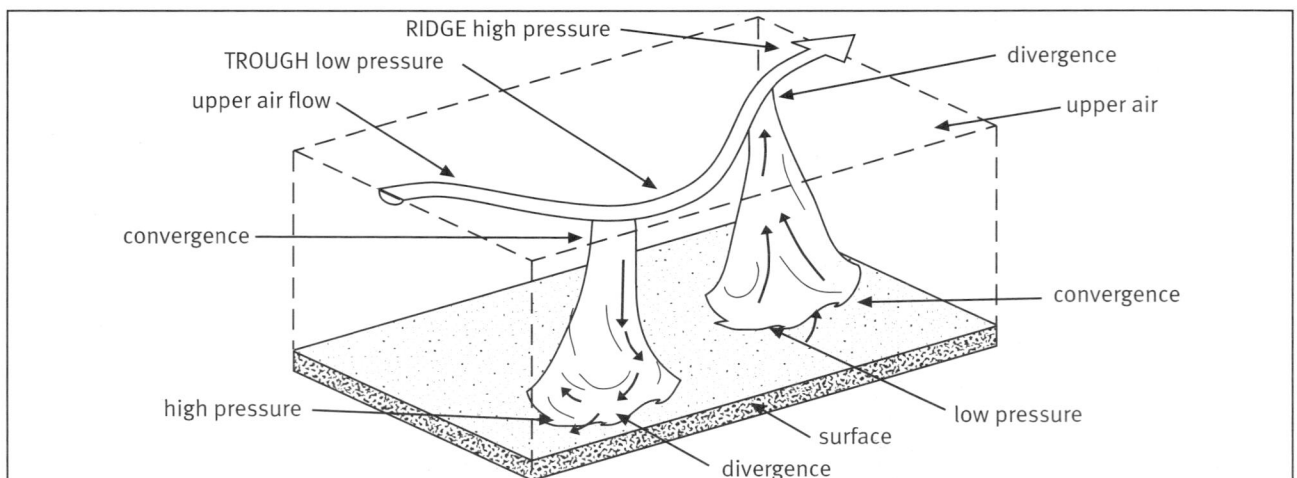

Figure 4.6 Jet streams, troughs and ridges and associated surface conditions

Pressure gradient force

steep pressure gradient – strong winds

gentle pressure gradient – light winds

988 992 996 1000 1004 1008 1012 1016

Geostrophic wind

High pressure

Geostrophic wind

1000 mb

c

996 mb

p

992 mb

988 mb

Low pressure

984 mb

980 mb

Coriolis force (c) cancels out the pressure gradient force (p) at an altitude above the surface where *friction* has no effect. Therefore, the geostrophic wind flows parallel to the isobars.

Coriolis force

North Pole

Rotation of the earth

Length of arrow is proportional to speed of rotation

(b)

(a)

equator

Air moving north (a) or south (b) carries the same momentum it starts with.

(a) is moving more quickly than the area of the earth it passes over and moves eastward faster than the ground underneath, that is, it is deflected eastwards or (in the northern hemsphere) to the right.

The opposite is the case with (b) which moves more slowly and is apparently deflected to the west (but still to the right) relative to the earth.

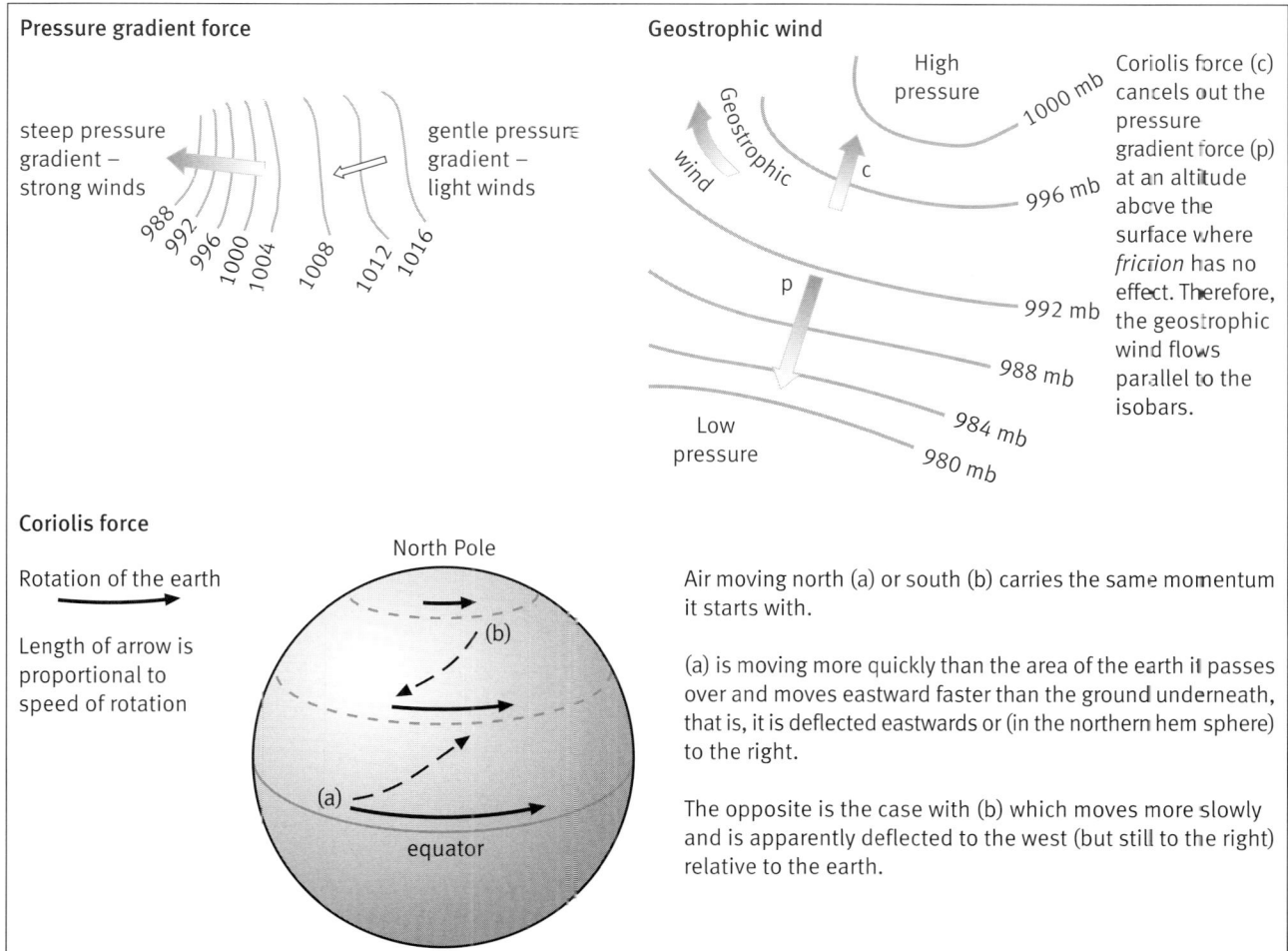

Figure 4.7 Atmospheric motion

The Sub-Tropical Jet Stream lies at about 25° to 30° and has much less pronounced waves and lower wind speeds, whereas the Easterly Equatorial Jet Stream is more seasonal and linked to the Indian subcontinent's monsoon.

Atmospheric motion

The horizontal movement of air is controlled by:

- **The pressure-gradient force.** Horizontal differences in pressure produce a pressure gradient with air moving away from areas of high pressure and towards areas of low pressure. The closer the isobars the greater the pressure gradient and the greater the wind speed (Figure 4.7).

- **Coriolis force.** This is an apparent deflecting force which causes a moving body on a rotating surface to follow a straight path in relation to outside objects but, when viewed in relation to the rotating surface itself, it moves to one side of its original line of movement (Figure 4.7).

- **Geostrophic wind.** At altitudes above the levels affected by surface friction winds blow more or less parallel to the isobars, that is, at right angles

to the pressure gradient. Under these circumstances the pressure gradient force and the Coriolis force are in balance (Figure 4.7).

- **Frictional forces.** Friction with the earth's surface reduces the wind speed which, in turn, reduces the deflection of movement due to the Coriolis force. The result is wind blowing obliquely across the isobars at 10° to 20° at the sea surface and 25° to 35° over land.

World precipitation patterns

Atmospheric circulation affects world precipitation patterns, and, at the same time, moisture in the atmosphere is important in the transfer of heat.

World precipitation patterns are the product of the general circulation of the atmosphere, seasonal variations and spatial differences. In the case of the last, the impact of mountain barriers and ocean currents is considerable. Mountain barriers produce orographic or relief rain with its associated rain-shadow effect on the leeward side of the barrier. Ocean currents influence precipitation over land by their nature: warm currents increase the temperature of adjacent air

Polar maritime air, cool, changeable, strong winds and showers.

Arctic maritime air, cold at all seasons, snow and sleet in winter especially in north and east.

Modified Polar maritime or Tropical maritime air, making the former warmer and less stable and cooling the latter. Depressions form along the Tm/Pm boundary bringing rainy, unsettled conditions.

Polar continental air brings very cold conditions in winter with snow or sleet in the east. Warm and settled for several days in summer.

Tropical maritime air, mild and damp in winter, warm in summer, thick cloud, associated with warm sectors of depressions so part of unsettled weather pattern.

Tropical continental air associated with high-pressure system from south. Stable conditions, warm and dry, but occasional thunderstorms.

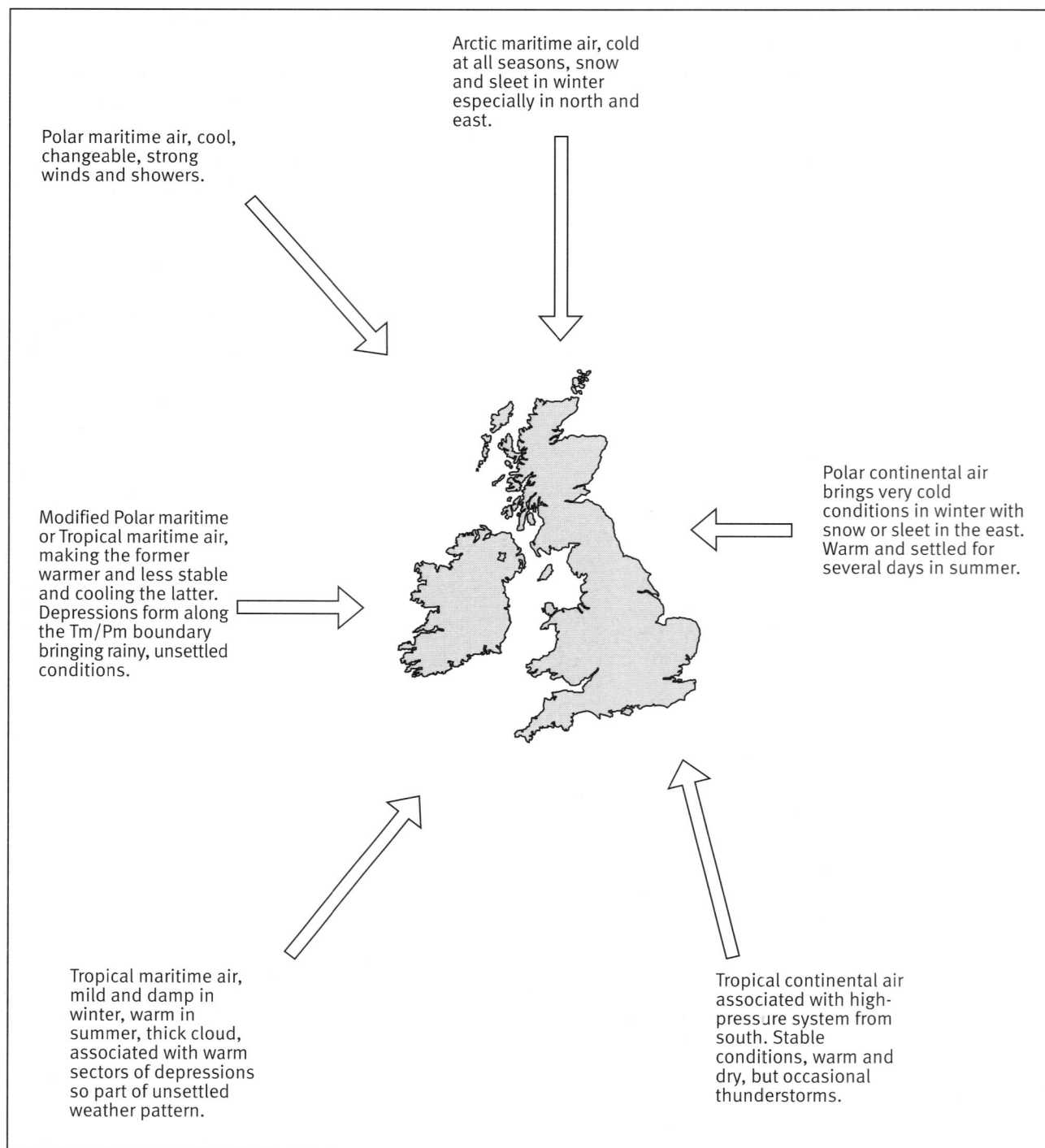

Figure 4.8 Air masses

masses and increase their ability to hold more moisture which provides increased precipitation over land; cold currents actually cause condensation and a loss of moisture from air moving across them.

Chapter 2 contains related material including the water cycle and water budgets.

Air masses and weather systems

Weather in mid-latitudes is extremely variable, due largely to the mixing of tropical and polar air

that takes place in the zone of westerly winds. The great thermal variations and the wave patterns of the westerlies, particularly the jet streams, are responsible for the depressions and anticyclones which are the chief characteristics of the weather.

Air masses

To talk about the mixing of polar and tropical air is an oversimplification. The weather is more specifically affected by the dominant **air mass** or masses. An air mass is 'a large body of air whose physical properties, especially temperature, moisture content and lapse rate, are more or less uniform for hundreds of kilometres'. Just how uniform an air mass is and in what way depends on:

- the characteristics of the source area
- the direction of movement of an air mass and the characteristics it acquires
- the age of the air mass

Review task

This is one to do over a period of time. Watch or listen to weather forecasts and reports. Note the different weather conditions that come from different directions. Build up your own simple map to show them, like in Figure 4.8, but add dates.

Precipitation	The range of forms in which condensed moisture occurs.
Condensation nuclei	Particles in the atmosphere on which condensation occurs.
Dew point	Temperature at which the air becomes saturated and water vapour condenses.
Evaporation	The change of state of water from a liquid to a gas (water vapour).
Humidity	The concentration of water vapour in the air.

Figure 4.9 Atmospheric moisture and definitions

Lapse rates, stability and instability

Air temperature at the surface reflects the characteristics of the prevailing air mass. Temperature decreases with height and this is called the **environmental lapse rate (ELR)**, which is roughly 6.5°C per 1000 m. Within this overall environment, pockets of warmer air rise and gradually expand and cool until they are in equilibrium with the surrounding air, the reverse applying to pockets of colder air which fall and gain temperature as they are compressed. On the broader scale, air may be forced to rise by mountain barriers or even by a mass of colder air. This expansion and cooling is known as **adiabatic cooling**.

Rising pockets of air cool at the **adiabatic lapse rate (ALR)**. If the cooling does not result in the air becoming

saturated it is known as the **dry adiabatic lapse rate (DALR)**, which is constant at 9.8°C per 1000 m. If saturation level is reached, condensation occurs, releasing heat as water vapour turns into water, compensating in part for the cooling with height. As a result, the **saturated adiabatic lapse rate (SALR)** is less than the DALR; it varies between about 6°C per 1000 m near the ground, where is there more moisture in the air, and 9°C per 1000 m at the top of the troposphere where there is very little moisture in the air.

Review task

Explain how the difference in the amount of moisture in the air at different altitudes (or in different air masses) affects the saturated adiabatic lapse rate.

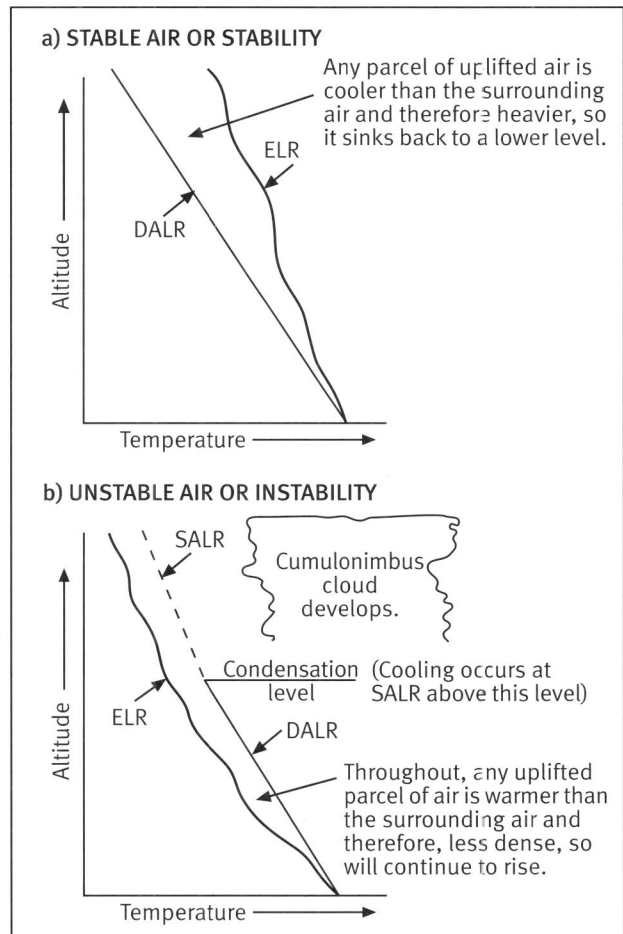

Figure 4.10 Stable and unstable air

Stability

Stable air is cooler and more dense than the surrounding air, in which case it has a tendency to sink. When it is warmer and less dense than the surrounding air, it has a tendency to rise making it **unstable** (Figure 4.10).

Conditional instability occurs frequently in the British Isles. As Figure 4.11 shows, the air is stable at low levels and stays stable while the lapse rate of any air forced to rise remains at the dry rate. If cooling with altitude results in the air becoming saturated then cooling continues at the saturated adiabatic lapse rate and the air becomes unstable. In the first case, air forced to rise remains cooler and denser than surrounding air and sinks back. In the second case, the air becomes warmer and less dense than the surrounding air and tends to rise, so the instability is *conditional* upon the air being forced to rise to the point where the air becomes saturated and therefore unstable. The cause could be a mountain range, a mass of cold air or a strong convection current.

Mid-latitude weather systems

Depressions

The boundaries between different air masses are marked by a distinct change in air mass properties, like temperature, humidity, speed and direction of movement. This zone of change is called a **front** and is a common feature on weather maps of the British Isles. **Depressions** are surface pressure systems which develop along the boundary between colder polar air and warmer tropical air, which is known as the **polar front**. They are areas of converging air within which the air spirals in an anticlockwise direction. Depressions are directly linked to the upper air flows as shown earlier in Figure 4.6. In Figure 4.12 a generalised model of a depression is shown and again you can see the link with upper air flows in the form of the jet stream. Notice that the front is named after the characteristics of the air *behind* the front.

> ### Review task
>
> **Use the information in Figure 4.12 and your own notes and textbooks to complete the table (Figure 4.13) to describe the weather changes as a depression passes over.**

Depression models

The **polar front model** of depressions described above was developed by Norwegian meteorologists in the 1920s, and despite recent ideas, is still very useful for examining the development and decline of a depression and the associated weather.

More recent models include the **conveyor-belt model** (Figure 4.14). This model explains importance of depressions in transferring heat from tropical to polar

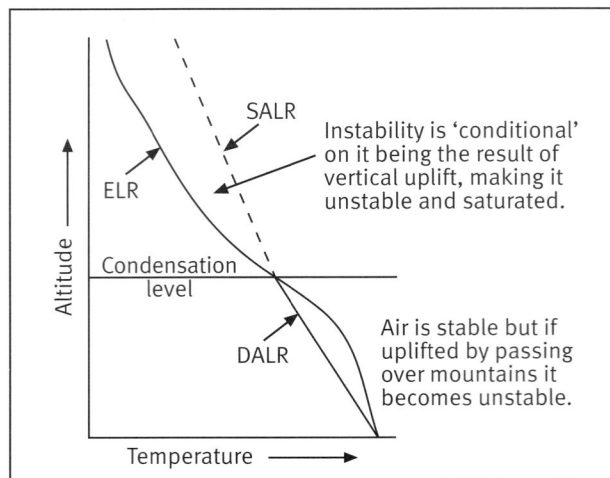

Figure 4.11 Conditional instability

areas and from lower to higher altitudes. Another development has been the recognition of different types of front. **Ana fronts** occur where the air rises in relation to the front, creating more unstable conditions and heavy rainfall, whereas **kata fronts** occur where the air is sinking in relation to the fronts, producing more stable conditions and only light rain or drizzle.

Anticyclones

Anticyclones are enclosed areas of high pressure, with the term **ridge** being used for the small areas of high pressure between depressions. Ridges of high pressure, like depressions, are linked to upper air flow patterns (see Figure 4.6). Sometimes the ridges develop to become enclosed anticyclones, in which case they form **blocking highs**. They block the paths of depressions and direct them further north or south than their normal path. This produces unusually long periods of settled weather in the area affected by the anticyclone and more depressions than normal in the areas to the north and south.

The other type of anticyclone is the large and permanent type of high-pressure area which is a part of the global pressure system pattern and is linked either to continental high-pressure systems or to the tropical high-pressure areas.

Anticyclones are areas of diverging air so winds spiral outwards in a clockwise direction in the northern hemisphere, and as the air is descending it warms adiabatically. This results in a decrease in relative humidity. Within the anticyclone, winds are generally light to calm due to the very small pressure gradient, although the edge of the anticyclone often has stronger winds. Conditions are generally stable and unchanging but do vary with the time of year.

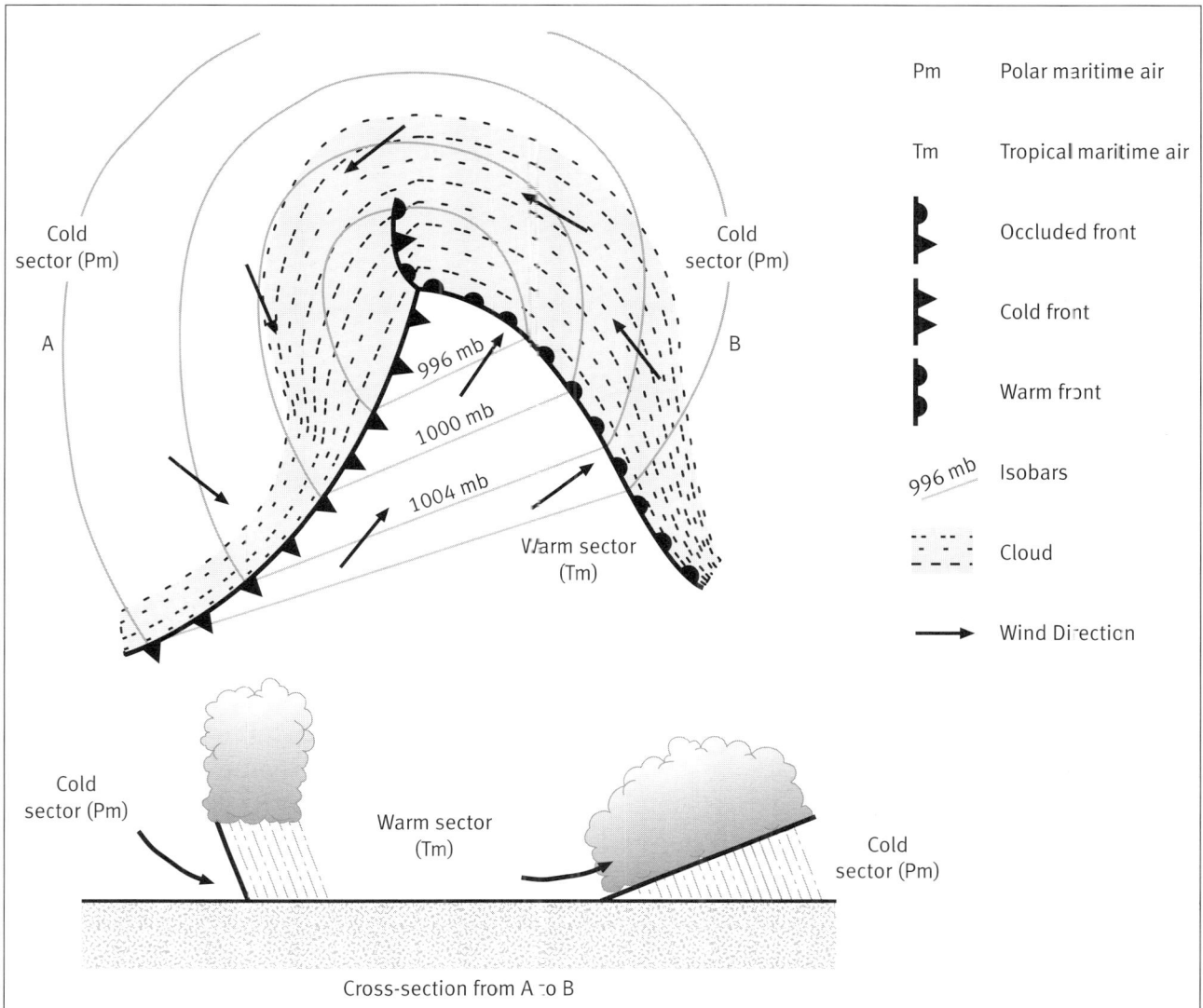

| | | Pm | Polar maritime air |
| | | Tm | Tropical maritime air |

Figure 4.12 **Weather and depression**

Figure 4.13 **Table for review task**

	Approaching depression	**Warm front**	**Warm sector**	**Cold front**	**Cold sector**
Pressure					
Wind					
Temperature					
Humidity					
Cloud cover					
Rainfall					
Visibility					

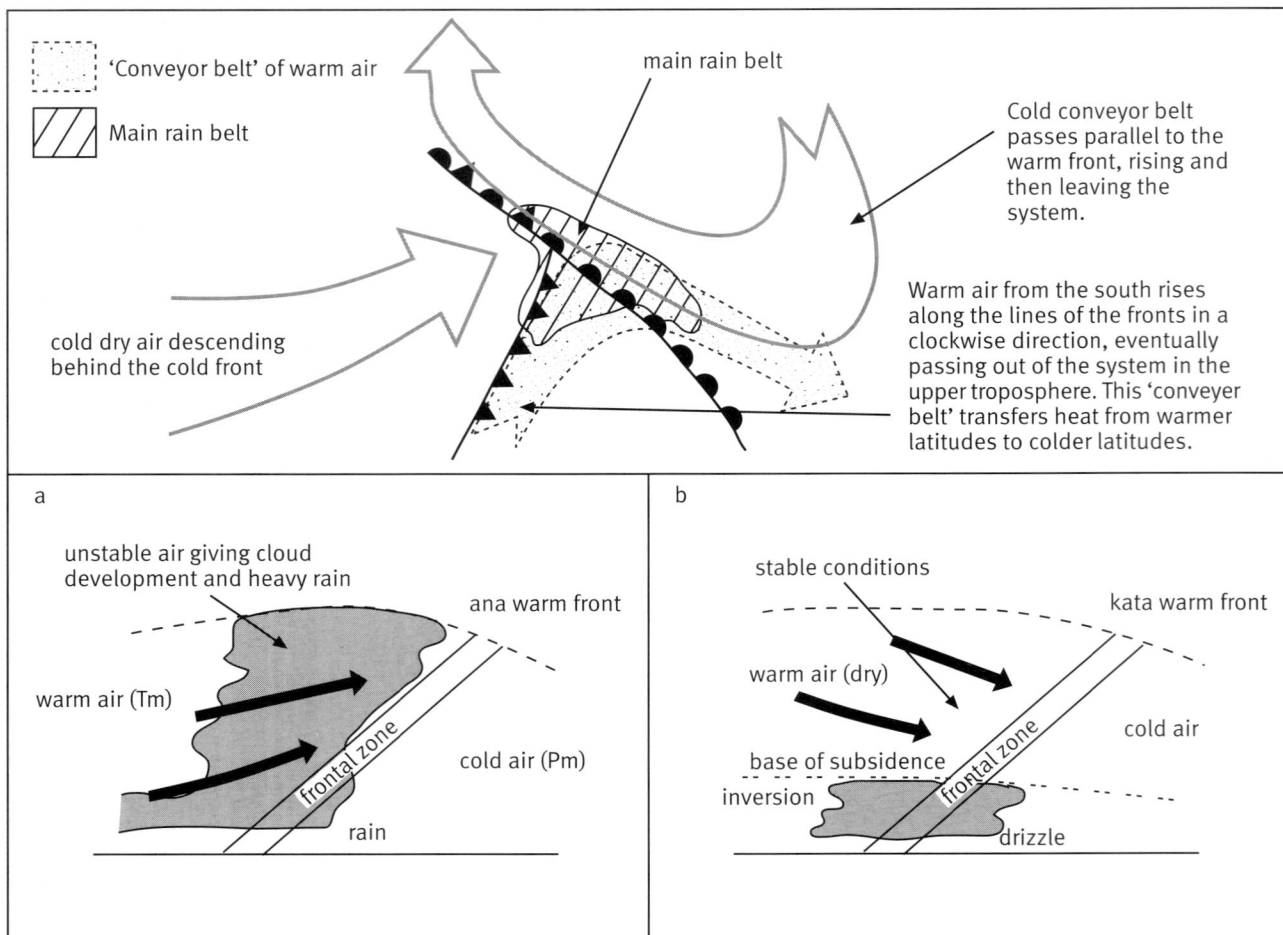

Figure 4.14 a) The conveyor-belt model of a depression, b) Ana-fronts and kata-fronts

Review task

1   Use what you know about air-mass origins, pressure systems, heat gain and loss and seasons to explain and link together the statements in Figure 4.15. Treat each column separately.

2   Do the same as you did with the review task on page 36 and keep an eye on weather forecasts and reports and note the weather conditions with particular anticyclones.

Winter	Summer
Often cloudy	Clear skies
Great heat loss when skies clear at night	Sunny High temperatures
Short days result in little heating during day	Low relative humidity Cooler nights
Low temperatures especially after clear night	Morning dew, radiation fogs or mist
Frost, black ice and freezing fog	Land and sea breezes along coasts

Figure 4.15 Seasonal anticyclonic weather variations

Local weather and climate

Local and regional winds

Lack of friction is one reason why **high-altitude winds** are much stronger than surface winds. However, a mountain barrier produces high-speed winds on or near the crest line due to vertical shrinking.

Mountain barriers also produce the **Föhn** effect when a warm, dry wind develops on the lee side of a mountain barrier as stable air is forced over it. As the air ascends over the barrier it cools and, if it cools to dew point, precipitation results and further cooling is at the saturated adiabatic lapse rate. The air then warms at the higher dry adiabatic lapse rate as it descends, giving rising temperatures and lower relative and absolute humidity (Figure 4.16).

Detailed relief affects wind speed and direction. Mountain barriers funnel the wind through valleys and

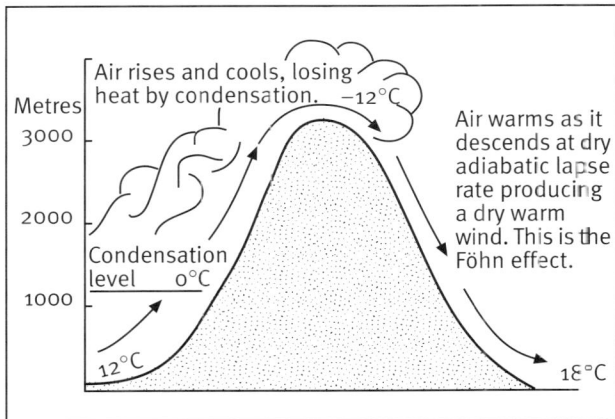

Figure 4.16 Föhn winds

across passes, producing strong winds as the air is forced to constrict and accelerate. The **mistral** wind shows the combined effect of orientation of valleys and the location of areas of high and low pressure. It occurs when there is high pressure to the north of the Alps and low pressure over the Mediterranean, forcing air flow to accelerate along the Rhône valley, producing a cold, strong wind.

Within mountain areas there are local mountain and valley winds which show a diurnal pattern. These are known as **anabatic** (upslope) and **katabatic** (downslope) winds. Anabatic winds occur in warm sunny conditions on the sides of valleys and, as a result of the heating of the valley sides, air flows up the valley sides. At the same time a general **valley wind** develops flowing along and up the valley. The situation is reversed at night as cooler, denser air flows downslope as a katabatic wind and, as the reverse of the valley wind, this dense air may flow along and down the valley as a **mountain wind** (Figure 4.17).

Katabatic air sometimes causes frost pockets or frost hollows as cold, dense air accumulates in valley

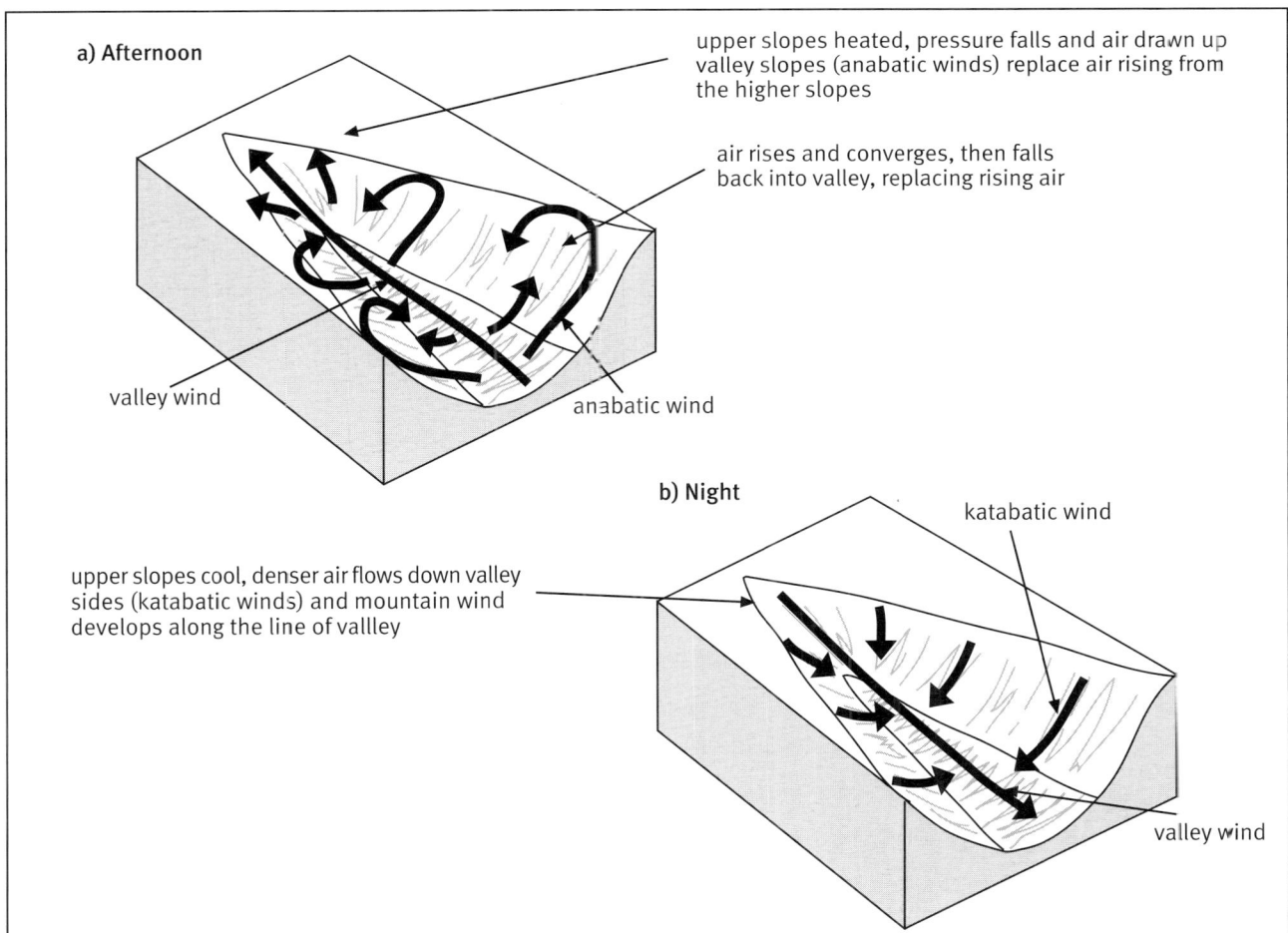

Figure 4.17 Mountain and valley winds

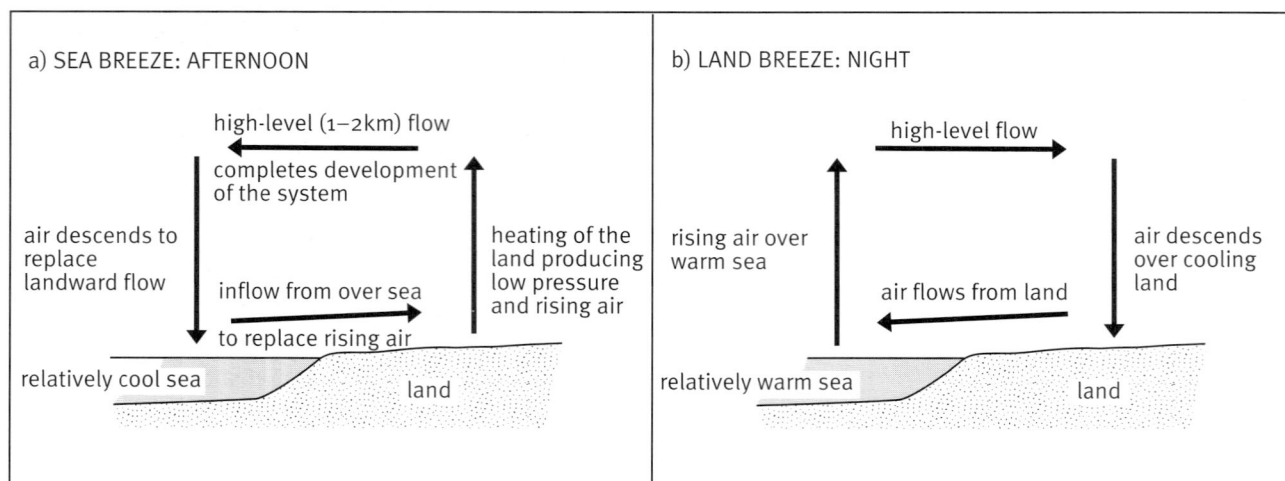

Figure 4.18 Land and sea breezes

bottoms. However, this is usually the result of nocturnal cooling giving a temperature inversion which often results in a **radiation fog** forming in the valley bottom as moisture in the air condenses. The fog disperses as temperatures rise during the day, but if the conditions persist over some time in winter then the fog may well not disperse as the inversion becomes intensified.

Land and sea breezes are generated by temperature-induced air movements in a somewhat similar way to the mountain winds (Figure 4.18).

Urban climate

Urban areas modify the local climate considerably so they contrast with rural areas in temperature, winds, visibility and precipitation.

Temperature – the urban heat island

Urban areas are generally warmer than surrounding rural areas, the effect being most noticeable after dark under calm anticyclonic conditions, when cooling in rural areas is much greater than in urban areas. Contributing factors to the generation of the heat island effect, which may give minimum temperatures 5° to 6°C more than in neighbouring rural areas are:

- heat loss from buildings which store heat from the sun or from fuel use
- size of cities
- density of buildings
- burning of fossil fuels
- heights of buildings trapping heat
- different reflective surfaces giving a complex pattern of re-reflection and trapping of heat

- rapid removal of surface water so less heat is used in evaporation
- lower wind speeds reducing heat loss
- temperature inversions trapping heat

The edge of the city is marked by a sharp difference in temperature, but windier conditions result in the heat island effect being felt to certain extent in downwind rural areas.

Precipitation

Cloud cover is greater over urban areas and the same is true of precipitation, both being about 10 per cent greater than rural areas. Both can be explained by convection currents created by the heat island effect and the greater concentration of condensation nuclei in the form of aerosol particles, which help to trigger thunderstorms.

Wind

Overall wind speeds are reduced in urban areas due to the roughness of the surface and increased friction, but tall buildings arranged in straight lines create a 'canyon' effect and funnel winds in the way they are funnelled along valleys.

Pollution

The atmosphere's composition is changed by aerosols (particles of salt, dust, organic matter and smoke), industrial gases and by car exhaust fumes which in combination bring about smogs, fogs and chemical changes as well as affecting incoming and outgoing radiation. The distribution of pollution is clearly marked around urban areas with a pollution dome forming under the temperature inversion that often forms. Light winds will displace the pollution so that plumes of pollution extend into adjacent rural areas downwind (see Figure 8.15).

Forest climates

Forests have a different kind of surface and modify the local climate. Their special climatic features vary according to:

- the size of forest
- the height of trees
- the density of trees
- the nature of the trees.

The effects of forests are:

- a wide range of albedos
- most solar energy is trapped at treetop level
- in dense forest little insolation reaches the forest floor
- wind speeds are reduced
- humidity levels are modified
- evapotranspiration may stimulate local rainfall
- small diurnal temperature range

Artificial modifications of local climates use shelter belts and wind breaks, the effects of which extend about 18 times the barrier's own height; and can be increased by introducing a series of barriers at a point before the shelter effect diminishes to the unprotected level.

The shelter belts also filters dust and fog droplets and cut the risk of frost damage on gentle slopes by reducing horizontal air movement.

Review task

1 Make a list of as many local-scale climate and weather modifications as you can. Then check through the sections above and your own notes to see what else can be added. Where possible name an example. Summarise it all in a Mind Map. Note that you can organise your Mind Maps in different ways: by type of location, such as city or forest; by feature, such as temperature or wind; by case study.

2 Brainstorm the entire chapter. This is an ideal task to do in a pair or small group. Write in the topics covered in the chapter under the appropriate column in the table below, together with as many of the important ideas as you can remember. Case studies can be included as well. Once you are satisfied that you have included as much as you can, you can check through the chapter to see what you have missed.

The atmospheric system	Air masses and weather systems	Local weather and climate

Now reorganise the contents of your table as a Mind Map; this gives you the flexibility to add other details when you review the topic.

PREVIEW

What you need to know:

- **How ecosystems function**
- **Vegetation succession; the polyclimax theory**
- **The characteristics of soils**
- **Soil forming processes**
- **Human impacts on ecosystems**
- **Contrasting examples of ecosystems**

Also, you will need to be able to:

- **analyse diagrams of ecosystem energy flows, chemical cycles and nutrient cycles;**
- **interpret diagrams of vegetation succession (seres);**
- **understand diagrams of soil profiles and catenas;**
- **relate your knowledge and understanding of vegetation and soils to examples of ecosystems in specific places.**

An **ecosystem** is a natural system in which plants and animals are linked to each other and to their environment. **Biogeography** is the study of spatial patterns of soils, vegetation and whole ecosystems. An understanding of **ecology** (the study of how organisms interact with each other and their environment) is a foundation of biogeography. The **environment** describes the conditions in which organisms live; it includes non-living or **abiotic** elements such as air, temperature, light or rocks and living or **biotic** elements such as other plants and animals. Ecosystems are maintained through energy flows and the cycling of materials. Within ecosystems relationships exist in a hierarchy of scales, from **micro-habitats** to **biomes** (Figure 5.1).

BIOME	Large-scale ecological unit, with its own characteristic flora and fauna: e.g. tropical rainforest
ZONE	Major unit within a biome: e.g. rainforest lower layer
HABITAT	Location within a specific localised environment and a related community: e.g. an area of swamp within a rainforest
MICRO-HABITAT	Very small specific location: e.g. under a fallen branch

Figure 5.1 Scales of ecosystems

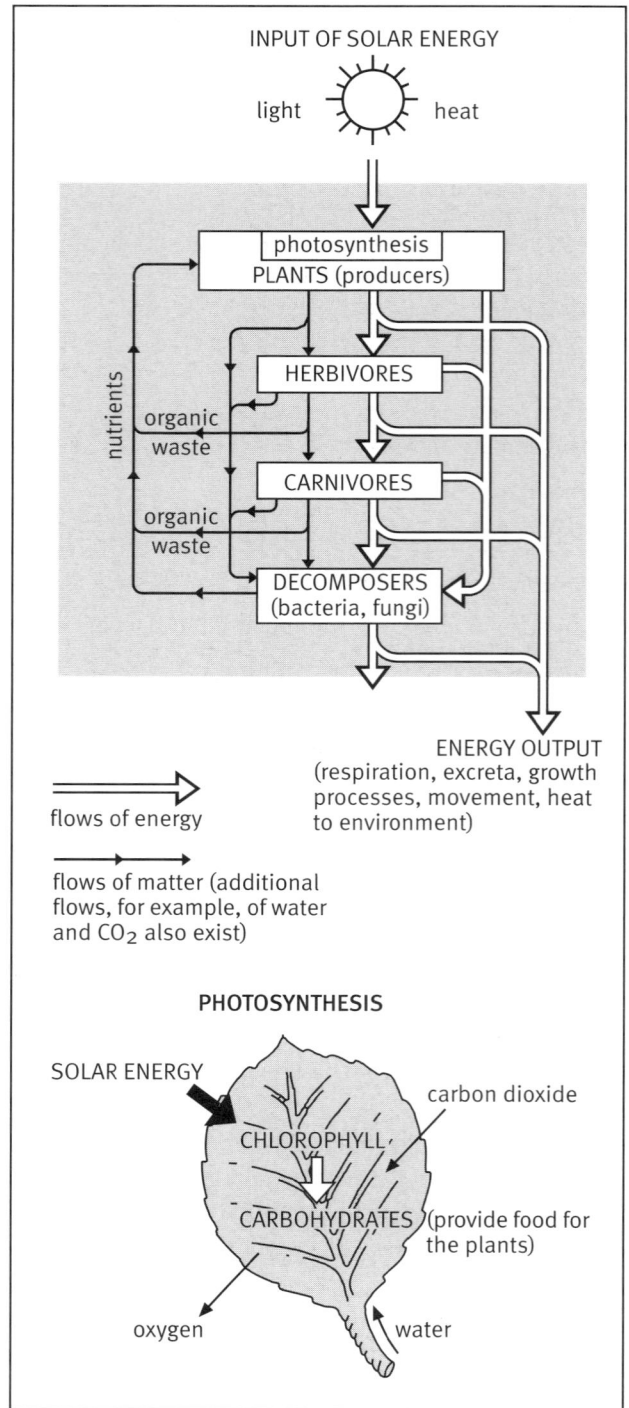

Figure 5.2 Energy flows in the biosphere

How ecosystems function

Energy flows

Figure 5.2 shows the flows of energy in the biosphere. These flows operate in an **open system,** as the input is solar energy (from outside the system) and energy is also returned to space. The key feature shown in Figure 5.2 is that solar energy in the form of light is trapped by chlorophyll in plants and so enables photosynthesis to take place. Carbon dioxide, water and heat energy from the sun are also required for photosynthesis. Figure 5.2 also shows that flows of matter accompany flows of energy; such material cycling occurs in **closed systems**.

Trophic levels

The green plants that produce their food (in the form of carbohydrates) by photosynthesis are known as **autotrophs** (in effect, self-feeders). They form the base of **food chains** and **trophic levels** (Figure 5.3). Each level in the chain feeds on and obtains energy from the level preceding it. These levels are occupied by consumers (or **heterotrophs**). Within trophic levels energy is lost through respiration, excreta and decay following death (X on diagram). At the points of energy transfer, energy is lost through heat (Y on diagram). This progressive loss of energy imposes limits on the quantity of living matter (**biomass**) which can exist at each trophic level; the result is a **trophic pyramid** (as Figure 5.3 shows) with a large biomass of plants or autotrophs and a very small biomass at the highest trophic level. In most ecosystems, there is a variety of plants and animals at each level so a complex **food web** exists rather than a simple food chain.

Productivity

The quantity of living matter or biomass in an ecosystem can be measured as a weight (for example, in kg/m²) or, because it represents stored energy, in calories. Over time, photosynthesis results in the creation of organic matter and thus the growth of the biomass. The growth of the biomass per unit of time is the gross **primary productivity** of an ecosystem. More precisely, **net primary productivity (NPP)** is a measure of added biomass (or energy fixed in photosynthesis) minus energy losses, such as through respiration (refer again to Figure 5.2). As the table in Figure 5.4 shows, NPP may be measured per unit area (g/m²/year) or on a world scale in tonnes per year. Figure 5.4 shows the very high productivity of the tropical rainforest compared with other ecosystems.

Material cycling

Figure 5.2 shows that flows of matter accompany flows of energy within ecosystems. Plants absorb chemical elements as soluble salts from the soil or as gases from the atmosphere. These elements are lost up the various trophic levels. As organisms at various levels die and decompose, the elements are returned to the soil or atmosphere. One of the most important chemical cycles is the carbon cycle, based on carbon dioxide gas. Thus, carbon dioxide passes from the atmosphere to the earth through processes such as photosynthesis by plants and from the earth to the atmosphere through, for example, respiration by organisms and the burning of forests and fossil fuels. Knowledge of this cycle is also important for understanding the process of global warming.

Nutrient cycles

Any atom, ion or molecule that an organism needs to live or grow is called a **nutrient**. P F Gersmehl

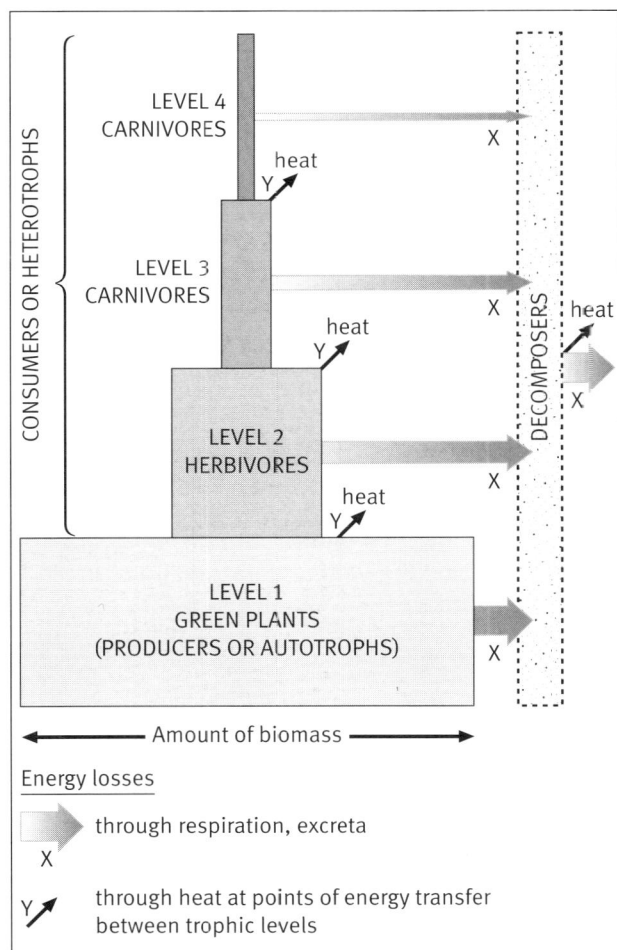

Figure 5.3 Food chain and trophic levels

Ecosystem	Net primary productivity (g/m²/year)	World net primary production (billion tonnes/year)
Tropical rainforest	2200	37.4
Desert scrub	90	1.6
Temperate deciduous forest	1200	8.4
Coniferous (boreal) forest	800	9.6

Figure 5.4 Productivity of different ecosystems

developed a model of **nutrient cycles**. The elements of nutrient cycles are three **stores** (biomass, litter and soil) and nutrient **transfers** between the stores (Figure 5.5). The cycling of nutrients is made up of parts of the various chemical cycles (carbon cycle, nitrogen cycle, etc.). Figure 5.5 shows nutrient cycles as they exist in different biomes. The size of the circles representing stores is proportional to the amount of nutrients stored. The width of the transfer arrows are proportional to the flow of nutrients.

Vegetation succession

The vegetation of an area will have developed in a number of stages over time from a bare surface. The first plants to colonise an area are a **pioneer community**. Each successive stage of vegetation, represented by a distinctive association of species, is known as a **sere**. A sequence of successive stages or seres is a **succession** (or **prisere**) and leads to a **climax community**. A climax community of plants that exists in a state of equilibrium with the climate of an area is called a **climatic climax vegetation**. On a global scale, climate is a major factor in determining large-scale groupings of plants (biomes) such as tropical rainforest or northern coniferous forest. Figure 5.6

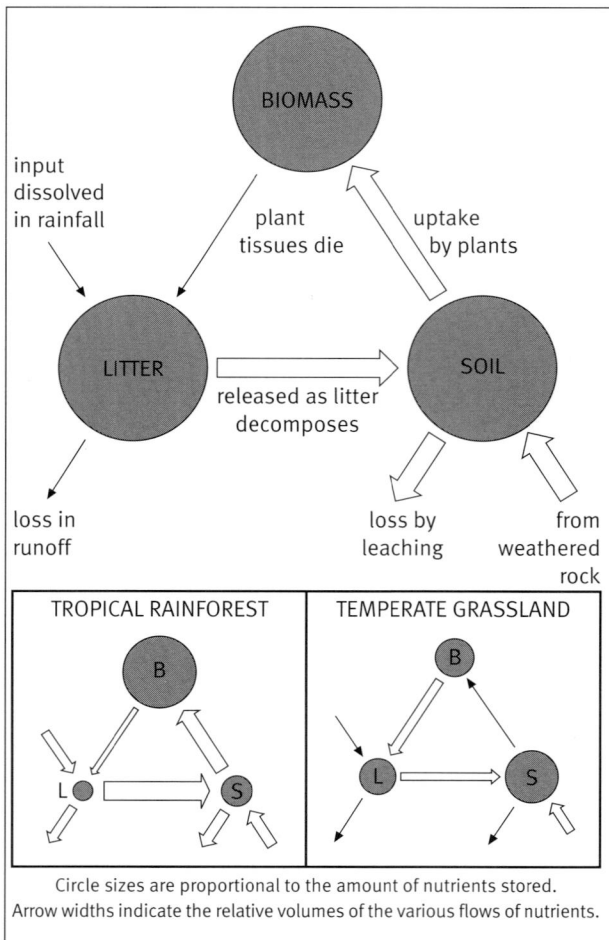

Circle sizes are proportional to the amount of nutrients stored.
Arrow widths indicate the relative volumes of the various flows of nutrients.

Figure 5.5 Ecosystem nutrient cycles

shows a model of succession developing towards the climatic climax of temperate broad-leaved deciduous woodland. Each sere is described by referring to the larger species in the community – the **dominant species**. This succession is developing from the infilling of a fresh water lake and is known as a **hydrosere**. Successions developing from other surfaces include a **lithosere** (on bare rock), a **psammosere** (on a sandy surface such as sand dunes) and a **halosere** (in a salt-water environment). Figure 6.10 in Chapter 6 describes a psammosere and a halosere.

Page 61

Few physical environments remain stable for sufficiently long for the climatic climax to be reached and maintained. There may be natural interruptions to the 'normal' or **primary succession**. An **edaphic control** (resulting from soils, usually because of limited fertility) or a **biotic control** (resulting from, for example, the grazing of wild animals) produces a **subclimax** vegetation. A **plagioclimax** occurs where a climatic climax community has been altered by human action such as burning vegetation or grazing domesticated animals. If the natural or human control on vegetation succession is removed, a **secondary succession** occurs as the vegetation develops towards the climax. Figure 5.7 summarises these successions, controls and climaxes diagrammatically (the **polyclimax theory**).

Review task

1 **Understanding ecosystems requires a precise understanding of many technical terms. Read through the preceding part of this chapter and, as you do so, construct a Mind Map or a check list to test yourself and then, with definitions added, use it as a late revision aid.**

2 **Figures 5.2, 5.3, and 5.5 include many flows. For each diagram follow through all the flows, checking your understanding of them and the items they connect. In Figure 5.5 make sure that you understand the varying volumes of flows shown.**

Soils

Rocks at the earth's surface are weathered to form a layer of broken material (the **regolith**). The weathered rock of the regolith becomes soil with the addition of water, air and organic matter (both live organisms and decayed organic matter or **humus**). Matter in the soil exists in solid, liquid and gaseous states.

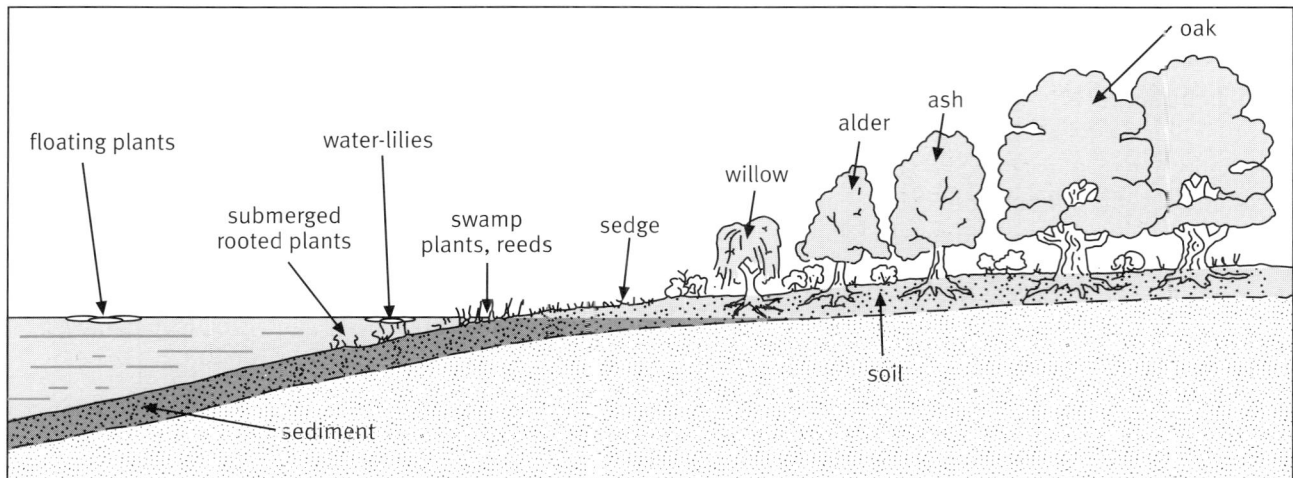

Figure 5.6 Primary succession from a lake to temperate deciduous woodland (a hydrosere)

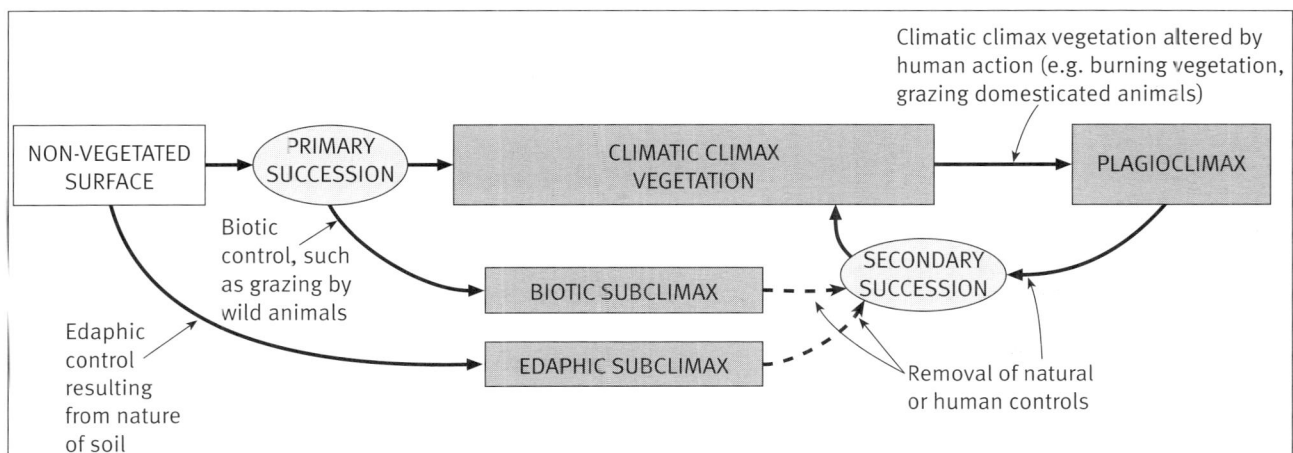

Figure 5.7 Monoclimax and polyclimax theories

Soil characteristics

Physical, chemical and biological processes interact between the four major components of soil (minerals from rocks, water, air and organic matter) and give rise to several soil characteristics or properties.

- **Inorganic (mineral) matter** originates from the weathering of rock (the soil's **parent material**). **Primary minerals**, such as quartz, resist chemical weathering and will stay present in the soil, for example as grains of sand. **Secondary minerals** have been altered by processes of chemical weathering such as oxidation and carbonation. They exist in the soil in the form of complex **clay minerals** or **sesquioxides** (the oxides of two common primary minerals, iron and aluminium).

- **Soil texture** refers to the size of solid particles in the soil and descriptions of soil texture are based on the proportion of sand, silt and clay particles. Figure 5.8a shows how particles are separated into these groups and how soils are classified in terms of texture by means of a triangular graph. Texture

controls the size and spacing of soil pores and therefore affects soil water content and flow. It also controls the retention of nutrients in the soil, as nutrients are absorbed into clay particles.

- **Soil structure** refers to the way in which particles combine to form a variety of different shapes known as **peds** (Figure 5.8b). Soil structure affects a soil's resistance to erosion, the movement of water and air, and the ease with which rocts may penetrate soils.

- **Soil moisture** is held within a soil or moves through it in a number of ways:
  - ✳ as **hygroscopic water,** a thin film held by tension around soil particles and unavailable to plants;
  - ✳ as **capillary water** which forms a film around the hygroscopic water. The tension with which it is held is less and so it is available to plant roots or can be drawn away through evaporation;
  - ✳ as **gravitational water** (or **free water**). This water has low surface tension and can drain away through gravity. Once free water is lost, the

remaining moisture that the soil can hold is its **field capacity**.

- **Organic matter** is largely derived from leaves and other plant remains. Earthworms and various insects bury, eat, excrete and so disperse leaf litter. Bacteria and fungi break down organic material to form humus. Humus is black in colour and can exist in solid, gel or liquid forms. Bacteria require oxygen, so if a soil is waterlogged, humus formation is checked and organic matter will remain unaltered (for example, as peat). If humus is well mixed with minerals in a crumbly layer it is known as **mull** (mild humus). If cold temperatures and excess water result in slow decomposition, an acidic humus develops, called **mor** (raw humus).

- **Soil nutrients** are chemical evidence in the soil which are essential for plant growth. Humus is the major source of nutrients and it combines with clay particles to form the **clay-humus complex**. Other sources are minerals from the parent rock, rainwater and artificially added fertiliser. The particles of clay and humus have negatively

charged ions on their surface (anions) which attract charged minerals (cations) as shown in Figure 5.9. The process of **cation exchange** takes place between the clay-humus particles and plant roots (also shown in Figure 5.9). Thus nutrients (Ca, Mg, K, Na) are supplied to plant roots while hydrogen (H+ cations) attached to the plant roots is released. A soil with much humus will have a high **cation exchange capacity** (CEC) and so will be more fertile than a very sandy soil (sands have a lower ability to retain nutrients) with a low CEC.

- **Soil acidity** is measured by the concentration of hydrogen cations in the soil and is indicated by the pH symbol. The pH scale is logarithmic and an acid soil is represented by a smaller pH than a basic or alkaline soil. Thus pH5 represents acidity ten times that of pH6. A pH value of 7 represents a neutral soil. A podsol is an example of an acid soil, while desert soils are alkaline.

Soil-forming processes

Many processes form soil and create structures within it. The presence or absence and relative importance of different processes contribute to the nature of the **soil profile**, made up of distinctive layers or **horizons**. Figure 5.10 locates the various processes within the soil and relates them to the profiles of two soils which are common in Britain: podsols and brown earths. There are several important processes:

- **Humification** is the decomposition of organic matter to form humus.

- **Chelation** and **cheluviation**. As organic matter decomposes, the organic acids attack clays, releasing iron and aluminium (chelation). The removal downwards of dissolved iron and aluminium sesquioxides is cheluviation.

- **Leaching** is the removal of soluble material (bases such as Ca and Mg) in solution from the upper A horizon making it more acid (as the material removed is replaced by H ions). The soluble material is deposited in the underlying B horizon. Leaching occurs where precipitation exceeds evapotranspiration and where a soil is well drained.

- **Podsolisation** is a more extreme form of leaching and involves percolating rainwater becoming sufficiently acidic (pH of below 4.5) to remove sesquioxides (of iron and aluminium, especially). In the B horizon, iron deposits can form a hardpan.

- **Gleying** occurs when a soil becomes waterlogged under conditions of very poor drainage. Pore spaces fill with stagnant, deoxygenised water; such an

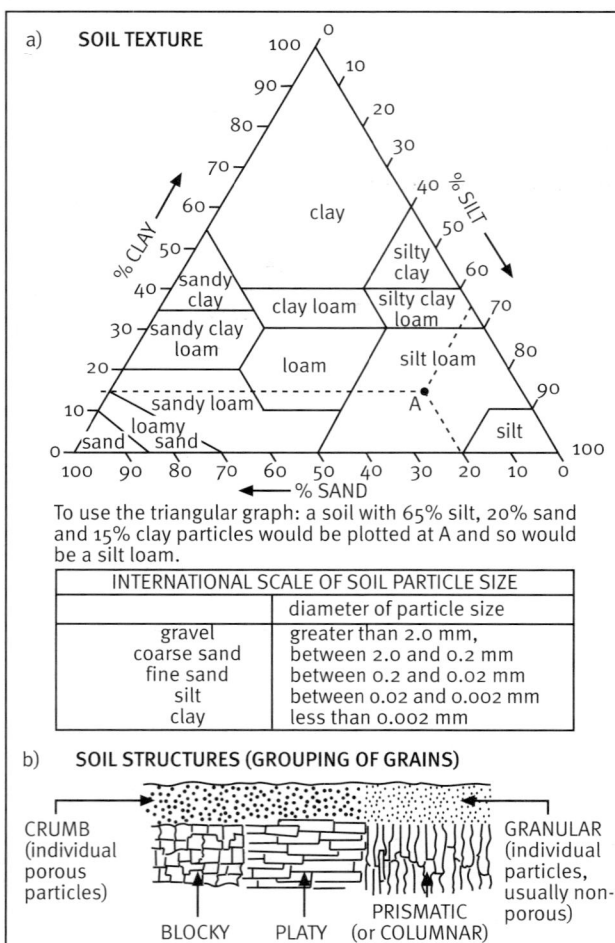

Figure 5.8 Soil texture and soil structure

1.

Ca^{++}
K$^+$ Na$^+$
CLAY-HUMUS PARTICLE.
Ca^{++} K$^+$
Na$^+$ Mg^{++}

PLANT ROOTLET
H$^+$
H$^+$
H$^+$
H$^+$
H$^+$
H$^+$

Cations are attracted to the clay-humus particle by the particle's negative charge. Cations originate from weathered parent rock or decay of organic matter.

H$^+$ cations attracted to plant rootlet.

2.

K$^+$
Ca^{++}
Na$^+$
H$^+$
H$^+$
H$^+$
H$^+$
H$^+$

Ca^{++}
Na$^+$
K$^+$
Mg^{++}

nutrients (cations) absorbed by plant rootlet

equal exchange of cations

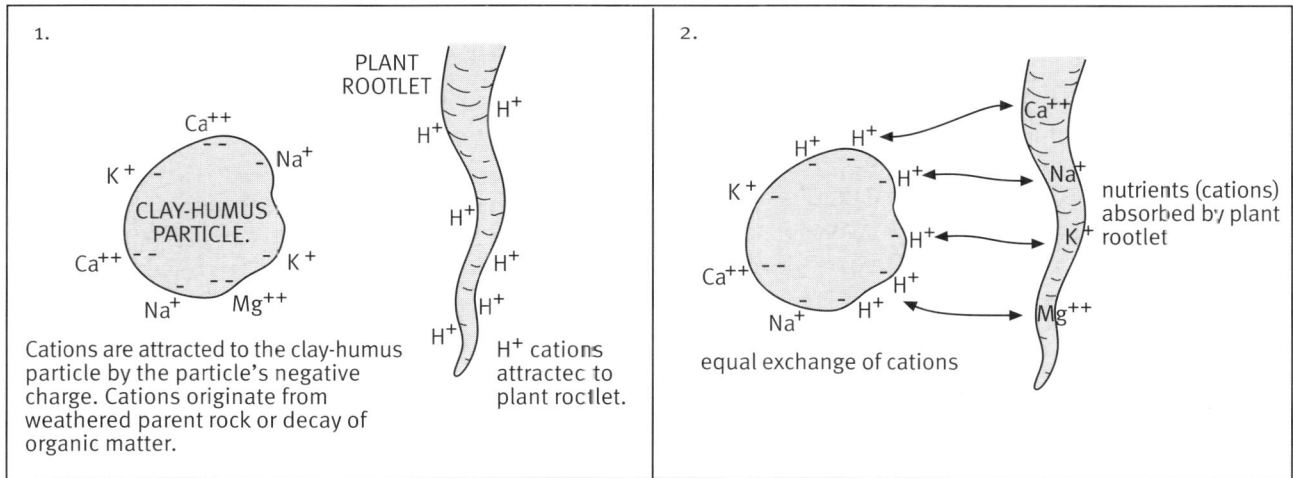

Figure 5.9 The cation exchange process

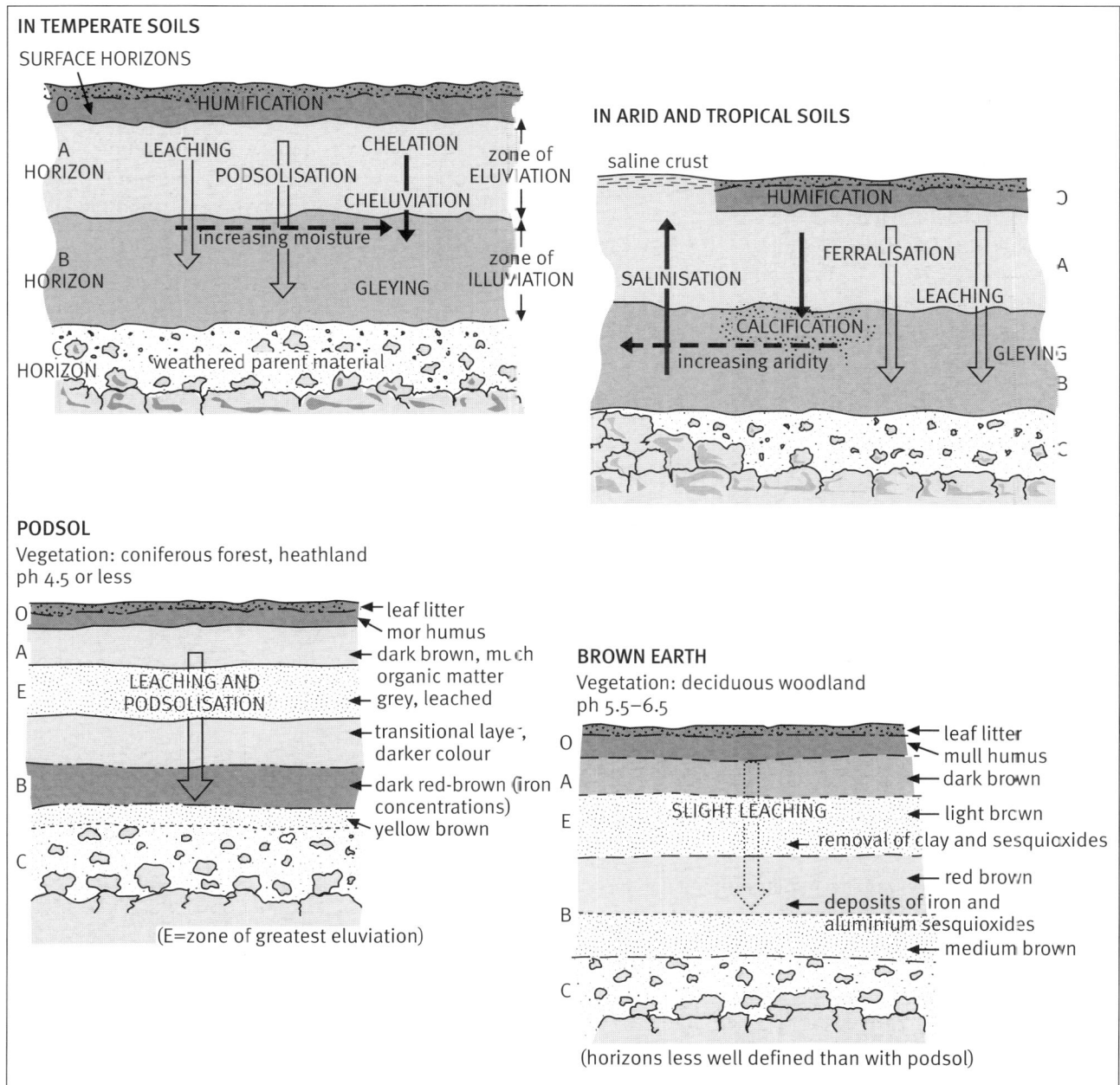

IN TEMPERATE SOILS

SURFACE HORIZONS

O   HUMIFICATION

A HORIZON   LEACHING   PODSOLISATION   CHELATION   zone of ELUVIATION

CHELUVIATION

increasing moisture

B HORIZON   GLEYING   zone of ILLUVIATION

C HORIZON   weathered parent material

IN ARID AND TROPICAL SOILS

saline crust

HUMIFICATION   O

FERRALISATION   A

SALINISATION   CALCIFICATION   LEACHING

increasing aridity   GLEYING   B

C

PODSOL

Vegetation: coniferous forest, heathland
ph 4.5 or less

O   leaf litter
mor humus
A   dark brown, much organic matter
E   LEACHING AND PODSOLISATION   grey, leached
transitional layer, darker colour
B   dark red-brown (iron concentrations)
yellow brown
C

(E=zone of greatest eluviation)

BROWN EARTH

Vegetation: deciduous woodland
ph 5.5–6.5

O   leaf litter
mull humus
A   dark brown
E   SLIGHT LEACHING   light brown
removal of clay and sesquioxides
red brown
B   deposits of iron and aluminium sesquioxides
medium brown
C

(horizons less well defined than with podsol)

Figure 5.10 Soil profiles and soil forming processes

anaerobic condition causes bacterial action to cease. Red oxidised iron (ferric iron) is chemically reduced to become grey-blue ferrous iron.

- **Ferralisation** is a process that occurs in tropical areas. High temperatures and rainfall lead to rapid chemical weathering of rock, producing silica and sesquioxides of Fe and Al. Vegetation also breaks down quickly and so the A horizon is not acidic. Rapid leaching removes silica and other bases (Na, K, Ca) leaving the sesquioxides of Fe and Al which give the soil its distinctive red colour.

- **Calcification** occurs in low rainfall areas. Leaching is limited and is unable to remove calcium which accumulates in the B horizon.

- **Salinisation** occurs where there is very low rainfall, high evapotranspiration *and* a high water table. Evaporation from the surface draws up soluble salts by capillary action. Thus a surface salt crust is formed.

Eluviation and **illuviation** result from processes such as leaching. Eluviation is the removal of soluble materials and clay particles and therefore distinguishes the A horizon. (Within the A horizon the main zone of eluviation is designated E – see Figure 5.10.) Illuviation is where the soluble materials and clays removed from the A horizon are redeposited in a zone of accumulation: the B horizon.

Factors affecting the formation and characteristics of soils

Climate

Temperature directly affects the rate of chemical and biological activity in soils. It also affects the rate of evaporation of soil moisture. Precipitation effectiveness (the extent to which precipitation exceeds potential evapotranspiration – see Figure 2.4 in Chapter 2) affects the extent of leaching and capillary action. Both temperature and precipitation have indirect effects on soil through their effect on vegetation (and thus the supply of humus) and on the rate of weathering of parent rock.

Page 11

As climate is such a key factor in determining soil type at a global scale, the most commonly used classification of soils is the **zonal classification,** so called because soils are related to climatic zones (Figure 5.11).

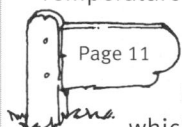

Geology

Rock type helps to determine soils' texture and structure (and therefore drainage), depth, colour and nutrient content. Rocks such as limestone or granite have such distinctive characteristics that rock type becomes the most important control on soil formation. **Rendzina** is a soil that develops on chalk or limestone with a grass vegetation cover. Chemical weathering leads to a continual release of calcium and few insoluble particles. The grasses produce a base-rich litter which rapid organic activity turns into a mull humus. The result is thin, alkaline but nutrient-rich soil. As rendzina occurs on a specific rock type within a climatic zone, on the zonal soil classification it is classified as an **intrazonal soil** (Figure 5.11).

Biotic factors

Biotic factors include the activity of micro-organisms which break down dead vegetation (bacteria and fungi), organisms which mix soils, vegetation (different types contribute different quantities and qualities of organic matter to the soil) and human action (see below).

Relief

Slope angle affects soil depth and soil drainage. Steeper slopes are more vulnerable to soil erosion. **Altitude** and **aspect** influence temperature and precipitation and thus affect soil formation, both directly and indirectly, through their effect on vegetation. The variation in soils along a slope due to changes in slope angle, drainage and local climate is illustrated by a **catena** (Figure 5.12). The soils at the various points along the catena are not developed in isolation; surface runoff and throughflow transport organic and mineral matter downslope.

Time

Soils take hundreds of years to form. As soils age, characteristics deriving from their parent rock become less important and characteristics derived from the addition of organic material and the working of the various soil-forming processes become more important. Many British soils started to develop under quite different climates and land uses than exist today. Soils which have had insufficient time to fully mature are known as azonal soils (Figure 5.11).

MAJOR <u>ZONAL</u> SOILS
and associated vegetation
types and climates

Note that this is a very simplified
outline. It does not cover all zonal
soils. Two transitional zonal soils
have been shown.

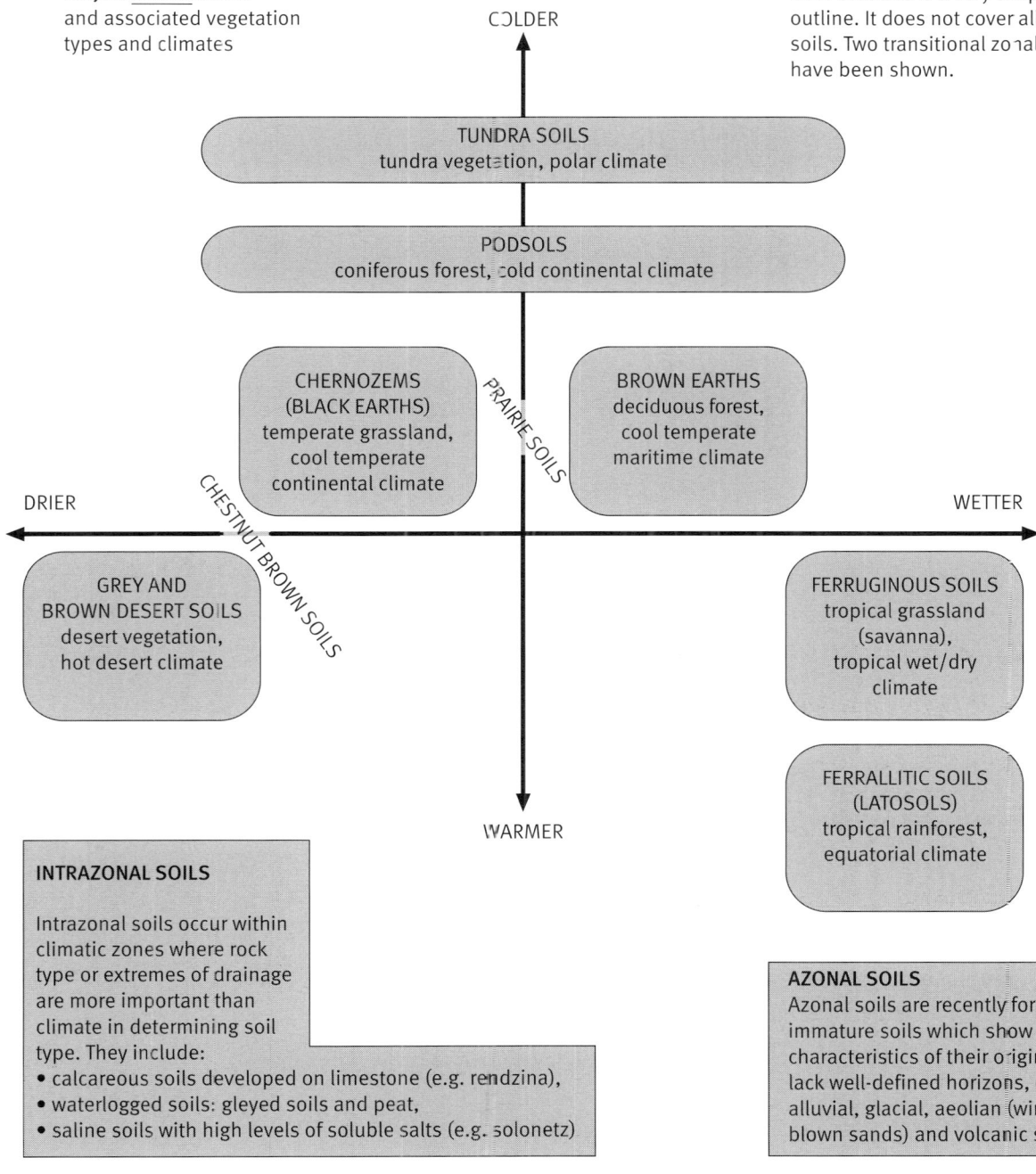

COLDER

TUNDRA SOILS
tundra vegetation, polar climate

PODSOLS
coniferous forest, cold continental climate

PRAIRIE SOILS

CHERNOZEMS
(BLACK EARTHS)
temperate grassland,
cool temperate
continental climate

BROWN EARTHS
deciduous forest,
cool temperate
maritime climate

DRIER

WETTER

CHESTNUT BROWN SOILS

GREY AND
BROWN DESERT SOILS
desert vegetation,
hot desert climate

FERRUGINOUS SOILS
tropical grassland
(savanna),
tropical wet/dry
climate

FERRALLITIC SOILS
(LATOSOLS)
tropical rainforest,
equatorial climate

WARMER

INTRAZONAL SOILS

Intrazonal soils occur within
climatic zones where rock
type or extremes of drainage
are more important than
climate in determining soil
type. They include:
• calcareous soils developed on limestone (e.g. rendzina),
• waterlogged soils: gleyed soils and peat,
• saline soils with high levels of soluble salts (e.g. solonetz)

AZONAL SOILS
Azonal soils are recently formed,
immature soils which show the
characteristics of their origin and
lack well-defined horizons, e.g.
alluvial, glacial, aeolian (wind-
blown sands) and volcanic soils

Figure 5.11 The zonal soil classification scheme

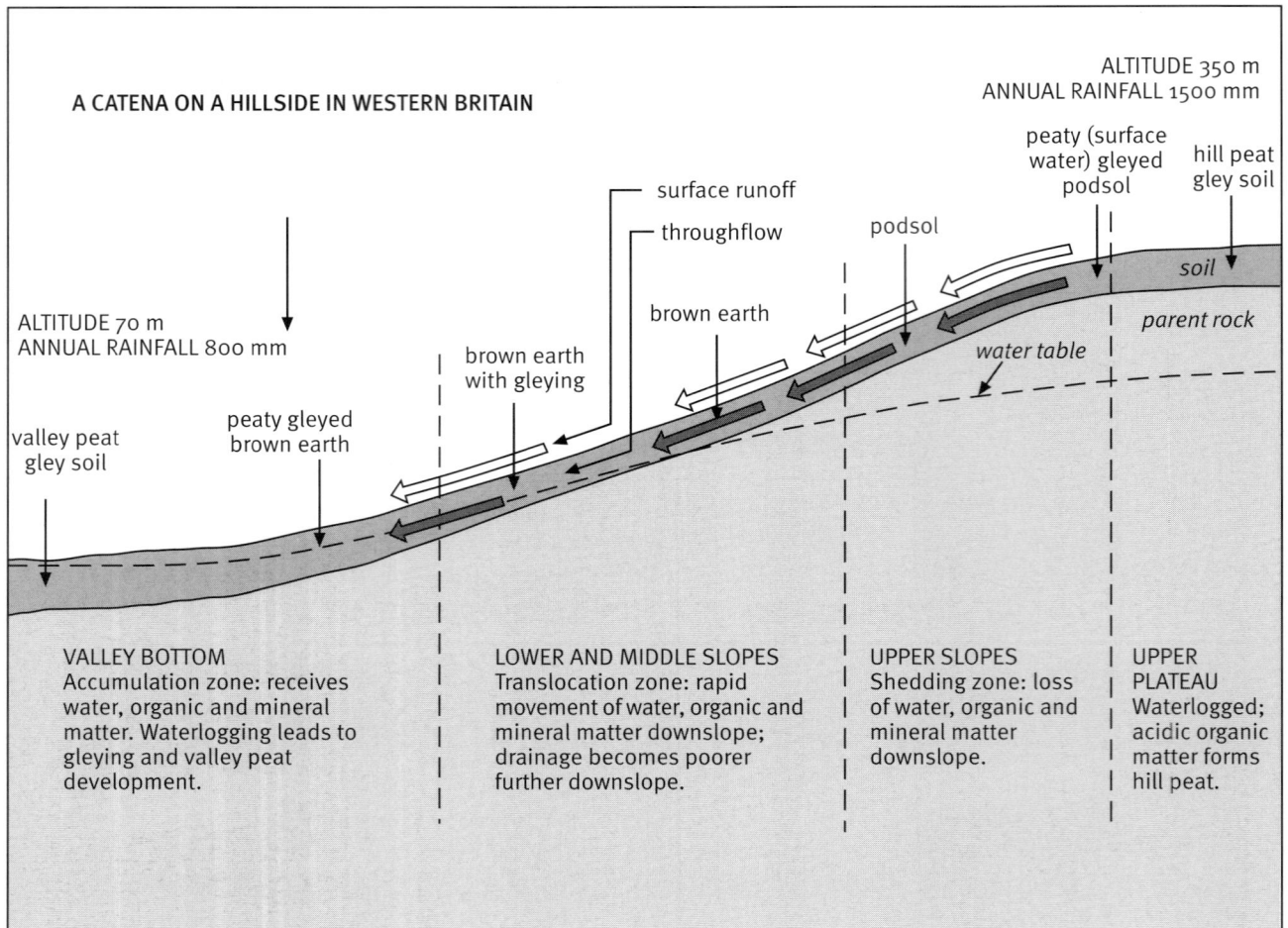

A CATENA ON A HILLSIDE IN WESTERN BRITAIN

ALTITUDE 350 m
ANNUAL RAINFALL 1500 mm

peaty (surface water) gleyed podsol

hill peat gley soil

surface runoff

throughflow

podsol

soil

parent rock

ALTITUDE 70 m
ANNUAL RAINFALL 800 mm

brown earth

brown earth with gleying

water table

valley peat gley soil

peaty gleyed brown earth

VALLEY BOTTOM
Accumulation zone: receives water, organic and mineral matter. Waterlogging leads to gleying and valley peat development.

LOWER AND MIDDLE SLOPES
Translocation zone: rapid movement of water, organic and mineral matter downslope; drainage becomes poorer further downslope.

UPPER SLOPES
Shedding zone: loss of water, organic and mineral matter downslope.

UPPER PLATEAU
Waterlogged; acidic organic matter forms hill peat.

Figure 5.12 A catena on a hillside in western Britain

Human impact on soils

Human action can easily disturb the equilibrium which exists between the formation of soil and the rate with which it is eroded or degraded naturally. Removal of vegetation, unwise cultivation methods, overgrazing, irrigation and the unwise use of farm chemicals can all lead to **soil degradation**. Soil management is required to tackle such problems. Figure 5.13 summarises the key points and relationships of this topic.

Review task

1   **Understanding soils requires a precise knowledge of characteristics, processes and influencing factors. The characteristics of soils can be checked by using a Mind Map or check list as in the review task on page 44. Your understanding of processes and influencing factors can be built up as follows. Divide an A4 page into three columns. In the left-hand column, list the various processes and factors. Head the second column 'podsols' and the third 'brown earths'. Then, by referring to the text,** Figure 5.10 and textbooks, make entries in these columns to link the processes and factors to these two soils. If a process does not apply, write in why it does not. If a factor is of only local importance (for example, geology or relief), note the influence it might have on a smaller area.

2   **Refer to Figure 5.13. Follow through the linkages shown on the diagram. Add references to examples you have studied of the various human impacts on soils.**

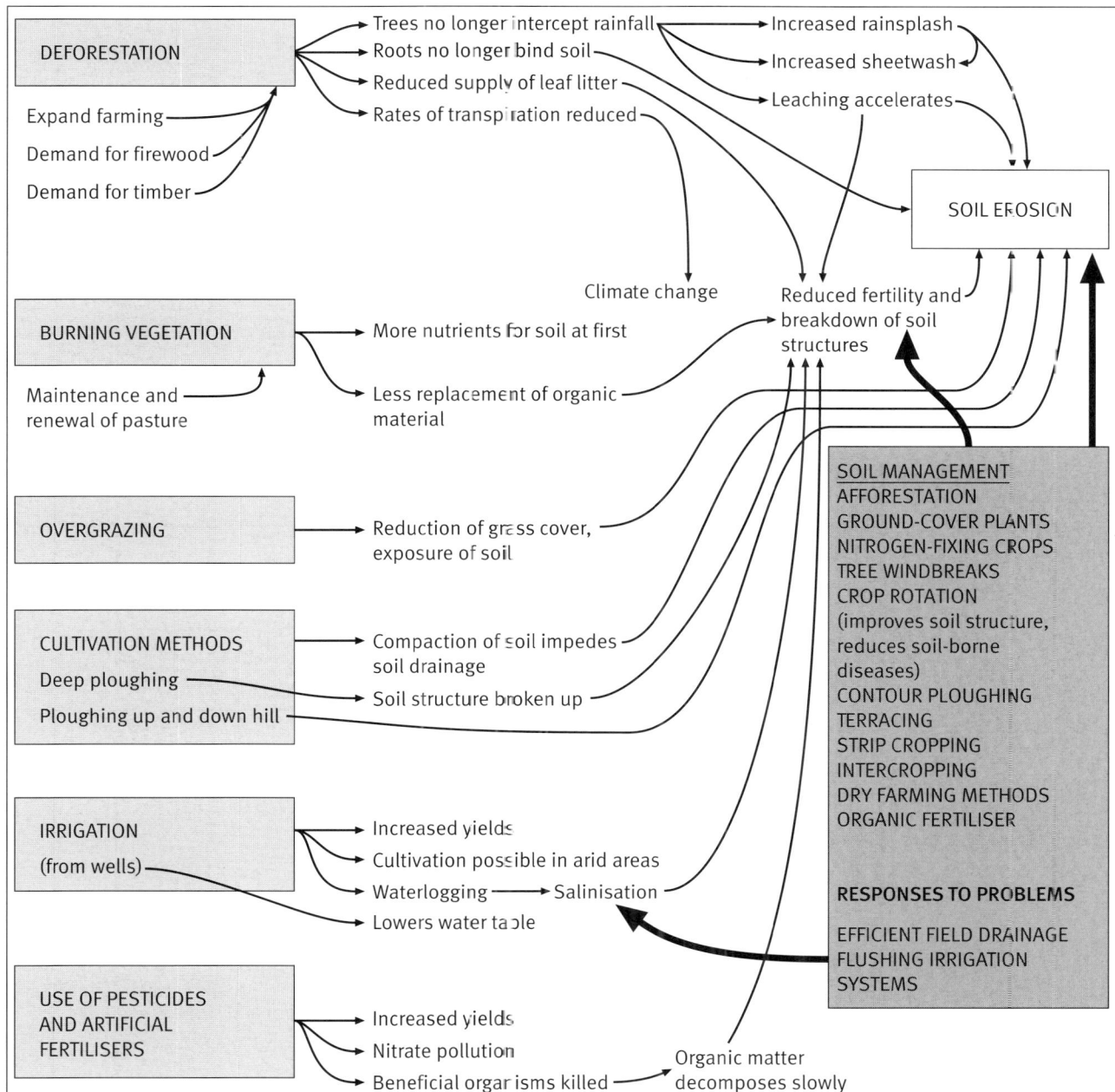

Figure 5.13 The impact on soils of human activity

Case studies of ecosystems

1 Broad-leaved deciduous forest (Figure 5.14)

The broad-leaved deciduous forest biome is a large-scale ecosystem associated with a cool temperate humid climate and brown earth soils. It is the major biome characteristic of north-west Europe including Britain. Most of the climatic climax vegetation has been cleared or has been greatly altered by human impact over long periods of time. Considerable regional and local variations of geology, relief and climate lead to variations in soil type or vegetation from the typical brown earth or deciduous woodland. Within this major biome, important more localised ecosystems exist: in

Britain the moorland ecosystem of the uplands is the most important. Most moorland vegetation is a plagioclimax, as burning and grazing prevent the vegetation progressing towards the climatic climax of deciduous woodland.

2 Coniferous (boreal) forest (Figure 5.15)

The coniferous (boreal) forest biome is a large-scale ecosystem associated with cold climates (long, cold winters and short summers) and podsol soils. In northern North America and Eurasia it is the major ecosystem, known as the **Taiga**. Coniferous forests extend into temperate latitudes in mountain areas and on the west coast of North America.

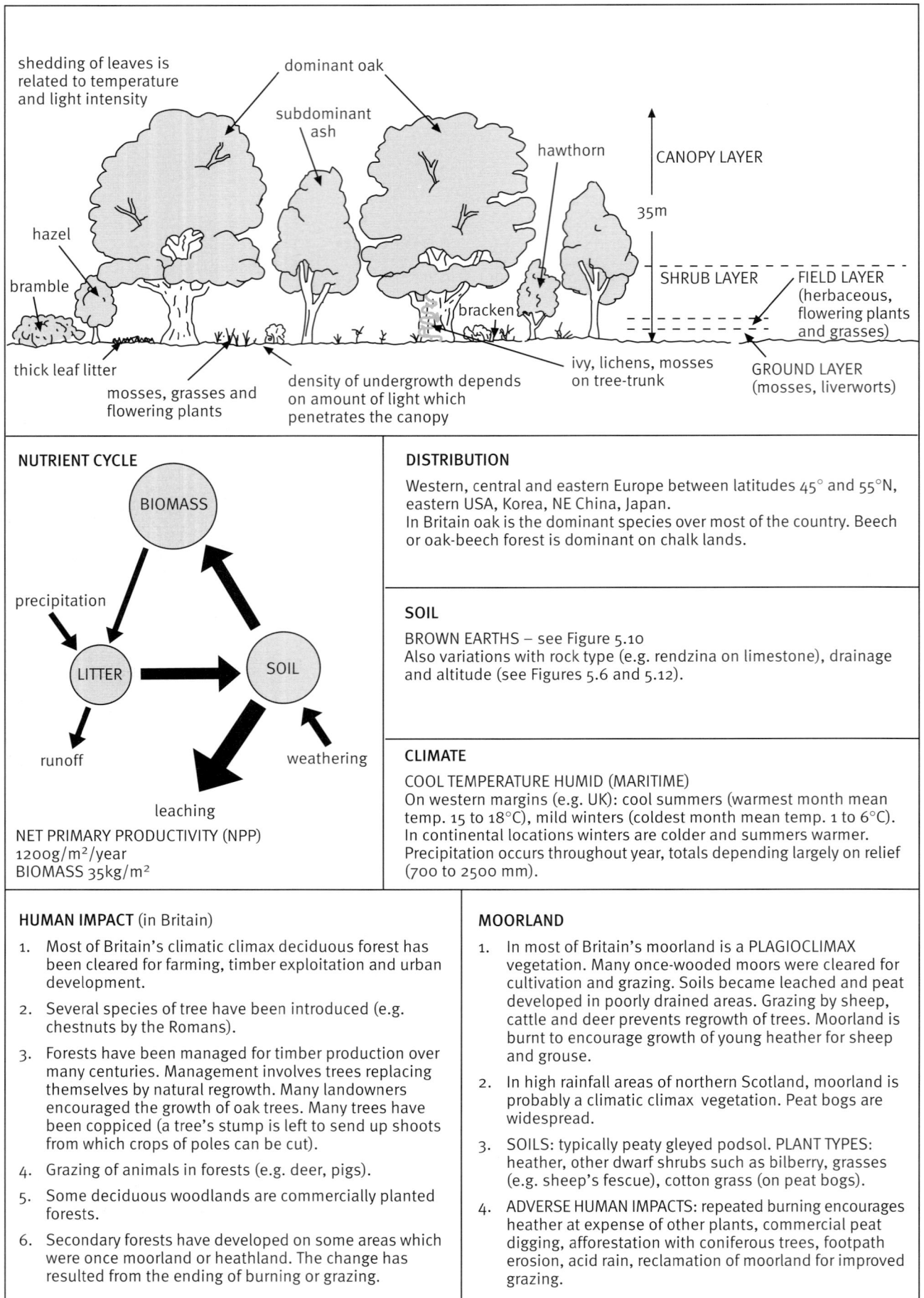

shedding of leaves is related to temperature and light intensity

dominant oak

subdominant ash

hawthorn

CANOPY LAYER

35m

hazel

bramble

SHRUB LAYER

FIELD LAYER (herbaceous, flowering plants and grasses)

bracken

thick leaf litter

mosses, grasses and flowering plants

density of undergrowth depends on amount of light which penetrates the canopy

ivy, lichens, mosses on tree-trunk

GROUND LAYER (mosses, liverworts)

NUTRIENT CYCLE

BIOMASS

precipitation

LITTER

SOIL

runoff

weathering

leaching

NET PRIMARY PRODUCTIVITY (NPP)
1200g/m²/year
BIOMASS 35kg/m²

DISTRIBUTION

Western, central and eastern Europe between latitudes 45° and 55°N, eastern USA, Korea, NE China, Japan.
In Britain oak is the dominant species over most of the country. Beech or oak-beech forest is dominant on chalk lands.

SOIL

BROWN EARTHS – see Figure 5.10
Also variations with rock type (e.g. rendzina on limestone), drainage and altitude (see Figures 5.6 and 5.12).

CLIMATE

COOL TEMPERATE HUMID (MARITIME)
On western margins (e.g. UK): cool summers (warmest month mean temp. 15 to 18°C), mild winters (coldest month mean temp. 1 to 6°C). In continental locations winters are colder and summers warmer. Precipitation occurs throughout year, totals depending largely on relief (700 to 2500 mm).

HUMAN IMPACT (in Britain)

1. Most of Britain's climatic climax deciduous forest has been cleared for farming, timber exploitation and urban development.

2. Several species of tree have been introduced (e.g. chestnuts by the Romans).

3. Forests have been managed for timber production over many centuries. Management involves trees replacing themselves by natural regrowth. Many landowners encouraged the growth of oak trees. Many trees have been coppiced (a tree's stump is left to send up shoots from which crops of poles can be cut).

4. Grazing of animals in forests (e.g. deer, pigs).

5. Some deciduous woodlands are commercially planted forests.

6. Secondary forests have developed on some areas which were once moorland or heathland. The change has resulted from the ending of burning or grazing.

MOORLAND

1. In most of Britain's moorland is a PLAGIOCLIMAX vegetation. Many once-wooded moors were cleared for cultivation and grazing. Soils became leached and peat developed in poorly drained areas. Grazing by sheep, cattle and deer prevents regrowth of trees. Moorland is burnt to encourage growth of young heather for sheep and grouse.

2. In high rainfall areas of northern Scotland, moorland is probably a climatic climax vegetation. Peat bogs are widespread.

3. SOILS: typically peaty gleyed podsol. PLANT TYPES: heather, other dwarf shrubs such as bilberry, grasses (e.g. sheep's fescue), cotton grass (on peat bogs).

4. ADVERSE HUMAN IMPACTS: repeated burning encourages heather at expense of other plants, commercial peat digging, afforestation with coniferous trees, footpath erosion, acid rain, reclamation of moorland for improved grazing.

Figure 5.14 The broad-leaved deciduous forest ecosystem

CHARACTERISTICS

1. Evergreen nature permits year-round photosynthesis. Coniferous larch forest is deciduous, an adaptation to extreme cold.

2. Needle leaves reduce transpiration.

3. Thick bark and seeds shielded by cones are protection against cold and fire.

4. Conical shape and downward-sloping branches shed snow.

5. Leaf litter is acid and decomposes slowly.

6. Forest floor cover is very limited (mosses, lichens) because of lack of sunlight and layer of undecomposed needle leaves.

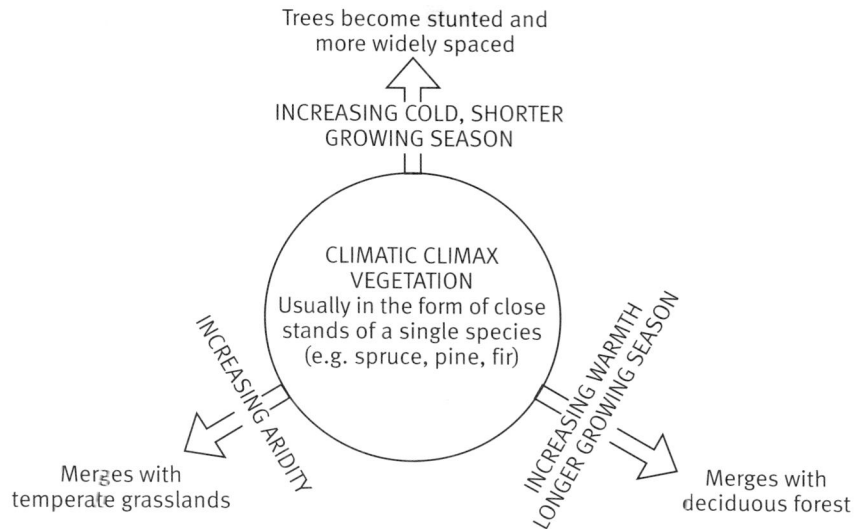

Trees become stunted and more widely spaced

INCREASING COLD, SHORTER GROWING SEASON

CLIMATIC CLIMAX VEGETATION
Usually in the form of close stands of a single species (e.g. spruce, pine, fir)

INCREASING ARIDITY

INCREASING WARMTH LONGER GROWING SEASON

Merges with temperate grasslands

Merges with deciduous forest

NUTRIENT CYCLE

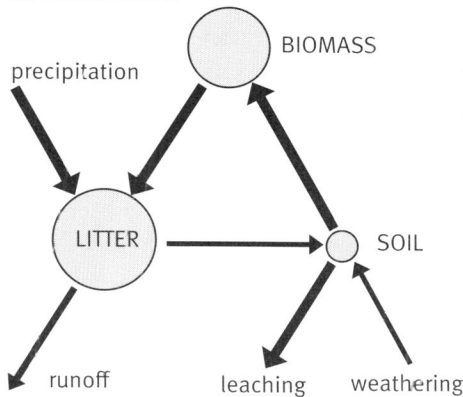

precipitation

BIOMASS

LITTER

SOIL

runoff

leaching   weathering

NET PRIMARY PRODUCTIVITY (NPP) 800 g/m²/year

BIOMASS 20 kg/m²

DISTRIBUTION

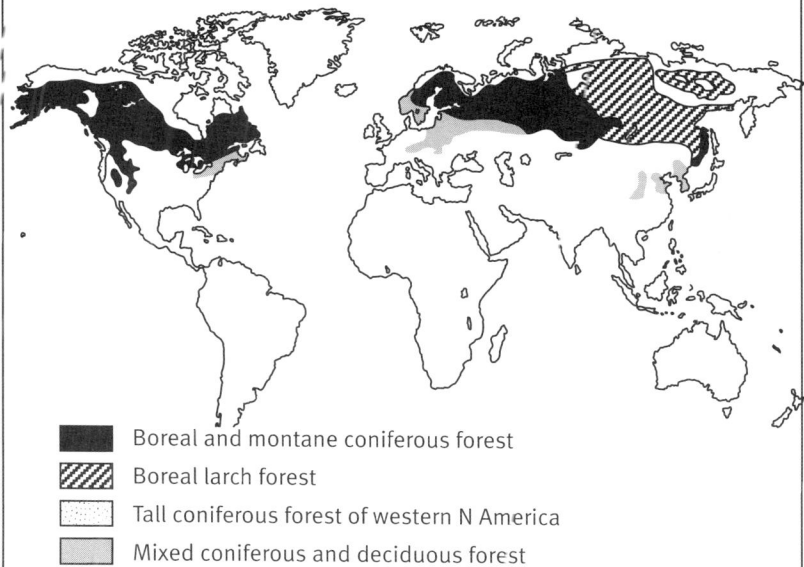

■ Boreal and montane coniferous forest

▨ Boreal larch forest

▦ Tall coniferous forest of western N America

▩ Mixed coniferous and deciduous forest

SOIL
PODSOLS – see Figure 5.10

CLIMATE
COLD CONTINENTAL: short cool summers (warmest month mean temp. 12 to 16°C). Cold or very cold winters (coldest month mean temp. −30 to −10°C). Low annual precipitation (400 to 600 mm) with summer maximum.

Figure 5.15 The coniferous (boreal) forest ecosystem

Review task

1   **Refer to Figures 5.14 and 5.15.**
    a) **Explain the relative sizes of the stores and flows in the nutrient cycling diagrams.**
    b) **Refer back to the soil profiles of brown earths and podsols (Figure 5.10). What relationships are there between climate, vegetation characteristics and soil characteristics?**

2   **Why is the deciduous woodland found in Britain not a true climatic climax 'natural' vegetation? Why is the moorland ecosystem a plagioclimax?**

3   **Make sure that you are able to link the key points of the ecosystems shown in Figures 5.14 and 5.15 to located examples. Note carefully the human influences on the ecosystem which apply to your examples.**

Waves

Waves are the main agents affecting coastal landforms. They are created by wind blowing over the ocean surface. A wave does *not* involve water moving across the ocean. As the wave passes across the surface it generates an orbital movement of water beneath it (Figure 6.1). The size of waves is affected by:

- the strength (velocity) of wind
- the length of time during which the wind blows
- the fetch – the distance of water over which the wind blows

Waves can travel much further than the winds that generate them. **Swell** describes waves that travel long distances, having been generated by distant winds; they tend to have long wavelengths and low height (see Figure 6.1 for wave terminology). In contrast, **storm waves** with short wavelengths and high height are generated more locally. The energy of a wave is proportional to its wavelength and the square of its height ($E \propto LH^2$). Thus a slight increase in height causes a large increase in energy and, therefore, destructive power.

As waves move into shallow water, friction with the sea bed increases. The orbital movement of water becomes elliptical, wavelength will shorten and wave height will increase. The wave height grows until it breaks at the **plunge line** where depth of water and wave height are about equal (Figure 6.2). The upward rush of water on a sloping beach is the **swash**, the following seawards movement of water is the **backwash**. The seaward movement may be concentrated in a **rip current**.

Waves breaking on a beach may be classified into constructive waves or destructive waves. **Constructive** waves, breaking on a beach with a low angle of slope, have a strong swash and a weak backwash and so build up material on the beach (Figure 6.2b). **Destructive** waves have a weak swash and strong backwash and so remove material from the beach (Figure 6.2a). Figures 6.2c and 6.2d show two other aspects of waves that affect coastlines:

1 **Wave refraction**, leading to a concentration of energy on a headland and a dispersal of energy in a bay.

2 **Longshore drift**, by which wave action is responsible for the transport of sand and shingle along the coastline.

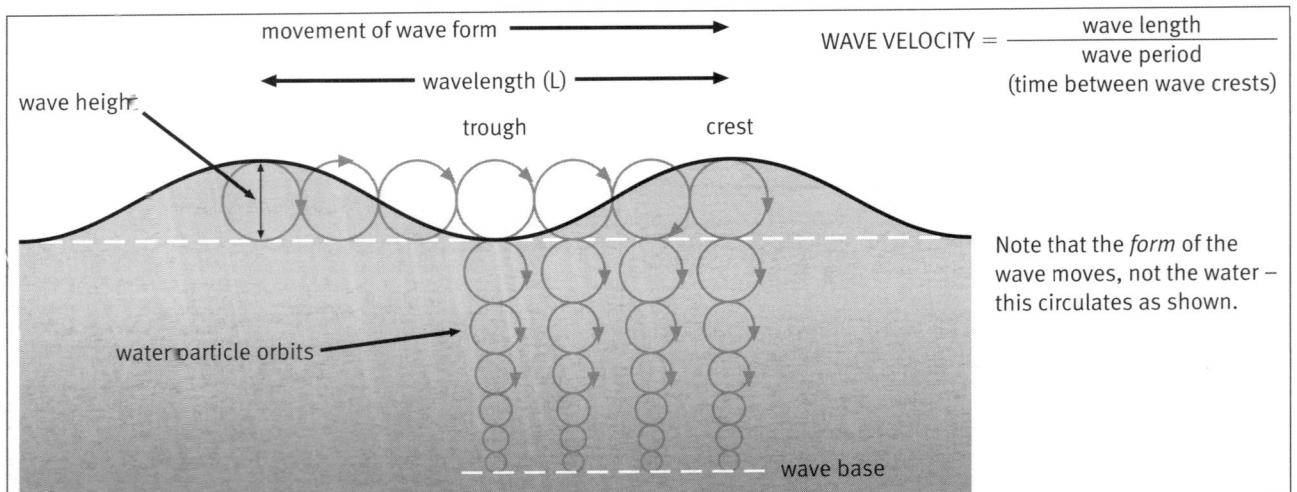

$$\text{WAVE VELOCITY} = \frac{\text{wave length}}{\text{wave period (time between wave crests)}}$$

Note that the *form* of the wave moves, not the water – this circulates as shown.

Figure 6.1 Terminology of waves

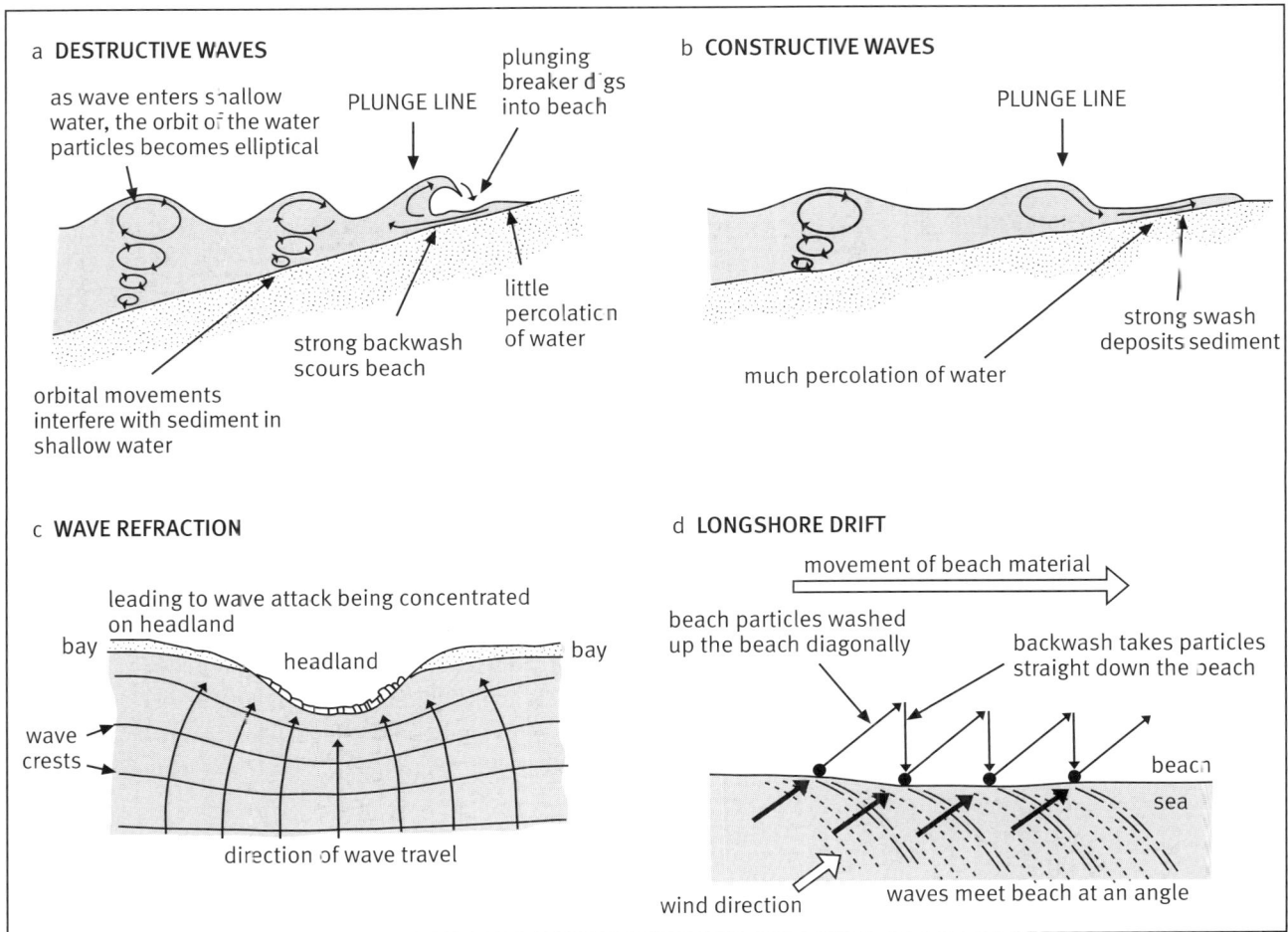

a **DESTRUCTIVE WAVES**

as wave enters shallow water, the orbit of the water particles becomes elliptical

PLUNGE LINE

plunging breaker digs into beach

strong backwash scours beach

little percolation of water

orbital movements interfere with sediment in shallow water

b **CONSTRUCTIVE WAVES**

PLUNGE LINE

strong swash deposits sediment

much percolation of water

c **WAVE REFRACTION**

leading to wave attack being concentrated on headland

bay

headland

bay

wave crests

direction of wave travel

d **LONGSHORE DRIFT**

movement of beach material

beach particles washed up the beach diagonally

backwash takes particles straight down the beach

beach

sea

wind direction

waves meet beach at an angle

Figure 6.2 Wave action

> ### Review task
>
> **Check your understanding of the various aspects of waves. Refer to Figures 6.2a and 6.2b; relate wavelength, wave height and steepness of beach to the two types of waves.**

Tides

Tides are regular movements of the sea's surface caused by the gravitational attraction of the moon and, to a lesser extent, the sun on the earth. Tidal range is affected by the shape of the coastline; in a funnel-shaped estuary, such as the Severn, incoming tides are forced into a narrowing space and so cause a particularly high tide. The impact of tides on coastal geomorphology includes the following influences:

- whereabouts waves break on beaches and the height of the zone of wave attack on cliffs;

- the extent of the inter-tidal zone on beaches, wave-cut platforms or in estuaries;

- the impact of **tidal currents**: where tidal ranges are great, strong flows of water in narrow estuaries or channels can move sediment and so affect the form of the coastline.

Coastal erosion

Erosion processes

- **Hydraulic action** has two aspects, first the severe pounding effect of waves which removes loose material and has a general weakening effect on solid rock, and, second, the compression of parcels of air into rock crevices which leads to sudden increases and decreases of pressure and therefore the break-up of rock.

- **Corrasion** (or **abrasion**) is the erosion of cliffs by shingle and boulders hurled against them by waves.

- **Solution** (or **corrosion**) includes the dissolving of limestone by carbonic acid in sea water and the effect of salt in corroding rocks (for example, attacking iron compounds).

- **Attrition** is the breaking down and rounding of boulders and pebbles already eroded from cliffs.

- **Sub-aerial processes** are non-marine processes. They include erosion by rain, surface runoff and wind, mass movement and weathering. Salt water from spray causes salt crystals to form, enlarge and cause weakening and disintegration of rock. Salt also corrodes various minerals.

Pages 1–6

Features of coastal erosion

Headlands and bays

On a relatively large scale, much of the plan of the coastline results from the **differential erosion** of different rock types. On the east side of the Isle of Purbeck, Dorset (Figure 6.3), less-resistant sands and clays form bays while more-resistant chalk and limestone form headlands – a **discordant coastline**. This pattern also reflects inland relief: the chalk forms a ridge with a considerably greater height of rock to be removed than where clay underlies a broad valley. Differential erosion on the southern Purbeck coast where there are contrasting bands of rock parallel to the coast has resulted in a **concordant coastline**.

Cliffs

On upland coasts, wave erosion creates cliffs (Figure 6.4). The rate of erosion and profile of cliffs depend on several factors:

- **Geology** exerts an influence through lithology (rock type), rock jointing and angle of dip (Figure 6.5). The influence of lithology is most clearly shown where the cliff is cut into beds of contrasting resistance. Unconsolidated deposits, such as boulder clay (forming the cliffs of Holderness), are eroded especially rapidly. Erosion at the base of boulder clay cliffs results in rotational slips, soil creep and mudflows (see page 57); the debris from these mass-movement processes is then removed by the waves. Wave erosion at the base of cliffs results in undercutting to form a **wave-cut notch**. Close jointing of rock, such as in a chalk cliff, facilitates cliff collapse above the notch while massive jointing (as in granite or Carboniferous limestone) permits large overhangs and results in a more irregular profile. The angle of dip of beds affects not only the cliff profile, as Figure 6.5 illustrates, but also the rate of erosion:

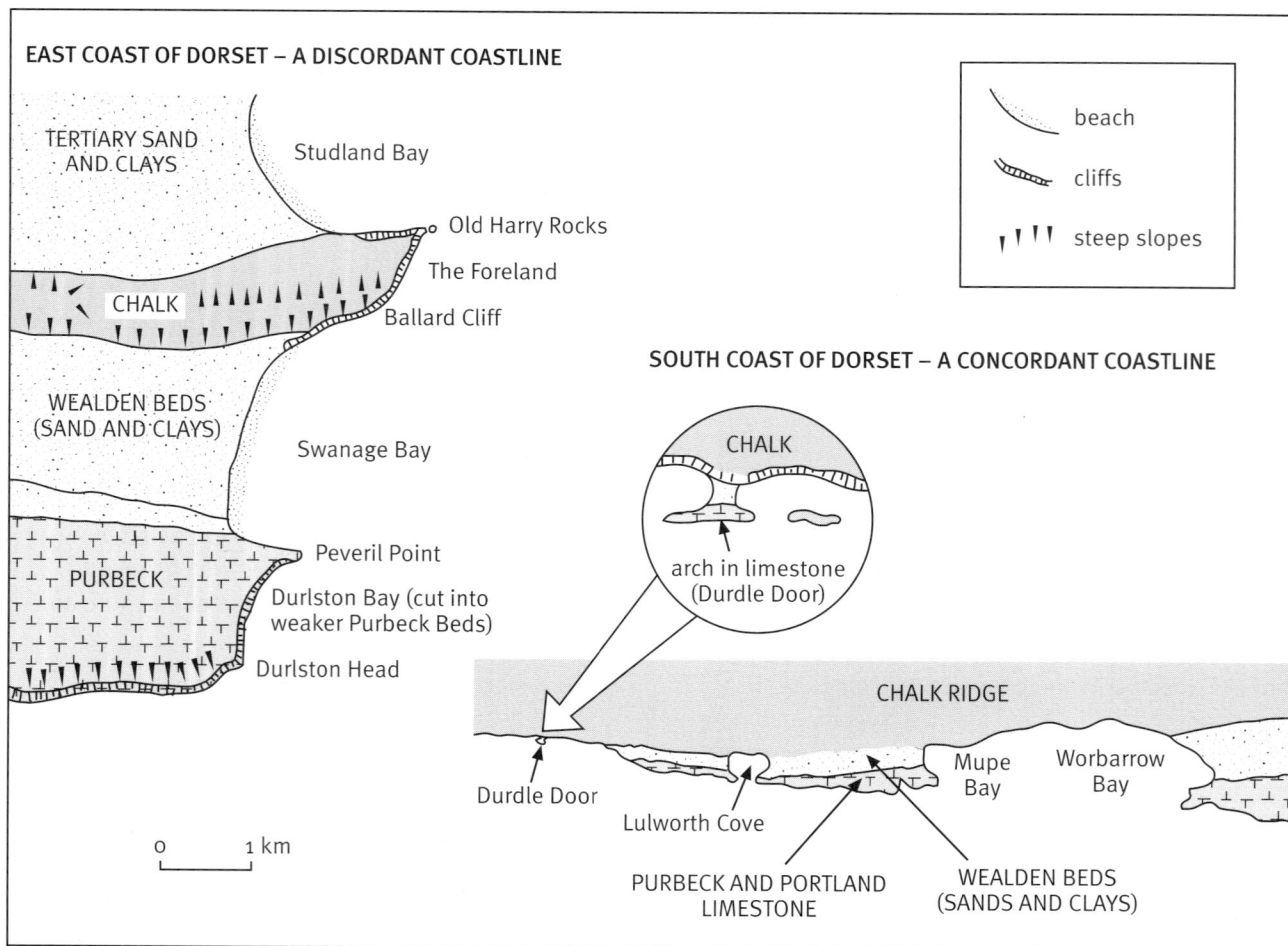

Figure 6.3 Dorset: discordant and concordant coastlines

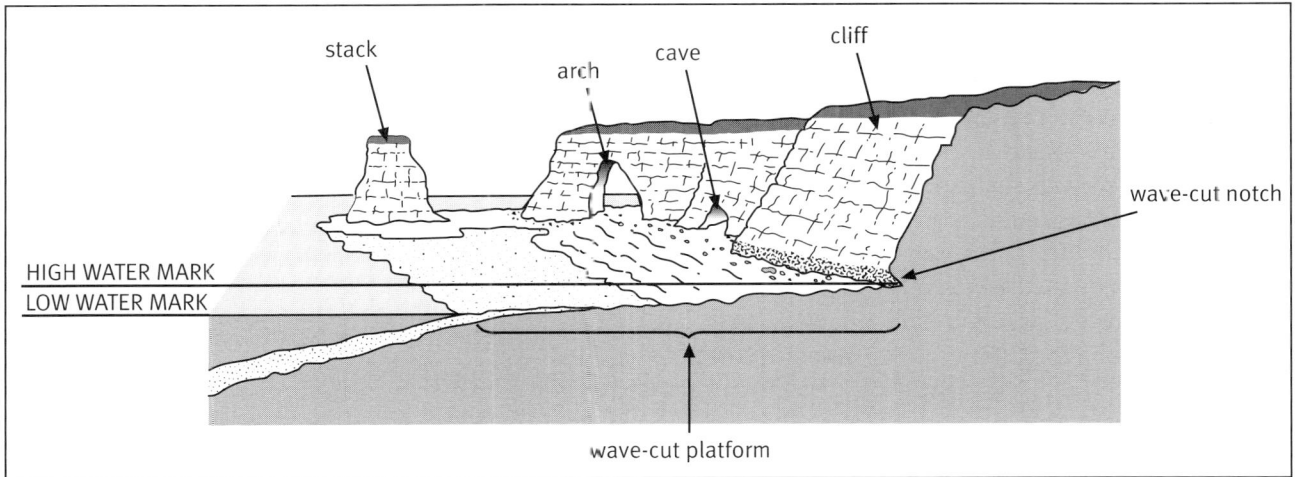

Figure 6.4 Features of coastal erosion

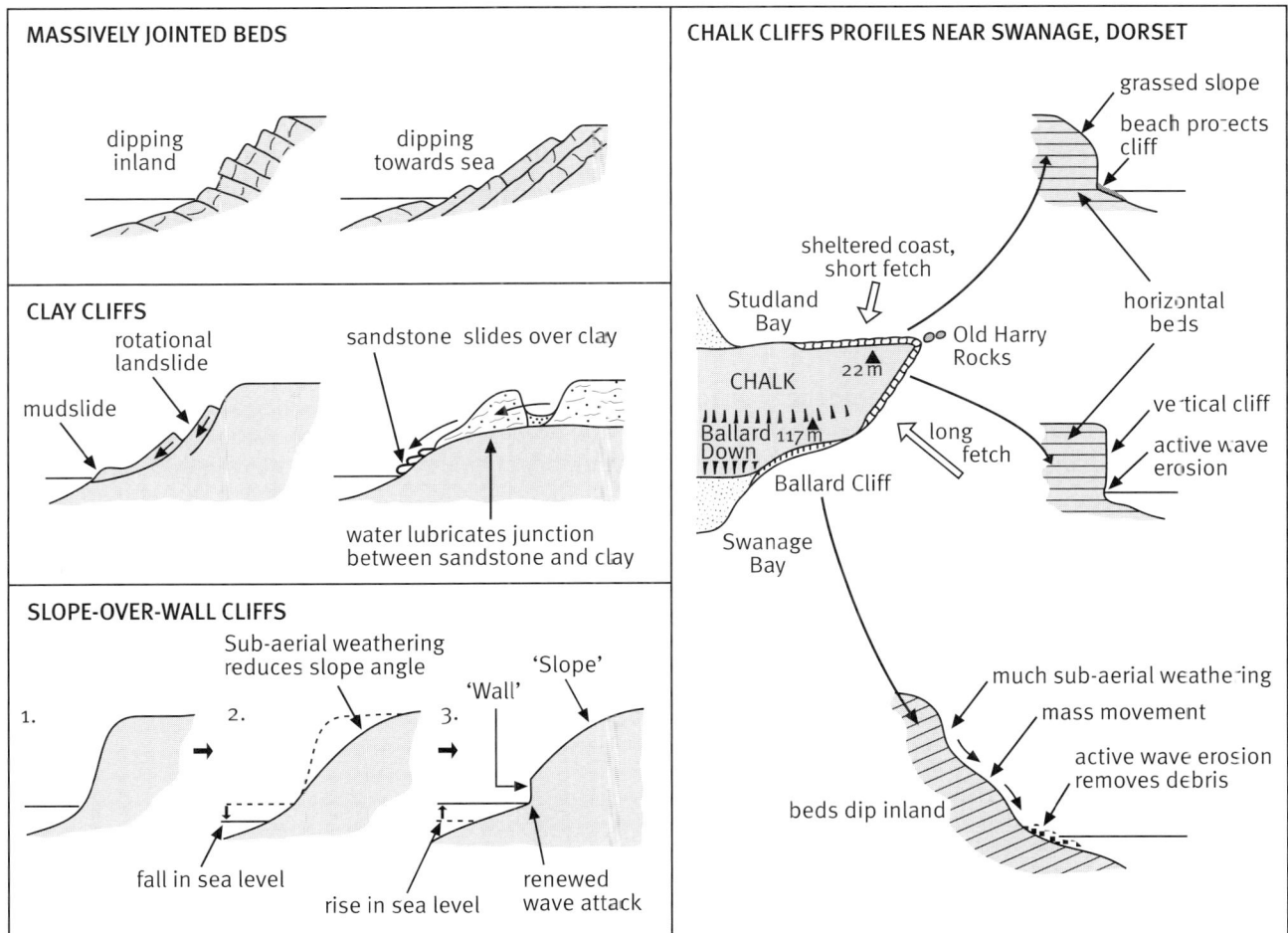

Figure 6.5 Cliff profiles

for example, beds dipping slightly seawards will collapse easily when undercut by wave erosion.

- **Length and direction of fetch.** A cliff attacked by waves developed over a long fetch will be eroded more rapidly than a cliff being attacked by waves which have less energy because of a short fetch (the Dorset chalk cliffs in Figure 6.5).

- **Protection of base of cliffs.** Wave attack on cliffs can be reduced in effectiveness by wide beaches, wide rock platforms or large accumulations of eroded debris. Human activity can also have an effect, reducing erosion by construction of sea defences or accelerating it by removing sand and shingle.

- **Geomorphological history.** Falling sea levels (or rising land) can result in wave action becoming less effective at the base of cliffs and eventually ceasing. The cliff then becomes degraded by sub-aerial processes. In much of Devon and Cornwall such a fall in sea level in Pleistocene (Ice Age) times resulted in periglacial processes (see Chapter 7) reducing the cliffs' slope. The subsequent rise in sea level has resulted in renewed wave attack and cliff formation to form **slope-over-wall cliffs**.

Pages 74–5

The detailed plan of a cliffed coast will also be influenced by many of the above factors. In particular, rocks of varying resistance, jointing and fault lines will result in an irregular plan. Narrow inlets called **geos** are formed by the sea easily eroding shattered rock along a fault.

Wave-cut platforms

As a cliff retreats under wave attack, it leaves at its base a gently sloping **wave-cut platform** (Figure 6.4). As waves cross the platform they use pebbles to abrade its surface. Rapid cliff retreat and a large tidal range lead to wide platforms.

Caves, arches and stacks

Erosion of lines of weakness in cliffs widens them to form deep recesses and **caves**. Where a cave is formed in a headland, it might cut right through to form an **arch** (Figure 6.4). The collapse of the arch leaves a **stack** which will eventually be worn down into a **stump**. All these features are found in the chalk cliffs of Dorset (Figure 6.5, Old Harry Rocks). Erosion into the roof of a cave may lead to a **blowhole** with an opening into the cliff top.

Review task

1 **Sub-aerial processes are often neglected as a set of processes important in coastal erosion. To summarise the relevant sub-aerial processes, construct a Mind Map. Start with three initial branches: erosion processes, mass movement and weathering. Do *not* include marine erosion processes (i.e. by waves). Refer back to Chapter 1 and to textbooks, if necessary, to add the relevant processes of mass movement and weathering.**

2 **You need to apply your knowledge of coastal erosion to a case study area. For your case study of a cliffed coastline, consider and summarise in turn: a) the factors influencing the general shape of the coastline; b) the nature of the cliffs (include sketch profiles, the main processes at work – marine and sub-aerial – together with the four main influences described above); c) detailed localised landforms. In addition, make sure that you are able to draw a simple sketch map of your coastline to show the major features and the location of important details.**

Coastal transportation and deposition

Beaches and other features of deposited sediment can be viewed as a sediment store system with a number of inputs and outputs (Figure 6.6). Where the accumulation of sediment exceeds its removal, net deposition takes place. Deposition occurs in sheltered areas with low-energy waves and where large supplies of sediment are available from offshore, along the coast or through river deposition. Stretches of coastline with large depositional

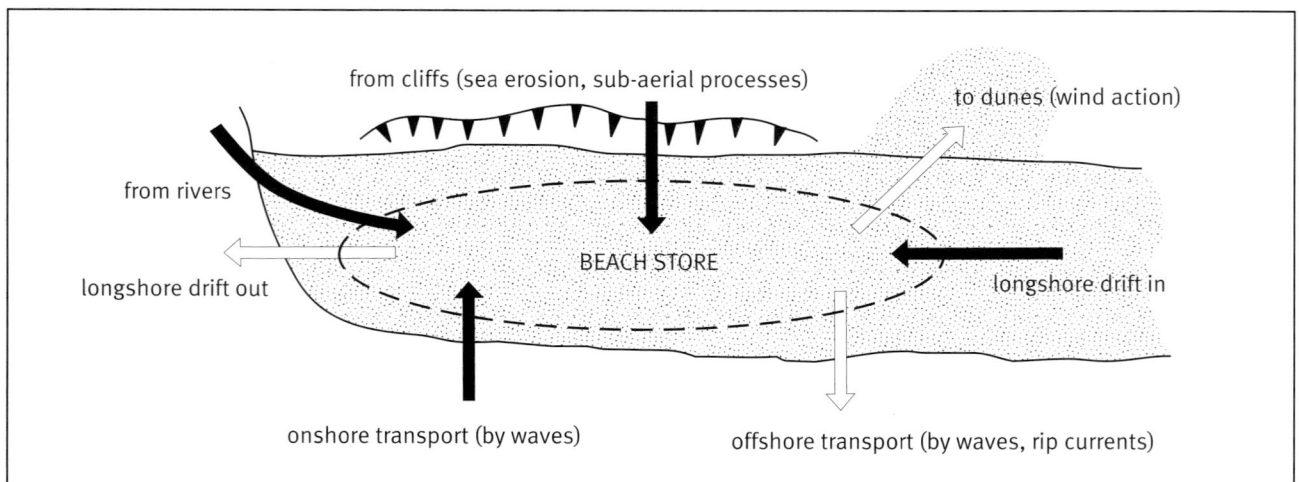

Figure 6.6 The marine sediment store system

landforms such as central Wales and the north Norfolk coast probably derive much of their material from offshore deposits dumped by the Pleistocene ice sheets. Many depositional landforms rely on a supply of sand and shingle transported along the coast by **longshore drift** (Figure 6.2). On relatively straight coastlines where there are no headlands to trap material, longshore drift may be the dominant process that creates the shape of the coastline.

Features of coastal transportation and deposition

Beaches

Beaches extend from the lowest level of low tide to the point where storm waves can throw material. Their profile varies over time dependent on whether storm or swell waves are dominant. Beach profile also varies with sediment type. Shingle supports steeper beaches partly because it has a higher angle of rest than sand. Also, shingle is more permeable; water from the swash quickly infiltrates, thus the effect of

the backwash to remove sediment towards the sea is reduced. Figure 6.7 shows the various detailed features of beaches.

Spits

Sand and shingle spits are formed where longshore drift continues the transport of sediment in a relatively constant direction, despite the alignment of the coast abruptly changing, for example, at a river estuary (Figures 6.8 and 6.9). However, the action of storm waves throwing material up above high water mark is also important. In the case of some spits and bars (see below) the action of storm waves may be the dominant process forming the landform; the feature so formed is described as being **swash-aligned** rather than the more common (longshore) **drift-aligned**. Many spits have **recurved ends**; these may be due to the refraction of waves as they reach the end of the spit or may be caused by storm waves from a different direction to the dominant waves (Figure 6.9). **Lateral ridges** of shingle represent former

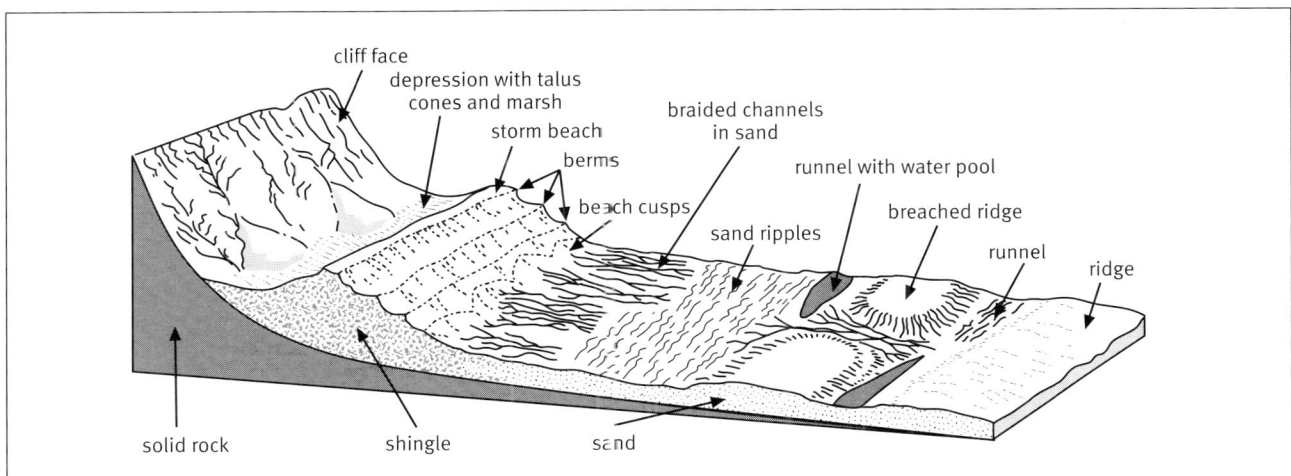

Figure 6.7 Features of beaches

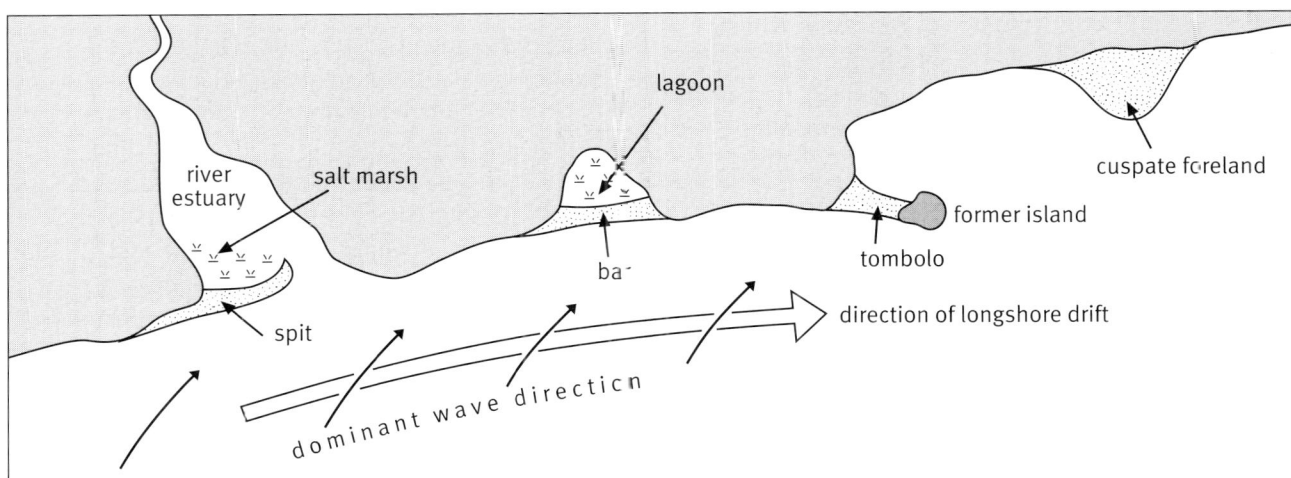

Figure 6.8 Features of coastal transportation and deposition

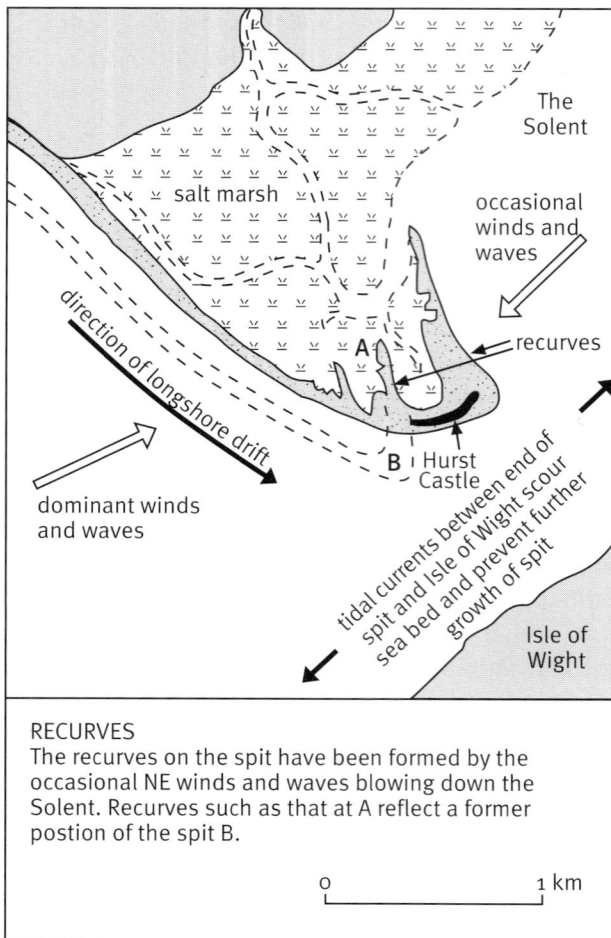

RECURVES
The recurves on the spit have been formed by the occasional NE winds and waves blowing down the Solent. Recurves such as that at A reflect a former postion of the spit B.

0   1 km

Figure 6.9 Hurst Castle Spit, Hampshire

recurved ends. **Sand dunes** develop on many spits and **salt marshes** develop in sheltered water behind them (see below).

Bars, tombolos and cuspate forelands (Figure 6.8)

If a spit is built across a bay (usually shallow and without a major river entering it) so as to reach the far side, it becomes a **bar**. An example is Slapton Ley, Devon. If a spit develops so as to join an island to the mainland, it becomes a **tombolo**. The part of Chesil Beach which links the Isle of Portland to the mainland is an example. A **cuspate foreland** is more complex. It may be caused by longshore drift from opposing directions or, as is believed to be the case with Dungeness in Kent, by an ancient shingle spit being realigned and modified by storm waves.

Barrier islands

Long and narrow sandy islands aligned roughly parallel to a coastline are known as **barrier islands**. Possible origins include: offshore bars of sand which are moved landwards into shallower water by storm waves (for example, along much of the coast of south-eastern USA); beach ridges and spits submerged by

rising sea levels and subsequently breached by storm waves (for example, the Frisian Islands of Netherlands and Germany).

Sand dunes

Areas of **sand dunes** are found on many spits and bars. Beach sand dries out at low tide and is then blown inland and above high water mark by the wind. The detailed form of sand dunes depends on the supply and movement of sand together with characteristic vegetation types which form a succession known as a **psammosere** (Figure 6.9).

Salt marshes

Salt marshes develop in sheltered water such as in an estuary behind a spit (Figure 6.9). They depend on a supply of fine sediment such as river alluvium. As the sediment is built up it is colonised by salt-tolerant plants (**halophytes**). The vegetation succession on a salt marsh is known as a **halosere** (Figure 6.10).

Review task

1 **Referring to your notes and textbooks, draw simple (learnable) sketch maps of examples of the various landforms of coastal transportation and deposition. Add labels to show details (for example, recurves on a spit) and relevant processes.**

2 **Revise your understanding of how sand dunes form and develop by referring to your notes or textbooks. Check your background knowledge of vegetation succession (seres) by referring to Chapter 5, page 44.**

Sea level changes

Sea level changes in recent geological history reflect the growth and decay of ice sheets. Worldwide changes so caused are **eustatic** changes. Local changes caused by thickening ice depressing the crust or the reduction in the weight of ice on the land (as it melts) leading to uplift are **isostatic** changes. Following the maximum extent of glaciation about 16 000 years ago there was a worldwide (eustatic) rise in sea level of about 110 m lasting about 12 000 years. The most recent isostatic rise of the land in northern Europe is a delayed response to the removal of the weight of the ice sheets. Landforms resulting from submergence (a rise in sea level) include **rias** (drowned river valleys) and **fjords** (drowned glaciated valleys). Landforms resulting from emergence include **raised beaches** and **abandoned cliff-lines**. Landforms resulting from both submergence and emergence may

PSAMMOSERE

← embryo dunes → ← fore dunes → ← main dunes → ← older dunes →

dune heath; some marram grass, red fescue, sea spurge, heather

gorse, bracken, heather, small shrubs

small trees   large trees
(e.g. pine, birch, oak)

marram grass

sea couch grass, some marram grass

increasing development of soil

sea   beach

water table   dune slack   cotton grass, juncus, reeds, rushes
(water table close to surface)

HALOSERE

non-halophytic plants

trees (e.g. alder, oak)

← mud flats → ← sward zone →

pioneer species

sea aster, sea lavender, grasses

shrubs

juncus, reeds

algae   marsh samphire (salicornia)   marsh cordgrass (spartina)

high water mark (spring tides)
high water mark (ordinary tides)
low water mark

creek   salt-pan depressions

increasing build-up of silt (plants such as spartina are very effective in trapping sediment)

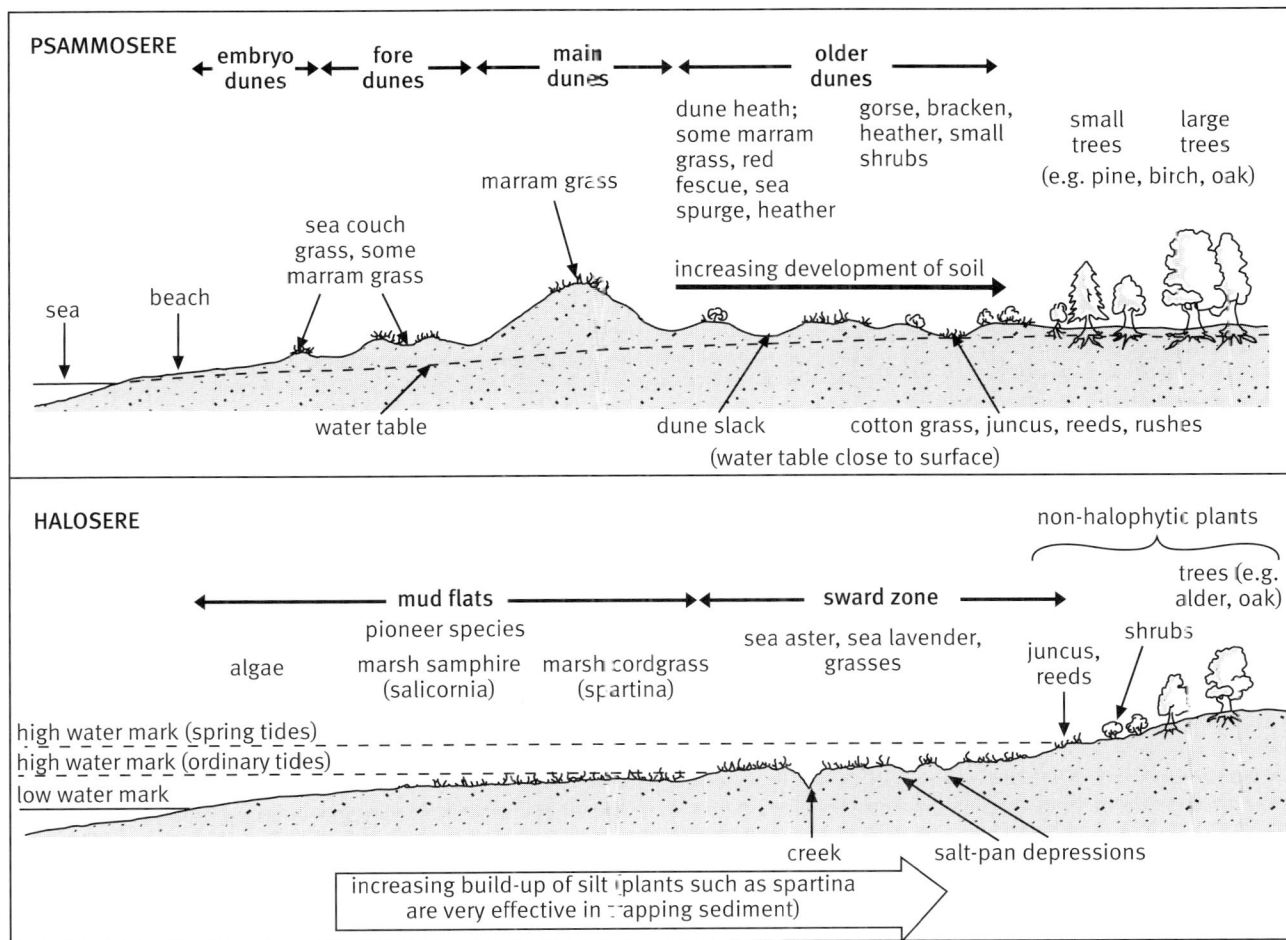

Figure 6.10 Coastal ecosystems

exist in the same area; for example, in western Scotland there are fjords (or sea lochs) resulting from the eustatic change of sea level and raised beaches resulting from the more recent isostatic change.

Global warming and rising sea levels

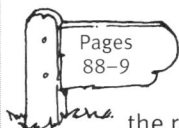

Pages 88–9

Since the 17th century, the world has been slowly warming. This trend has accelerated in recent years because of the release of carbon dioxide and other 'greenhouse' gases and is predicted to continue. Rising temperatures cause rising sea levels, primarily because as ocean water heats, it expands. The extent to which global warming will melt glaciers and polar ice sheets and so cause rises in sea level is more uncertain. Figure 6.11 shows the likely impacts.

Slight increase in sea level

- increased coastal erosion
- salinisation of coastal ecosystems
- greater threats from storm surges and marine flooding

An increase in sea level of 1 metre

- loss of low-lying agricultural land (e.g. Nile Delta, Ganges Delta)
- flooding threat to many major ports and conurbations, including London, Rotterdam, Tokyo, Bangkok and Calcutta
- destruction of coastal ecosystems
- total submergence of low-lying island countries (e.g. Maldive Islands)

Figure 6.11 The impact of rising sea level

Management of coastlines

Managing coastal erosion

People have settled on coastlines and so have to deal with the impact of coastal flooding and erosion. Human activity (such as removing beach material) may make the threat from the sea worse. The response over the past century has been to build various sea-defence works. Many defence works are coming to the end of their useful life. Sea defences are very costly to build and to improve. Contrasting approaches to management are:

- **Cost-benefit analysis** in which a decision to build sea defences or not is based on balancing the value of land and property saved against the cost of works.

- **Managed retreat** in which the defence of some coastal areas may be abandoned. The principle of adapting and supplementing natural processes would be followed.

Figure 6.12 shows recent approaches to managing the coastline of Christchurch Bay. Notice how defence measures at one location may have a negative impact further along the coast. For example, the construction of the Long Groyne (A in Figure 6.12) at Hengistbury Head in 1938 prevented the movement of sediment from the west into Christchurch Bay. The groynes built by Christchurch council also trapped sediment. The reduced sediment supply meant that the cliffs at Barton-on-Sea had less protective beach beneath them and so became more vulnerable to erosion. Also, further east, the supply of shingle to Hurst Castle Spit was reduced.

Managing human activities

Coastlines require management for other reasons as well. Various, sometimes conflicting, land uses put pressures on coastlines. Thus the coast of the Isle of Purbeck and Poole Harbour in Dorset (Figure 6.13) faces pressures from recreation, resource development, urban growth and other human activities in an area with sensitive environments and attractive coastal scenery. Many of the human activities listed in Figure 6.13 are not compatible with each other. Thus on the water, sailing and canoeing conflict with powerboating and water-skiing; commercial shipping does not coexist easily with recreational activities. Various human land uses and pressures for new development impact upon the scenery and wildlife habitats of this area. In Poole Harbour the Wych Farm oil field is located in a sensitive marsh ecosystem and extends southwards into a Site of Special Scientific Interest (SSSI) on the Dorset heathlands. In the Isle of Purbeck facilities for visitors such as caravan parks may cause visual pollution. Large numbers of visitors cause footpath erosion, especially on the coastal footpath, disturb wildlife and can be a nuisance to farmers.

To resolve these conflicts and manage human activities, much of this area of Dorset, as Figure 6.13 shows, is protected by various conservation measures. In some locations organisations such as the National Trust and RSPB have purchased land so as to preserve landscape and protect wildlife. Most of the area has been designated an Area of Outstanding Natural Beauty (AONB) in order to tighten planning controls and particular locations have been given the special status of Sites of Special Scientific Interest so as to preserve habitats through strict controls. On the water, various recreational activities are zoned so as to avoid conflict, with the southern part of Poole Harbour being designated a quiet zone where motorised craft are prohibited. In addition, the structure plans and local plans of local authorities aim to balance the various pressures and plan for the future.

Review task

1 **Either for the coastline of Christchurch Bay (Figure 6.12) or another area you have studied, a) list the reasons why management is needed – consider both the physical nature of the coast and its human use, b) list the protection measures attempted, giving reasons for their construction, c) list the likely benefits and costs, including possible negative impacts elsewhere on the coast. If your case study has contrasting examples within it, as with Barton Cliffs and Hurst Castle Spit in Figure 6.12, then work through this task separately for each localised example. You may find it convenient to use diagrams to illustrate, or even replace, your lists.**

2 **For a coastal area you have studied, list the various human activities that impose pressures. Also, list locations where conservation is important, briefly noting reasons. Where conflicts exist between activities and between an activity and a conservation location, join the items concerned with a line, labelled to note the nature of the conflict. List separately the attempts which have been made to manage your case study area. Link them to the relevant conflicts.**

A

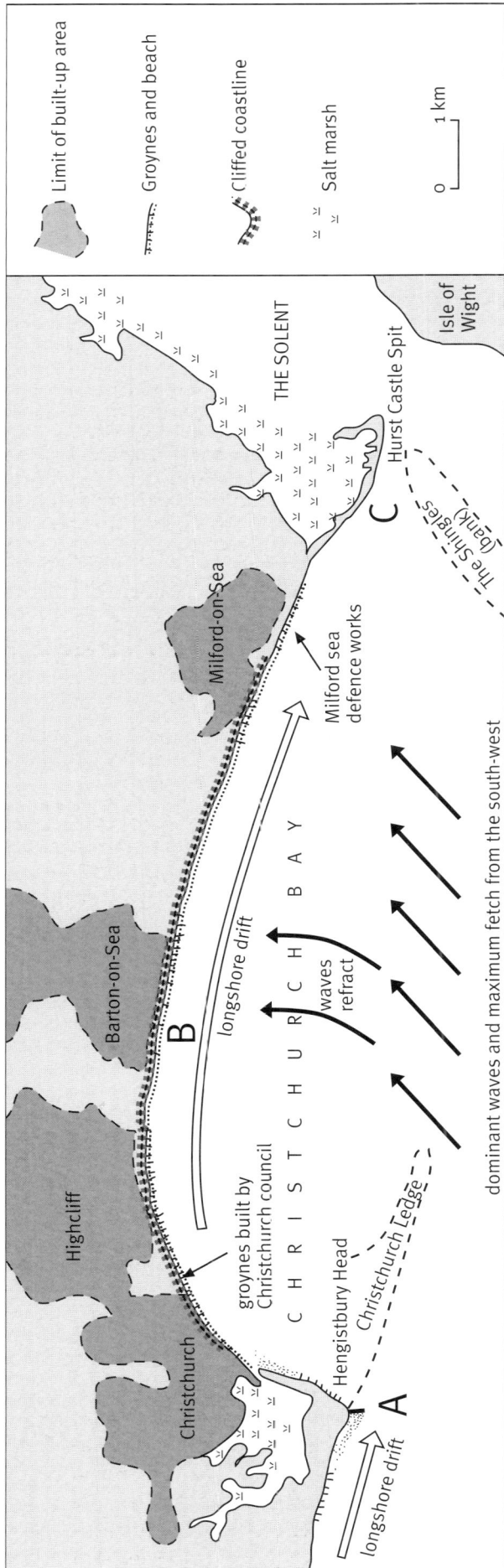

A LONG GROYNE – designed to protect Hengistbury Head from erosion; prevents movement of sediment into Christchurch Bay (constructed in 1938)

C HURST CASTLE SPIT
The spit protects the low-lying Solent coast from erosion. It has been breached by storms several times. Sea defence measures elsewhere have reduced the supply of shingle by longshore drift.

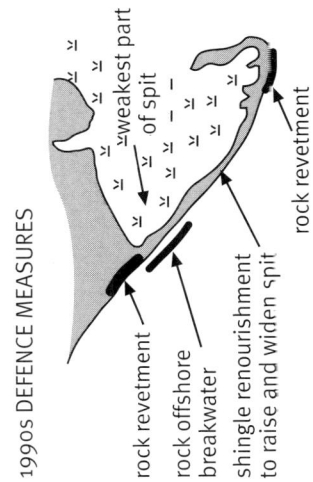

B BARTON-ON-SEA DEFENCE WORKS (CONSTRUCTED 1990s)
The problem:
1. Geology – easily eroded cliffs.
2. Housing close to cliff top.
3. Defences built in 1960s breaking up.
4. Reduced supply of sediment to beach from further west because of Long Groyne and groynes at Christchurch.

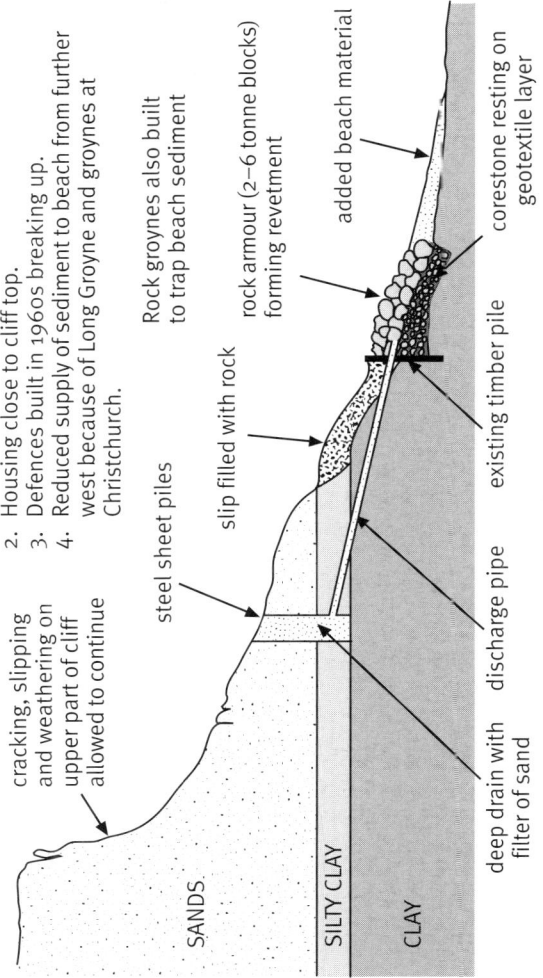

Figure 6.12 Christchurch Bay: management of coastal erosion

PRESSURE FROM HUMAN ACTIVITIES

Royal Marines Base

Poole

Bournemouth

Poole Bay

Poole Harbour

Wareham

Wych Farm oil field

Ferry services to France and cargo ships

Isle of Purbeck

Studland

Swanage   Swanage Bay

Durlston Head

St Aldhelm's Head

Legend:
- 🏠 Built-up areas
- Ⓕ Fishing ports (inshore fishing mainly for crab and lobster)
- ⒮ Spoil dumping grounds (for harbour dredgers)
- Ⓠ Quarrying
- 〰 Sand beaches
- ⏚ Resort development
- ◎ Caravan park
- Ⓡ Rock climbing on cliffs
- ▲ Boat marinas

OTHER RECREATIONAL ACTIVITIES

On land: walking, mountain biking, bird watching.

On water: sailing, canoeing, powerboating, waterskiing, boardsailing, jetskiing, diving, fishing.

0 ——————— 10 km

ENVIRONMENTAL PROTECTION AND PLANNING CONTROLS

- ⤸ Marine recreation areas (in Poole Harbour, different activities occupy their own zones)
- Ⓡₑₛ Nature reserve
- ⓃⓉ National Trust land
- ⒸⓅ Country Park

PLANNING AND WILDLIFE PROTECTION

AREA OF OUTSTANDING NATURAL BEAUTY (AONB)

– all of coastal area and much of inland area

HERITAGE COAST – all the coast south of Poole Harbour

SITES OF SPECIAL SCIENTIFIC INTEREST (SSSI)

– all the coast except built-up areas. Also, Poole Harbour and Dorset Heathlands

SPECIAL PROTECTION AREAS (for birds)

– proposed for Poole Harbour and Dorset Heathlands

ALSO: National Nature Reserves, National Trust owned land, RSPB reserves, Dorset Trust for Nature Conservation

SENSITIVE MARINE AREA (SMA)

– proposed for most of coastal waters of Isle of Purbeck

Poole

Bournemouth

Poole Bay

Poole Harbour

Wareham

Dorset Heathlands

Power boat racing area

Isle of Purbeck

Studland

Studland Bay

Swanage

Swanage Bay

St Aldhelm's Head

Durlston Head

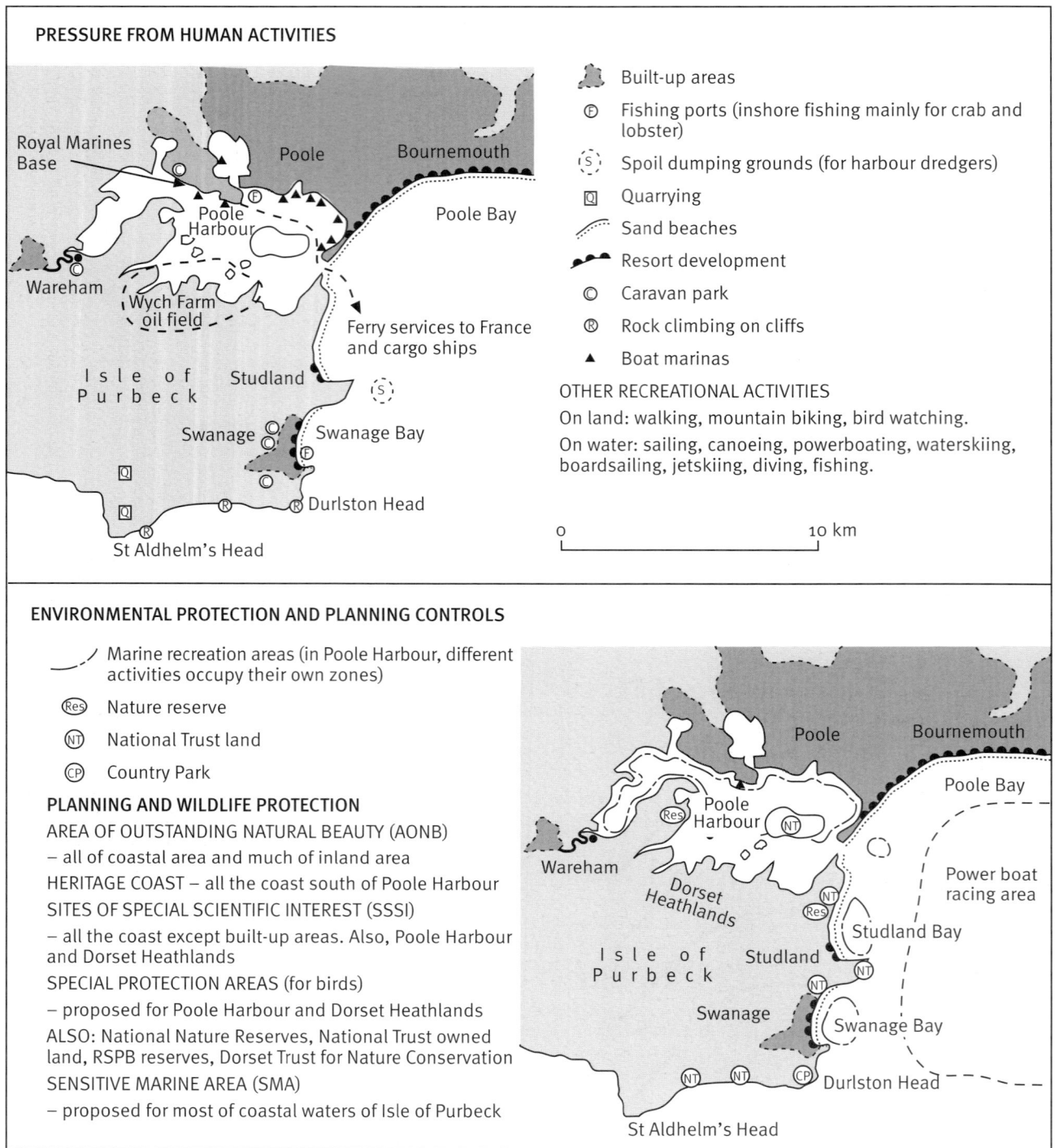

Figure 6.13 Isle of Purbeck and Poole Harbour: human activities and coastal management

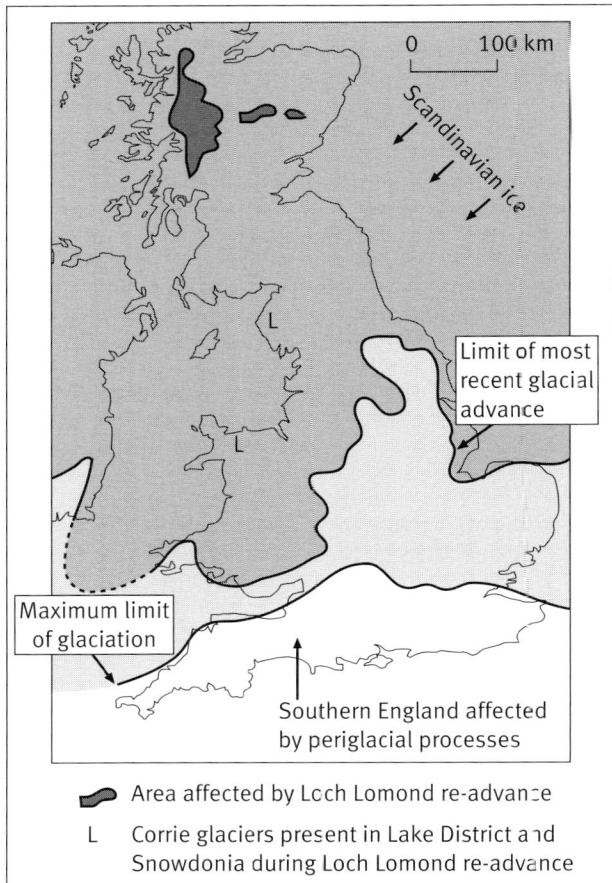

Ice ages

Although there have been many ice ages and periods of glaciation in the earth's geological history, only the most recent one during the **Pleistocene** period of the **Quaternary** era (Figure 7.1) has had a significant effect on the present-day landscape. The Pleistocene Ice Age was marked by cold periods of advancing ice (**glacials**) separated by warmer **interglacials**. Glaciologists disagree as to the number of glacials; the table in Figure 7.1 shows a very simplified picture and highlights the periods of most significance in shaping the landscape and relates them to the map showing the spatial pattern of glaciation in Britain. At

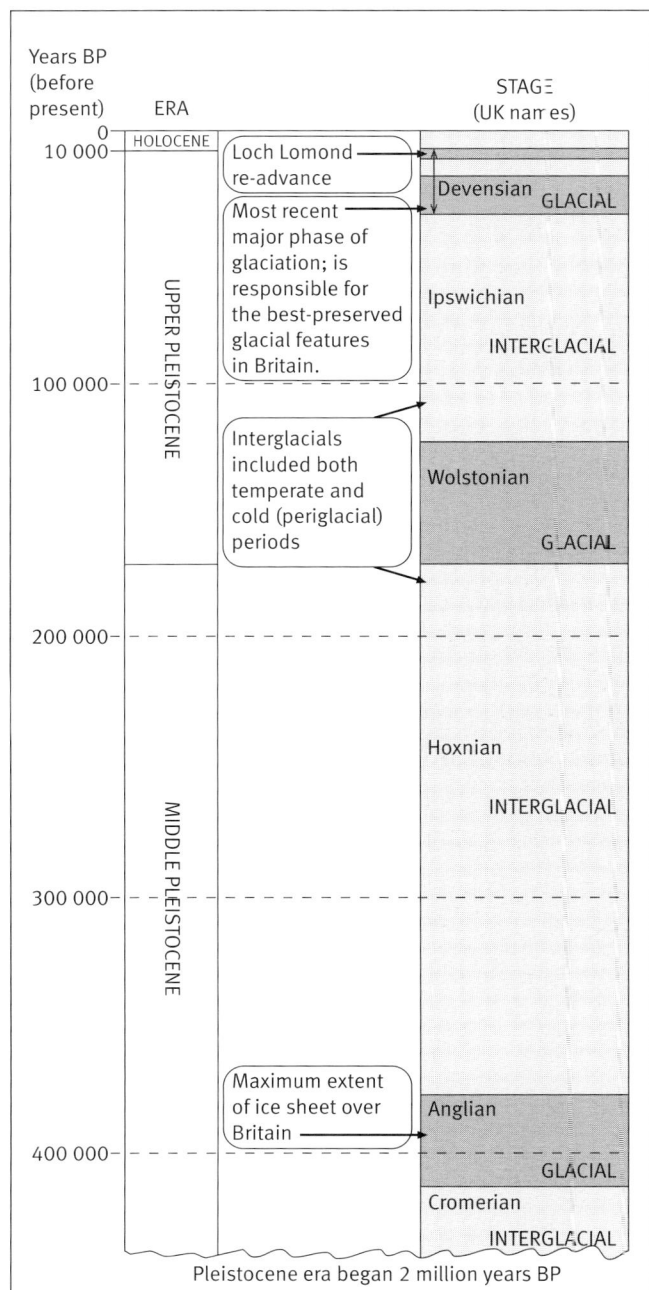

Figure 7.1 Ice Age chronology

its maximum, ice covered about 30 per cent of the earth's land surface compared with about 10 per cent today.

The Ice Age also had other significant effects:

- The pattern of global climate and vegetation belts changed. Thus areas fringing the ice sheets experienced tundra conditions and the impact of **periglacial** processes (intense freeze-thaw action and the various processes associated with conditions of permafrost).

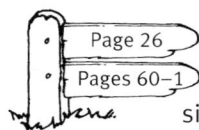

- Periglacial processes affected those areas covered by the ice sheet in the periods before and after glaciation. (Consider why this is of significance.)

Page 26
Pages 60–1

- Advances and retreats of ice caused changing sea levels which in turn affected the geomorphology of river valleys (see Chapter 3) and coastlines (see Chapter 6).

What causes ice ages? Many different causes of climatic changes leading to ice ages have been suggested: sunspot activity affecting incoming solar energy; changes in the earth's orbit or the tilt of the axis; changes in the earth's surface caused by plate tectonics (in particular the distribution and relative positioning of land and sea); a reduction of carbon dioxide in the atmosphere and changes in the pattern of ocean currents.

Ice sheets and glaciers

Classification of types

Figure 7.2 shows how ice sheets and glaciers may be classified according to their size and shape. They may also be classified thermally into temperate and polar glaciers. Temperate (**warm based**) glaciers occur in areas with mild summers (for example, the Alps) and summer melting leads to the presence of much water which reduces friction and allows the glacier to move freely. Polar (**cold based**) glaciers are found in Arctic areas (for example, Greenland); no melting occurs beneath the surface and so movement and erosion will be more limited.

Glaciers as systems

As Figure 7.3 shows, the glacial system is more than an input of snow leading to ice which eventually melts. The major input is snow. As snow is compacted by the weight of later snowfalls it becomes **firn** or **névé**. Further pressure squeezes out air and the firn becomes glacier ice (at a depth of around 30 m). Most of the outputs are losses due to **ablation**, the melting of ice. The upper part of a glacier has inputs exceeding outputs and is known as the **zone of accumulation** in contrast to the lower part where outputs exceed inputs – the **zone of ablation** (Figure 7.3). A **line of equilibrium** (equivalent to the snow line) separates the two zones. Continued accumulation above this line and ablation below it would in theory lead to the glacier becoming top heavy. In practice this does not happen as a transfer of ice from head to toe occurs through glacier movement, so maintaining the equilibrium.

Inputs and outputs also vary over time. The occurrence of much snow and little melting in winter balanced by little accumulation and much melting in summer, give the annual balance (Figure 7.4). If the positive balance of summer continues to exceed the negative balance of winter (i.e. ablation exceeds accumulation), over a period of years a glacier retreats. If the opposite occurs, a glacier advances. Most Alpine glaciers have retreated over the past hundred years.

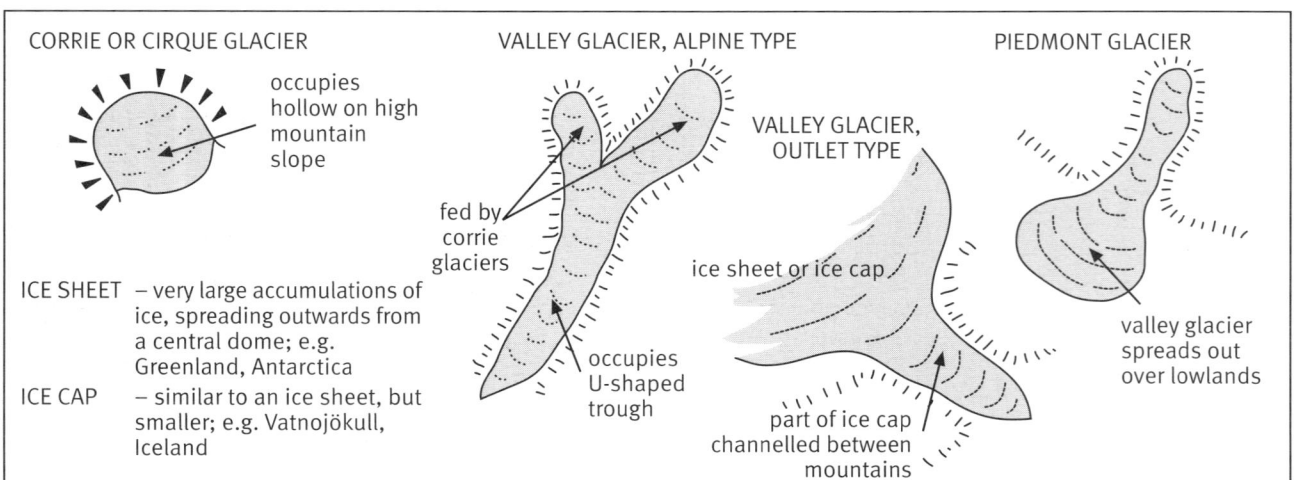

CORRIE OR CIRQUE GLACIER — occupies hollow on high mountain slope

VALLEY GLACIER, ALPINE TYPE — fed by corrie glaciers; occupies U-shaped trough

VALLEY GLACIER, OUTLET TYPE — ice sheet or ice cap; part of ice cap channelled between mountains

PIEDMONT GLACIER — valley glacier spreads out over lowlands

ICE SHEET — very large accumulations of ice, spreading outwards from a central dome; e.g. Greenland, Antarctica

ICE CAP — similar to an ice sheet, but smaller; e.g. Vatnojökull, Iceland

Figure 7.2 Glaciers and ice masses

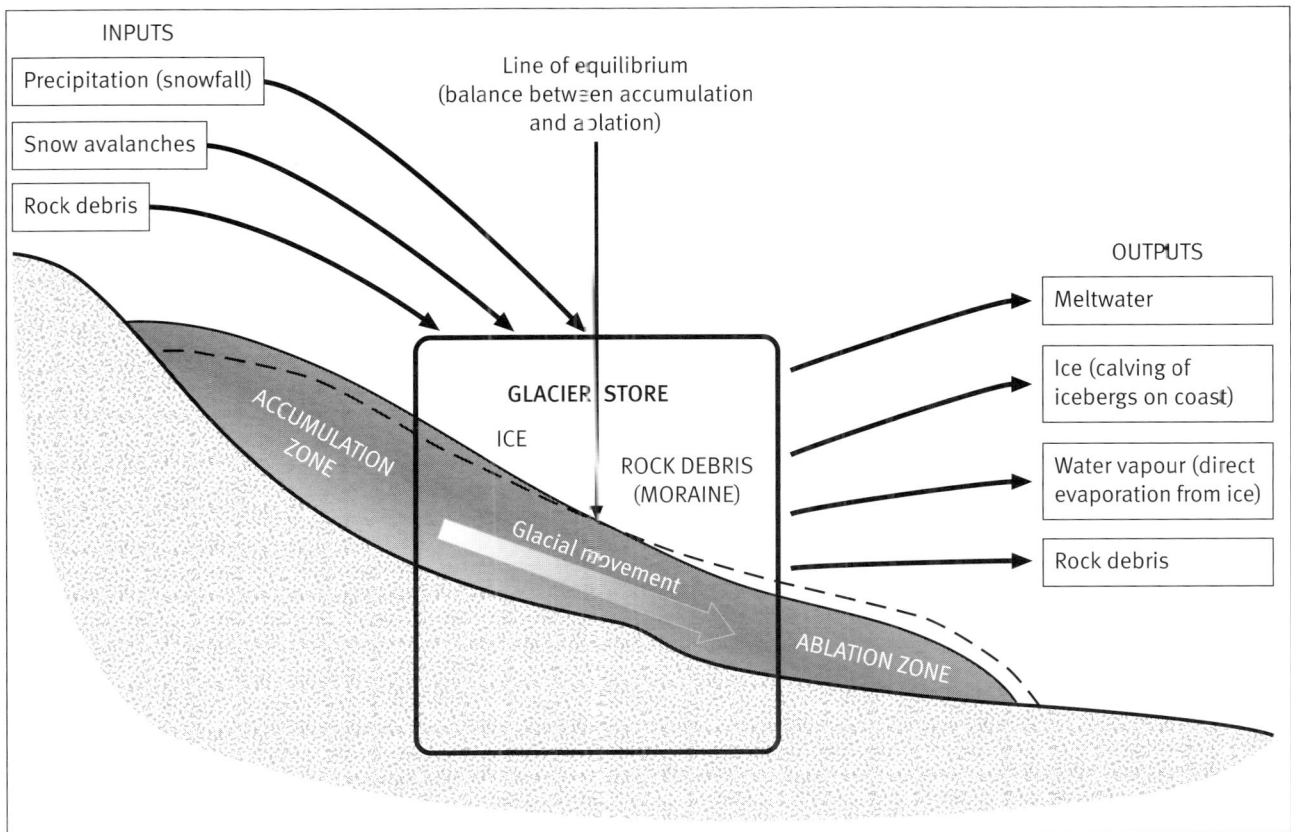

Figure 7.3 The glacial system

Ice movement

Two basic processes contribute to the downslope flow of glacier ice:

1 **Basal sliding** is the sliding of ice over bedrock at the base of temperate glaciers. It is made possible by the presence of meltwater, which acts as a lubricant. Much of the meltwater occurs because local increases of pressure at obstacles lead to a rise in temperature and thus a melting of ice.

2 **Internal deformation** is a combination of various processes occurring within the ice. The pull of gravity downslope and the weight of ice above exerts stress on ice crystals so that they slide over each other and deform within themselves; the result is **creep**. If stress is very great, ice crystals may **fracture** and cause slippage along the fracture plains.

The speed of glacial flow is affected by gradient, the temperature of the ice (at its base especially), the thickness of the glacier and the mass of the ice upslope. Across a glacier the rate of flow varies because of friction with the valley sides and bottom. Transverse crevasses therefore become curved and great differences in flow may lead to

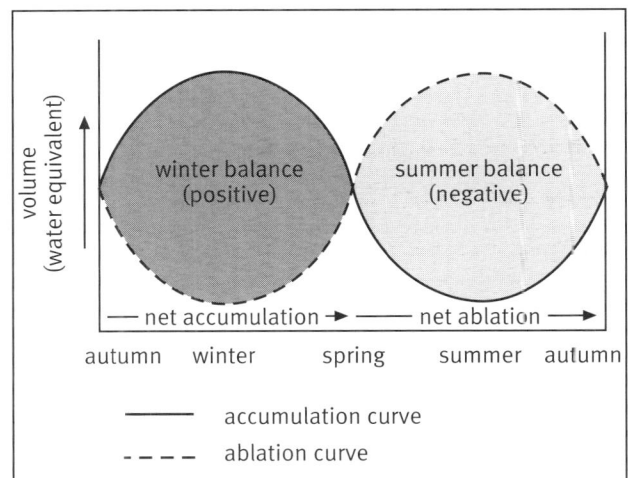

Figure 7.4 The annual balance (or budget) of a glacier

the development of longitudinal crevasses (Figure 7.5). Along the course of a glacier, variations in rates of flow have dramatic effects. On a steepening gradient ice accelerates, and therefore stretches and thins out. This is called **extending flow**. On a gentle gradient, ice decelerates, compresses and thickens; this is **compressing flow**. Note in Figure 7.5 how slip planes develop with both types of flow and result in the development and closing up of transverse crevasses. Extreme extending flow over a rock step may lead to crevassing forming an **icefall**.

Ice transport

The rock debris carried by a glacier originates from the weathering and landsliding of slopes above the glacier and from material eroded by the glacier itself. The rock debris is classified into:

- **Supraglacial debris** on top of the glacier, mostly in the form of **lateral** and **medial moraines**. (A lateral moraine is a band of debris at the edge of a glacier. A medial moraine is formed by lateral moraines joining as glaciers merge together.)

- **Englacial debris** within the body of the glacier. It may fall from the surface down crevasses or be carried down by meltwater.

- **Subglacial debris** at the base of a glacier, mostly originating from the glacier's erosion of the bedrock.

Near the glacier's snout, ablation uncovers much of the englacial debris. Compressing flow (see Figure 7.5) also brings much debris to the surface. The entire surface of the glacier may be covered with debris.

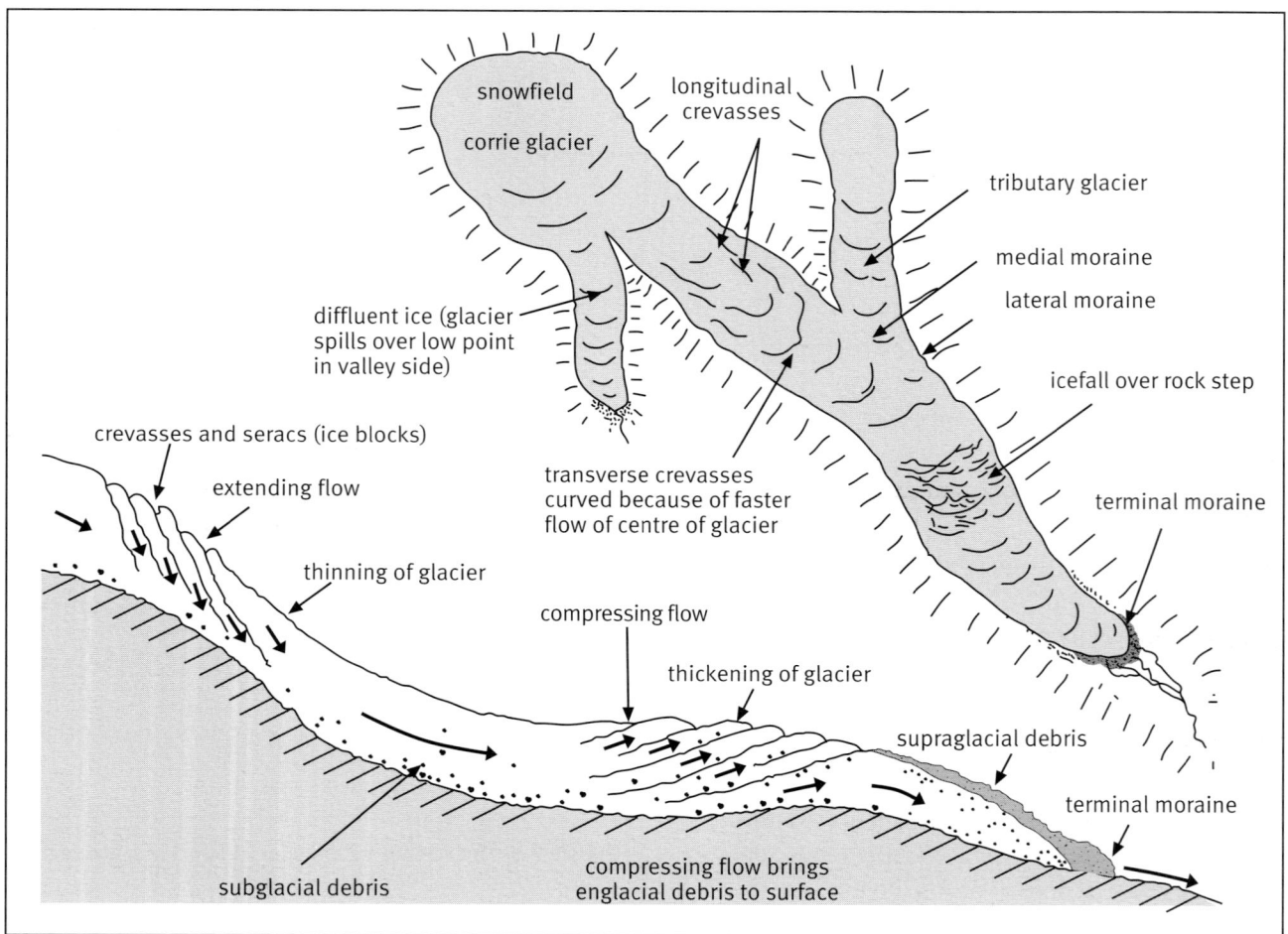

Figure 7.5 Features of valley glaciers

Glacial erosion

Erosion processes

- **Abrasion** (or **corrasion**) is the glacier's use of angular debris, held by the ice, to scrape away at the underlying rock. Evidence of this process are scratches on rock known as **striations**.

- **Plucking** involves the glacier freezing on to blocks of rock and pulling them away. It probably occurs in this simple form with previously loosened rock.

Meltwater will assist in this plucking process: pressure builds up behind a protrusion of rock and so causes melting. The meltwater penetrates any cracks and freezes around the rock (**regelation**) which is then pulled out by the glacier.

In Figure 7.6 these processes are applied to the erosion of a corrie. The **rotational movement** of the corrie glacier adds power to the erosional processes and overdeepens the corrie. Extending flow and compressing flow also have their effect on the effectiveness of erosion (consider how and why).

Processes that weather or weaken rock also contribute to the effectiveness of erosion:

- The weathering of rock by **freeze-thaw action** (or **frost shattering**) may break up rock in periglacial conditions before glaciers advance. At the early stage of corrie formation freeze-thaw action and possibly chemical weathering will weather rock beneath the accumulating patch of snow (the process of **nivation**). During glaciation, meltwater will give rise to freeze-thaw action at the base of the glacier.

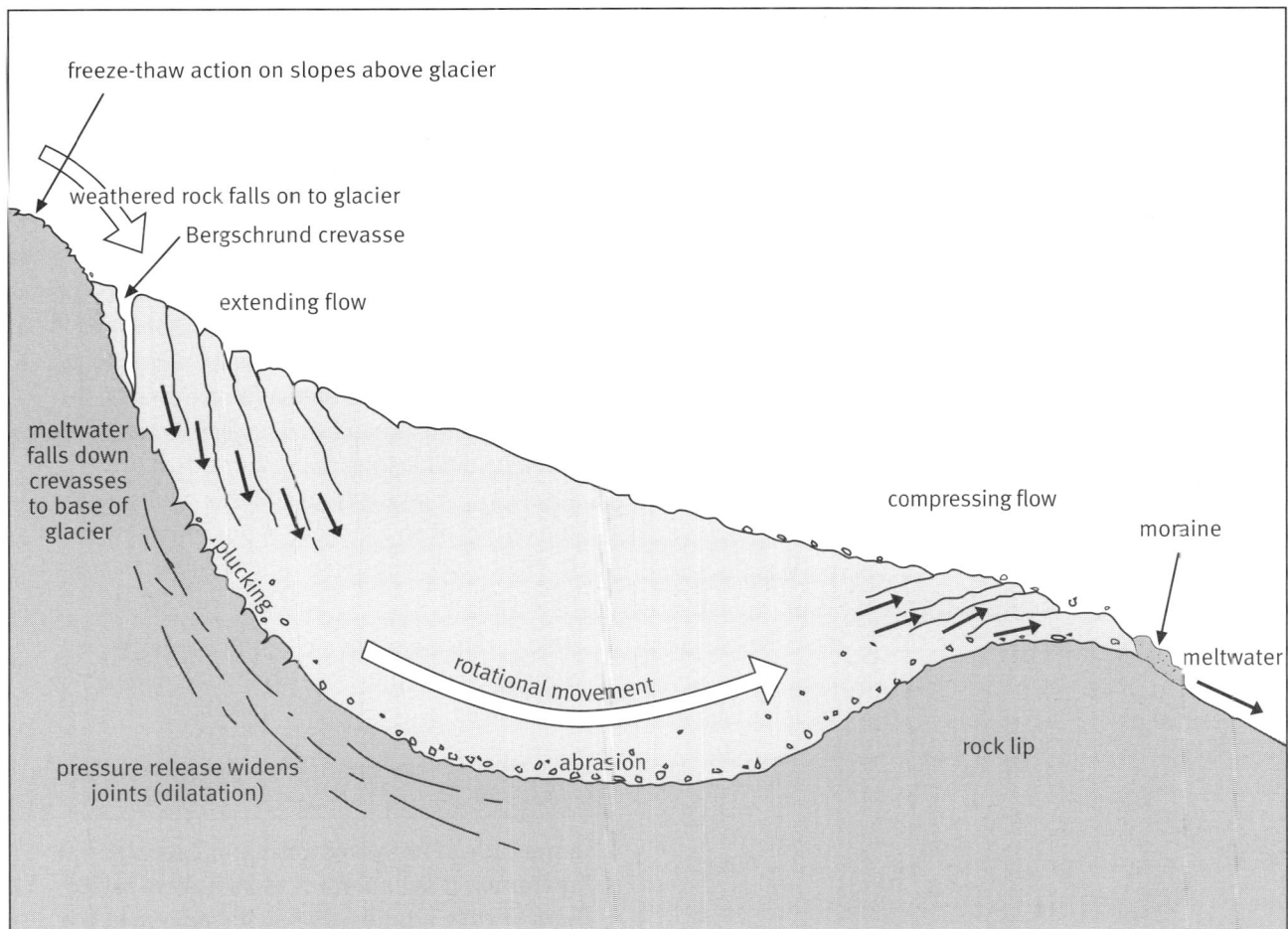

Figure 7.6 A corrie glacier: erosion processes

- Removal of layers of rock by glacial erosion causes a release of pressure in underlying rocks and the expansion of joints. This phenomenon, known as **dilatation**, serves to weaken the rock and make it more vulnerable to erosion.

Landforms produced by glacial erosion

- **Corries** (also known as **cirques** or **cwms**) – see Figure 7.6.

- **Glacial troughs** are typically U-shaped in cross section and are formed by valley glaciers which follow preglacial river valleys. In contrast to a river, a glacier occupies all of its valley and so widens, deepens and straightens it into an efficient channel for its ice. Overdeepening is a feature of many troughs; theories explaining this are either geological (for example, the presence of weaker rocks) or involve localised greater erosive power (e.g. caused by the confluence of two glaciers). **Ribbon lakes** may occupy overdeepened sections.

- Erosive features associated with glacial troughs: **rock steps** (between overdeepened basins), **roches moutonnées** (Figure 7.7) (ice-smoothed rocks with irregular steep faces pointing down-valley), **hanging valleys**, **truncated spurs**, **cols** formed by **diffluent ice** (ice breaks away from the main glacier through a low point in the valley side which it then enlarges).

- Features separating glacial troughs or corries: **arêtes** (knife-edge ridges) and **pyramidal peaks**.

- A **crag and tail** is formed by ice-sheet erosion. A crag of resistant rock protects in its lee an area of less resistant rock from erosion, so forming the tail.

- Large expanses of lowland severely eroded by ice sheets may produce a **knock and lochan**

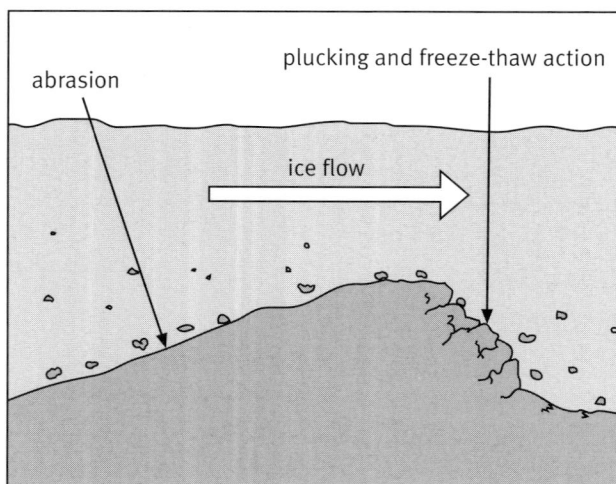

Figure 7.7 Roche moutonnée

landscape as in the coastal areas of north-west Scotland. Bare rock, hummocky terrain and a confused drainage pattern result.

Glacial deposition

As it melts, ice deposits the material it has transported. Such debris varies greatly in size and composition and, in contrast to fluvial deposits, is unsorted. Meltwater will transport ice-deposited debris further and produce fluvioglacial deposits of sorted material. The term **drift** includes both glacial and fluvioglacial deposits. Glacial deposits are termed **till**. **Lodgement till** is laid down beneath moving ice and **ablation till** beneath relatively static, melting ice.

Landforms produced by glacial deposition

- **Till plains** occur where wide expanses of lowland had been covered by glacial deposits (for example, much of East Anglia). The till deposited, also known as **boulder clay**, consists of sub-angular fragments and reflects the nature of rocks over which the ice has passed – thus in East Anglia the till is chalky.

- **Drumlins** are egg-shaped mounds of till smoothed by later ice movement.

- **Erratics** are boulders deposited by ice; 'erratic' because their rock type will differ from the underlying geology.

- Moraines are landforms built of glacial debris. **Lateral moraines** and **medial moraines** are found on the surface of glaciers (see above), their distinctive forms largely disappearing when the ice melts. At the snout of a glacier, the combination of the ice's forward movement and subsequent ablation results in the piling up of debris to form a **terminal** (or **end**) **moraine**. With ice sheets, the term terminal moraine is usually used for the large moraine marking the maximum extent of glaciation. A **recessional moraine** marks a pause in the retreat of an ice sheet. A **push moraine** results from a re-advance of ice which bulldozes material into a ridge.

- When static dead ice melts, the overlying supraglacial debris is dumped to form **dead ice hummocks**.

> **E X A M   T I P**
>
> In the case of both erosional and depositional landforms, it is important to be able to relate them to case studies, as has been done in the case of Easedale in Figure 7.8.

1 Work through the lists of landforms produced by glacial erosion and deposition. For each landform, refer to your notes or textbooks to check that you are familiar with how it has been formed.

2 For either Easedale (Figure 7.8) or a glaciated valley you have studied, construct a flow diagram to summarise the landforms and processes as follows. On the left-hand side of your page, list the landforms of the valley, naming major examples. At the bottom of the page list other influences on the valley's landforms (for example, pre-glacial drainage pattern, geology, human activity). On the right-hand side of your page, list the various processes: those of glacial erosion and deposition, weathering, fluvioglacial, post-glacial fluvial action. Link the various processes and influences to the landforms by lines with arrows. Thicken a line where the factor is of major importance. Add explanatory notes to lines where relevant.

Fluvioglacial processes and landforms

Fluvioglacial deposits are sorted vertically and horizontally and consist of particles that are more rounded than glacial deposits (because of attrition in flowing water). Material is deposited by meltwater because of a loss of energy, for example when water enters a lake or when there is a sudden drop in discharge. Figure 7.9 shows the resulting landforms.

The most important features formed by fluvioglacial erosion are **meltwater channels**. Such channels may be proglacial (immediately in front of an ice sheet), marginal (alongside an ice sheet), subglacial or overflow channels (see below).

Proglacial lakes, overflow channels and glacial diversions of drainage

As the ice sheet moved southwards over Britain during the Ice Age, rivers flowing northwards were blocked. **Proglacial lakes** built up between the ice sheet front and higher ground. Eventually a lake's level would rise high enough to overflow at the lowest point of its rim. As the water flowed out rapid erosion

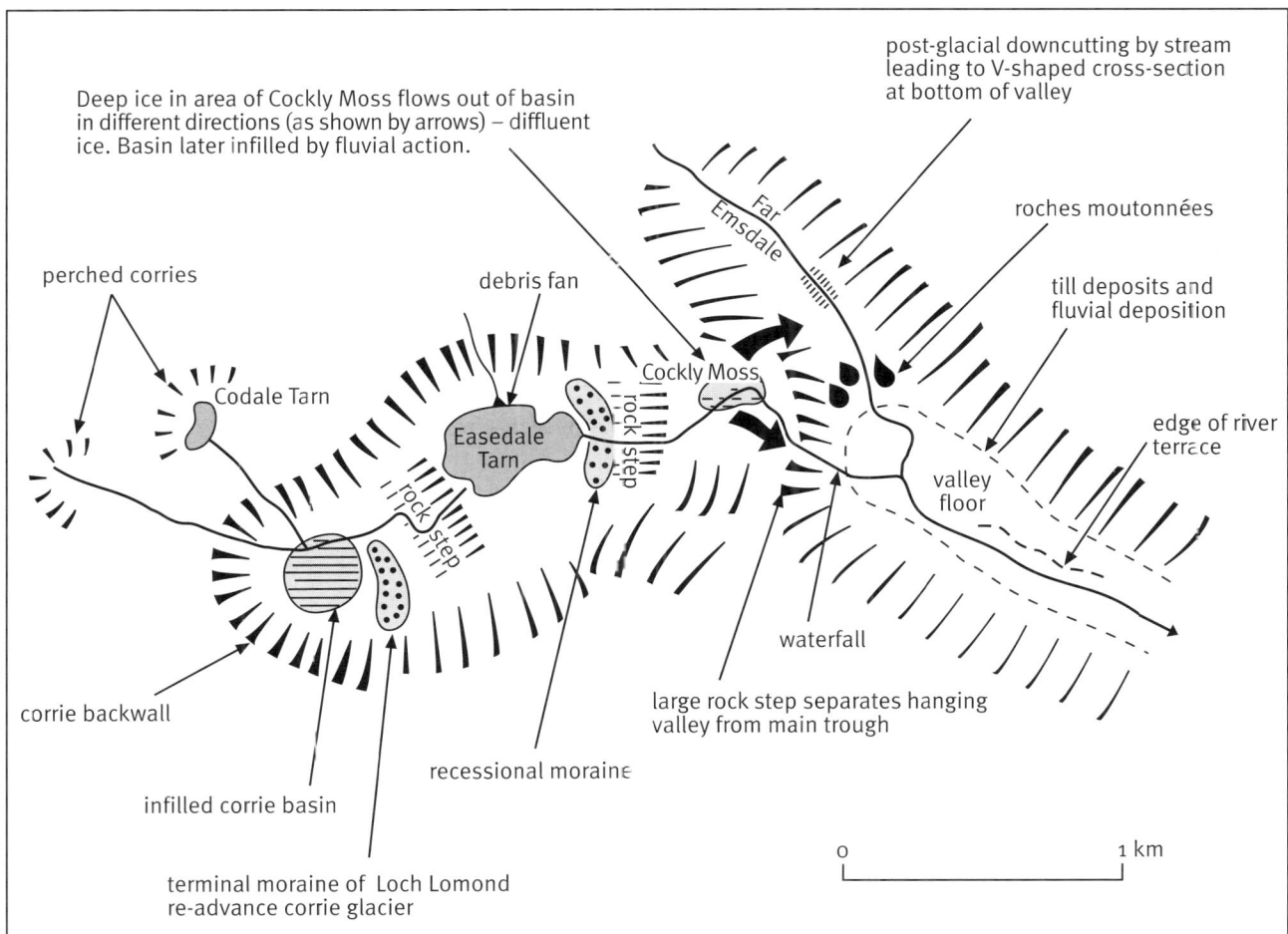

Figure 7.8 Easedale, Cumbria: a glacial valley

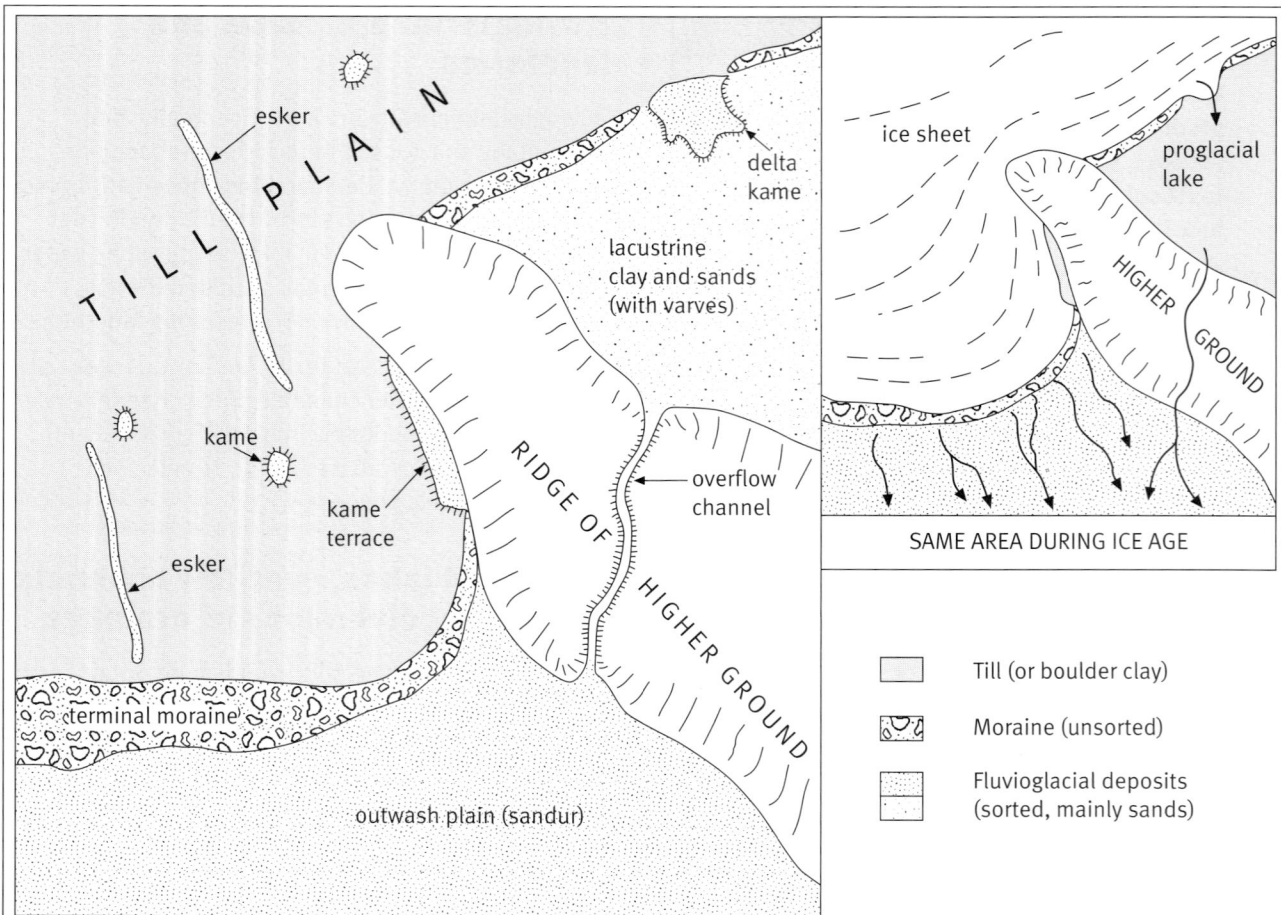

FLUVIOGLACIAL LANDFORMS

1. *Outwash plain*

2. *Esker* – ridge of sands and gravels formed in a sub-glacial tunnel.

3. *Kame* – irregular mound of sands and gravels deposited by meltwater on or at edge of ice sheet.

4. *Kame terrace* – deposited by meltwater between unglaciated higher ground and edge of ice sheet.

5. *Kame delta* – delta built into proglacial lake.

6. *Lacustrine clays and sands* – deposited on bottom of proglacial lake. Deposits show *varves*, bands of alternating coarse and fine material. (Each varve represents one year's deposition.)

Figure 7.9 Fluvioglacial landforms

would cut an **overflow channel**. Figure 7.10 shows how this happened in North Yorkshire. Water was dammed by the ice sheet to form a series of lakes in the Eskdale Valley, linked to each other by marginal channels. As the water overflowed to the south the large overflow channel of Newtondale was cut, leading to a further proglacial lake: Lake Pickering. This in turn overflowed through another overflow channel: Kirkham Abbey Gorge. Besides the meltwater channels, this sequence of events has had other impacts: delta deposits at Pickering, lacustrine deposits marking the extent of Lake Pickering and,

most significantly, a **glacial diversion of drainage** (see the inset map in Figure 7.10).

Glaciation and human activity

The impact of glaciation on human activity has been considerable both in lowland and highland areas:

- Highland areas, both those glaciated in the past and those with glaciers remaining today, are major areas of tourism because of the spectacular scenery and their suitability for winter sports. Glacial lakes are important for recreational use.

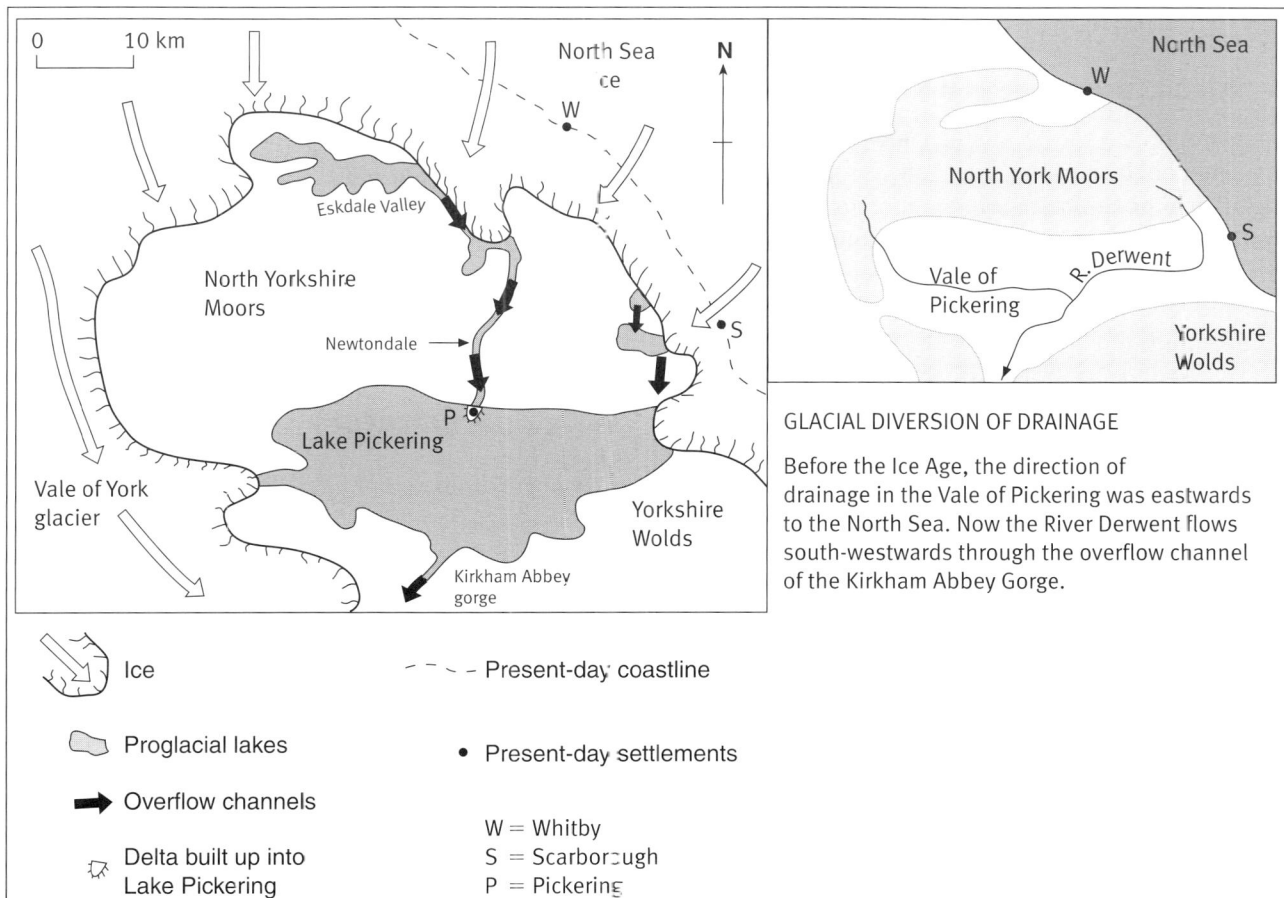

GLACIAL DIVERSION OF DRAINAGE

Before the Ice Age, the direction of drainage in the Vale of Pickering was eastwards to the North Sea. Now the River Derwent flows south-westwards through the overflow channel of the Kirkham Abbey Gorge.

Figure 7.10 North Yorkshire: overflow channels and glacial diversion of drainage

- Glacial troughs provide transport routeways.
- The marked relief of glacial highlands and the presence of lakes at high altitude makes them suitable for the development of hydro-electric power.
- Glaciatied highlands with high snowfalls present hazards such as avalanches. As the pressure of development increases, as in skiing areas, the impact of avalanches on people and settlements becomes greater.
- Glacial deposits provide aggregates (sand, gravel, etc.) for construction purposes.
- In lowlands the nature of drift has a major impact on agriculture. Thus in East Anglia fertile clay and loam soils, developed on chalky till, contrast with the infertile outwash sands of Breckland.

Review task

1 In the same way that features of ice erosion and deposition are linked to the example of Easedale (Figure 7.8), the fluvioglacial features in Figure 7.9 need to be linked to examples. Your information could be summarised in a table with three columns, headed 'Landform', 'Process' and 'Examples'. In the 'Examples' column you could include simple sketch maps and diagrams where appropriate, or references to your notes on this subject.

2 Again, examples are important in your knowledge of the impact of glaciation on human activity. Work through the list of impacts given above and make sure that you have knowledge of relevant examples in areas which have been glaciated or are undergoing glaciation at the present time. A Mind Map would be a useful means of approaching this task.

Periglaciation

Periglacial areas

Areas around ice sheets and glaciers are described as being **periglacial**. They are characterised by the presence of **permafrost**, permanently frozen ground. Continuous and widespread discontinuous permafrost covers about 20 per cent of the world's land surface. Large areas occur in Alaska, northern Canada, northern and eastern Russia and the Tibetan Plateau. Periglacial processes also occur in many high mountain areas, where temperatures are reduced by altitude, but permafrost does not exist. Periglacial features are also found in areas which were on the fringes of the Pleistocene ice sheets. Thus at the maximum extent of glaciation over Britain, southern England was affected by periglaciation.

Permafrost and the active layer

Permafrost varies greatly in thickness, up to 1500 m in Siberia. On its surface is a layer which is frozen in winter but thaws in summer: the **active layer**. Key features of the active layer are:

- When melting occurs, because the frozen ground beneath is impermeable, the active layer becomes waterlogged.

- Flows become major processes in this mobile layer.

- With the onset of winter, freezing progresses from the surface downwards. Unfrozen mobile materials are therefore trapped under increasing pressure and become contorted.

- Melting ice and snow lead to high stream discharges in the short summer.

Periglacial processes

- **Frost shattering** (or **freeze-thaw action**) is the most important weathering process in periglacial areas. Frost action also contributes to the process of **nivation** under snow patches.

Page 1
Page 69

- Groundwater is unevenly distributed, so when it freezes, as small crystals or ice lenses, the resultant expansion of the ground is uneven, but mainly upwards: this is **frost heave**. Sideways movement through freezing and expansion is **frost thrust**. Heaving/thrusting movements under pressure result in churning of materials – **cryoturbation**.

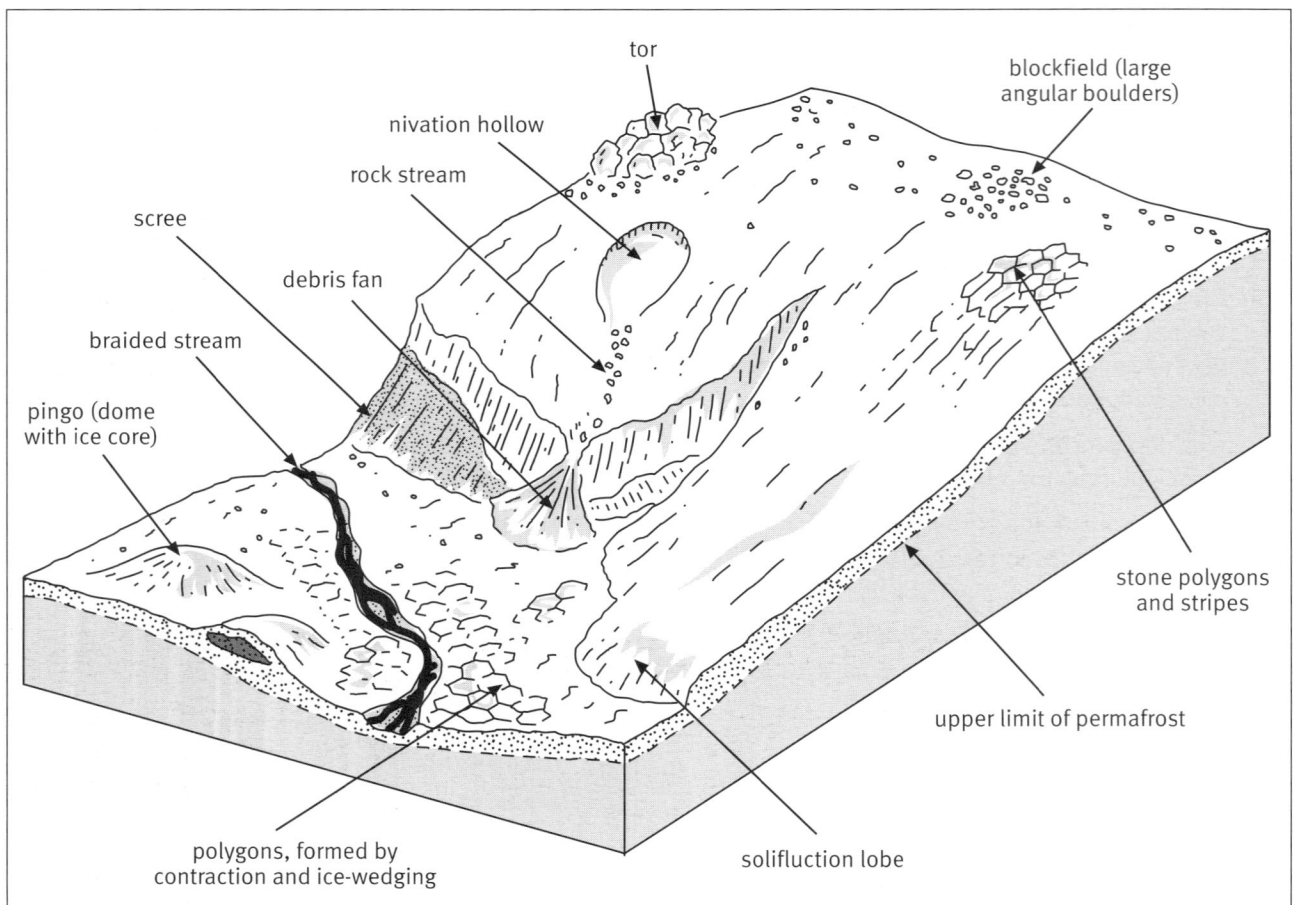

Figure 7.11 Periglacial landforms

- **Solifluction** is the slow downslope flow of a saturated, seasonally thawed layer. It can occur in cold environments without permafrost. When it is associated with the active layer on top of permafrost it is termed **gelifluction**.

- **Ground contraction** takes place under conditions of severe cold, leading to a polygonal pattern of cracks. Water later collects in the cracks and, when it freezes, wedges them further open.

- **Fluvial action** is of major importance in the short summers.

- **Wind action:** strong winds easily pick up fine glacial and fluvioglacial sediments in areas with little vegetation. The material may then be deposited as layers of **loess**.

Periglacial landforms

The various periglacial processes, often interacting with each other, produce very distinctive landforms (Figure 7.11). The different types of movement produce various forms of **hummock** and **patterned ground**. Solifluction tends to even out relief. Long-term melting of permafrost also produces a distinctive topography: melting of groundwater causes subsidence and because groundwater is unevenly distributed, the subsidence is uneven. The result is a landscape with a hummocky surface and water-filled depressions – a landscape known as **thermokarst**.

Periglacial areas and human activity

Human settlement in periglacial areas leads to a localised melting of permafrost and an expansion of thermokarst conditions. Low temperatures limit decomposition and so periglacial areas are very vulnerable to pollution. Permafrost causes considerable problems for building. They add to the problems of living in a severe climate. Figure 7.12 lists some of the problems and the measures adopted to remedy them.

Problems of living in cold environments

- Long winters, with prolonged snow cover, low temperatures, blizzards and icing, hinder human activity in general, prevent agriculture and hinder transport.
- Lack of light in winter hinders human activity.
- Freezing of rivers and coastlines hinders transport (although rivers may be used as winter highways).
- Disturbance of environment, such as felling vegetation, increases the effect of summer heating and so lowers permafrost and increases the effect of thermokarst.
- Permafrost gives rise to problems of constructing buildings and roads. Heat from buildings causes thawing and subsidence. Roads insulate underlying ground from summer heating and so the permafrost rises underneath. Construction of long-distance pipelines (e.g. for oil) is a major problem.
- Periglacial processes, such as frost heave, solifluction and thermokarst subsidence, cause difficulties for settlement.

- Services requiring pipes, such as water supply and sewage, cannot be laid underground because of freezing temperatures and their heating effect on frozen ground.
- Pollution has a major impact in cold environments because low temperatures limit decomposition.

Human responses to problems

- Buildings are constructed on stilts or on concrete rafts laid on top of 2-m thick gravel pads.
- Roads and aircraft runways are built on gravel pads.
- Insulation of buildings includes triple glazing.
- Water supply and sewage disposal pipes run through **utilidors**, insulated ducting heated by steam pipes and running above ground.
- Heating systems are provided for oil storage tanks and parked vehicles have engine heaters which are plugged into the electricity supply.
- Long-distance oil pipelines are insulated and run above ground, mounted on stilts.

Figure 7.12 Living in cold environments

Review task

1 **It is important that you are able to relate periglacial processes to the characteristic landforms. You can do this with a Mind Map or a flow diagram, starting with the landforms and then continuing with the processes. Refer to Figure 7.11, your notes and textbooks as necessary.**

2 **Use the information in Figure 7.12 as a check list or framework in your revision of a case study of human activities in a periglacial area.**

What you need to know:

This varies enormously from course to course, but even where there is no section entitled hazards there are often hazards included in other topics. For example, hurricanes and tornadoes may be in a weather and climate module, and major global environmental issues might well have their own module.

Also, you will need to be able to:

- **understand the physical and human processes involved with any hazard or group of hazards;**
- **be aware of the various ways of defining and classifying hazards and the way that responses to hazards at all scales depends on perception of risk.**

Definitions

In a narrow sense, **environmental hazards** include natural events (**natural hazards**) which are seen as a threat to life and property; they are called **natural disasters** when they actually occur. In a broader sense, environmental hazards include natural hazards, such as flooding, which may be increased by human activities, as well as longer term problems caused by human changes to the environment, such as acid rain and global warming. The topic also includes the effects of and responses to hazards and disasters.

The factors involved in environmental hazards are:

- **natural processes** which produce the possibility of an event (e.g. a hurricane)
- **social and spatial factors** (the distribution of population) which determine whether populated areas will be affected
- **technological factors** (e.g. weather forecasting, disaster planning) play a major role in determining how great the disaster becomes.

With hazards on a global or regional scale, social and technological factors may be far more significant than the natural factors.

Classification of hazards

Hazards can be classified by:

- cause – geological and geomorphological hazards, atmospheric hazards, biological hazards, hazards caused or triggered by human activity

- level of interaction – human influence, technological developments and natural processes overlap to give social hazards, technological hazards, natural hazards and hybrid or mixed hazards; quasi-natural hazards are apparently natural hazards but have been precipitated by human activity

- effects – long- or short-term hazards, widespread or localised hazards, damage, deaths, disruption

- vulnerability

- location, level of poverty, warning time

- risk perception and response.

The **perception** of risk and the **response** of people to the risk varies for a variety of reasons:

- good or poor level of understanding of processes

- immediate or a delayed impact

- familiar or unfamiliar hazard

- common or rare hazard

- many or few deaths

- voluntary or involuntary hazard

- controllable or uncontrollable hazard

- deaths spread over time or concentrated in a clearly identifiable incident

- deaths spread over space or clustered in one locality.

Perception is coloured by the emphasis given to events in the media, but increased knowledge of hazards leads to **adjustments** involving:

- making changes to minimise the likelihood or scale of the event, such as flood control measures or water storage;

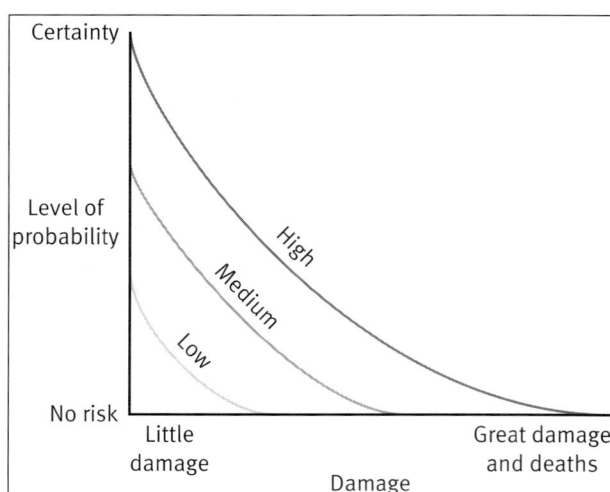

Figure 8.1 Hazard risk graph

- reducing levels of vulnerability by, for example, controls on buildings or establishment of warning systems;
- rearranging losses, such as by planning for emergency relief and insuring against losses.

Review task

Make a short list of activities, ranging from extremely hazardous, to less hazardous, to not at all hazardous, and a list of natural hazards of different types. Mark them on a copy of the hazard risk graph (Figure 8.1).

Living in hazardous locations

Some of the most densely populated parts of the world are also perceived as, and are, at great risk from various natural hazards. Although hazards can be predicted in terms of probability in particular areas, it is difficult to be precise about the location and scale of particular events.

In LEDCs in particular, people have little alternative but to remain in high-risk areas for a wide range of social, economic and political reasons. Some hazardous environments have specific advantages, like the fertile soils of flood plains and volcanic areas. In MEDCs some hazards can be overcome by expensive technological 'fixes' but there are many examples of apparently illogical developments, such as dense population concentrations in earthquake prone areas of Japan and of California. In both cases a **cost-benefit analysis** strategy is at work in which the risks of remaining are set against the benefits of remaining.

Climatic hazards

This section includes some very short-term events (hurricanes and tornadoes) and some medium- to long-term events (drought and El Niño/ENSO). Other related hazards are found elsewhere – see flooding in Figure 2.13 in Chapter 2.

Review task

Spend five minutes making a list of climatic hazards. Name the specific dangers associated with each one and the difficulties of coping with the danger.

Hurricanes

Hurricanes are weather events notorious for their scale and ferocity and for the damage they cause (Figure 8.4), but they form only one of a range of tropical weather systems which are all low-pressure systems drawing in warm moist air (Figure 8.2):

- Small-scale wave disturbances may grow into major systems the further away they are from the equator.

- About 650 km in diameter, often larger in the western Pacific.
- Central air pressure at sea level below 950 mb or even 900 mb.
- Considerable vertical development with the tops of cumulonimbus clouds reaching 12 000 m.
- Warm core of dense cumulonimbus maintained by condensation and release of latent heat intensifying the high-level high pressure, and enhancing the drawing in of heat and moisture at low levels.
- Innermost part of the hurricane (the eye) has a diameter of 30 to 50 km and is calm with clear skies or broken cloud as it is an area of descending air.
- Solid wall of cumulonimbus cloud surrounds the eye. Towards the edge of the hurricane the clouds become lower and begin to fragment.

Figure 8.2 Typical features of a hurricane system

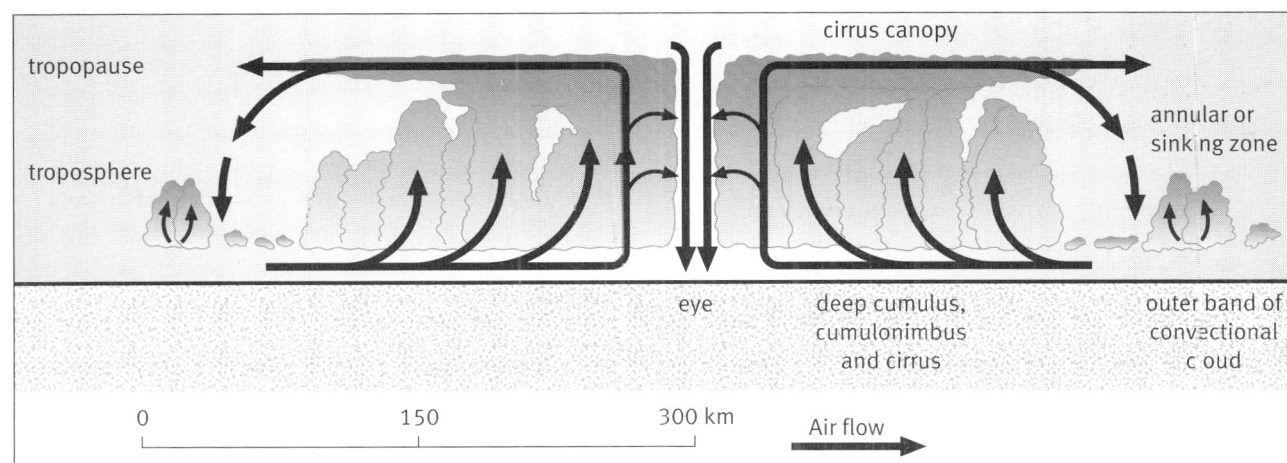

Figure 8.3 Structure of a hurricane

Physical factors

- **Wind** – intensity, distance to hurricane path where the winds are strongest, distance from the sea as wind speed tends to decrease away from the sea as the hurricane slowly dies, speed of movement of the hurricane, a slow-moving hurricane causes extra damage as it moves over an area.
- **Rainfall** – flooding, landslides.
- **Relief** – low relief and flooding from rain, coastal flooding from tidal surge.
- **Tide timings** – high tide coinciding with a tidal surge magnifies flooding.
- **Storm surge or tidal surge** – reduction in atmospheric pressure over the sea produces a sea surface rise of about 1 m for every 100 mb drop in pressure. Effect made worse as strong winds and decreasing water depth near the coast pile up the water so the surge across low-lying islands may be over 6 m high.
- **Longer term changes from global warming** – increase in sea temperatures in tropical areas will increase the formation area for hurricanes and intensify conditions in existing areas. Rising sea levels will increase areas at risk of flooding.

Human factors

- **Population density and distribution** – affect numbers at risk and procedures, like evacuation in vulnerable areas.
- **Living standards and infrastructure** – poorer countries suffer because of the lack of sea defences and emergency services and suffer greater numbers of deaths. Wealthier countries can evacuate people but lose enormous amounts in money terms. Note that measures used undervalue homes and welfare of people in LEDCs, and usually ignore longer term problems, like interest payments, loss of exports and tourists.
- **Levels of precautionary measures** – forecasting and prediction are generally good enough to prevent major disasters in MEDCs; maintenance of disaster centres and emergency services is straightforward for wealthy countries; in LEDCs dissemination of warnings is difficult and evacuation possibilities limited.

Figure 8.4 Hurricane hazards

- Tropical depressions are rotating wind and cloud systems with wind speeds up to 61 km/h.
- Tropical storms rotate more quickly and have wind speeds of 61 to 118 km/h. (These are sometimes called tropical cyclones.)
- **Hurricanes** have wind speeds in excess of 119 km/h. (They are called hurricanes in the Atlantic, typhoons in the western North Pacific and cyclones elsewhere.)

Conditions for development

- Hurricanes develop in both hemispheres in late summer and autumn when the Inter-Tropical Discontinuity is at its greatest distance from the equator. (Close to the equator the Coriolis force is too weak for a rotating air flow to grow, and low-pressure areas fill quickly.)
- Hurricanes need a sea surface temperature of over 27°C to provide the necessary heat energy and moisture to raise relative humidity so that condensation will release that latent energy and produce intense convection (Figure 8.3).
- The friction-free sea surface allows the system to move and continually provides a warm surface.
- In the upper atmosphere, waves in the tropical easterly winds spread air outwards and this allows strong upwards convection to develop.
- On the western sides of the oceans humid and unstable air favours the development of large-scale convection.

When any of these conditions ceases to exist the hurricanes begin to decay.

Review task

Refer to a case study of a hurricane, for example Hurricane Floyd (Atlantic, September 1999). Apply the ideas above and in Figures 8.2, 8.3 and 8.4 to it.

Tornadoes

Tornadoes are **small, short-lived and extremely violent storms**. They consist of a funnel of rapidly circling air reaching down from the base of a cumulonimbus cloud and, because the air pressure in the centre of the funnel is very low the wind speeds are typically 150 to 200 km/h and occasionally 400 km/h. The funnel is normally no more than 600 m across although some are larger, and it moves across country at 30 to 70 km/h. Most travel only a few kilometres but exceptionally they go for several hundred kilometres and these tend to be particularly erratic in their path and to do most damage. The damage is partly done by the wind itself, partly by blown debris and partly by the intense low pressure in the centre of the tornado which is sufficiently low to lift off house roofs.

Tornadoes occur most frequently in north-west Europe and the Midwest plains of the USA. In the USA the states of Oklahoma, Indiana, Iowa and Kansas are the highest risk states in what is called 'tornado alley'.

The **conditions for the development** of tornadoes are still not fully understood. In 'tornado alley' they occur most frequently in the spring when warm, humid and unstable air from the Gulf of Mexico meets colder, drier air from the north. Intense frontal activity at the boundary with extreme instability and convection gives rise to thunderstorms. Convergence of air masses below the cumulonimbus clouds together with complex vertical movements sets off a rapid upwards movement of air causing a sudden drop in air pressure at the ground. This is when the funnel becomes visible as air entering the vortex reaches saturation.

The **occurrence** of tornadoes in the USA is about 750 per year with an average of 100 deaths per year, the majority of deaths from the few longer lasting tornadoes. When a tornado is detected, a tornado warning is issued alerting the public and emergency services via television and radio, with sirens in places where a tornado is imminent. Advice to take shelter, in a specially constructed tornado shelter if possible, combined with the forecast and warning system has led to a significant decline in deaths in recent years. Damage to property of over $100 million has been recorded after several tornadoes in the USA, but little can be done to reduce this aspect of the hazard.

Drought

Drought is a hazard with a **delayed and indirect impact** since areas which experience it the most suffer from famine brought about by crop failure due to lack of water and perhaps extensive soil degradation.

The word drought has no one specific meaning. In general terms it includes some or all of the following: a period with insufficient precipitation to maintain soil moisture levels as a result of evapotranspiration, or to maintain stream flow levels or, over a longer period, to maintain or replenish groundwater levels, all of which have an impact on crops and natural vegetation and human activity.

Droughts tend to be most common under the following circumstances:

- Where sub-tropical high-pressure cells increase and persist so the extension of the Azores high to the east and south has been linked to droughts in the Sahel.

- Irregularities in the summer monsoon, delaying the arrival of maritime tropical air.

- Changes in ocean currents or upwellings producing cooler ocean surface temperatures.

- Displacement of the tracks of mid-latitude depressions by the development of persistent blocking high-pressure areas.

- El Niño events are associated with erratic weather conditions in many parts of the world and drought is one of these (see pages 80–1).

Drought is not classed as a hazard in true desert areas since it is a normal condition, but rather in areas where fluctuations from year to year or for several years at a time may bring about drought conditions. Marginal areas, where small variations can take an area into drought are the most vulnerable, although almost all areas of the world experience some temporary and irregular periods of drought.

Desertification

This a large-scale problem in dryland areas of the world and the amount of land and number of people affected is increasing. It is a problem associated with drought but the causes of desertification are much more complex than long-term drought alone. The UN definition of desertification is 'the diminution or destruction of the biological potential of the land ... leading ultimately to desert-like conditions', but this definition fails to include social, Page 104 economic and political factors. Generally human action is considered to be the major cause (see also Chapter 10).

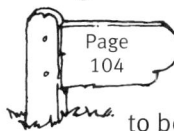

Problems are greatest in semi-arid areas where there is enough precipitation to support cultivation or pastoral activity, and enough to cause severe and rapid erosion of unprotected areas of soil. During periods of normal or wetter-than-normal rainfall, cultivation and grazing is extended into areas normally too dry. These areas are particularly susceptible to erosion (or soil degradation) when drier conditions return, and when longer term droughts occur the problems are multiplied.

The fundamental causes of desertification

- Use of the land beyond its natural carrying capacity.
- Fragile resource systems.
- Adverse climatic conditions including recurring droughts.

The direct causes of desertification

- Faulty agricultural policies:
  - ✳ overgrazing
  - ✳ overcultivation
  - ✳ cultivation and grazing of bare slopes
  - ✳ removal of tree cover for firewood
  - ✳ waterlogging and salinisation (salt deposition) as a result of large-scale irrigation
  - ✳ lack of advice services
  - ✳ little attention to subsistence farming.
- External factors:
  - ✳ the international trade system
  - ✳ foreign debts.
    Both encourage the production of cash crops for export and, as these tend to occupy the more favoured areas and use more of the available irrigation water, there is increased pressure on other areas.
- Population: especially as the dry countries tend to have higher population growth rates.

Local-scale problems resulting from desertification

- Disruption of local farming communities as a result of poor water supplies and land degradation.
- Need for relief aid.
- Out-migration of population.
- Problems in host areas for environmental refugees.

Global problems

- Reduction of the world's food-producing capacity.
- Reductions in biodiversity.
- Effects on climate of increased atmospheric dust and increased surface albedo.

> ### Review task
>
> **Summarise the section on desertification as a table or diagram. Add examples throughout, making sure they cover a range of scales.**

Measures to control and reverse desertification

- Conservation.
- Reclamation.
- Development of national resources in a sustainable manner.

Competition for water

Drought conditions in MEDCs produce serious effects even though the drought may not have a great direct effect on peoples' lives. The effects are aggravated by poor water conservation methods and wasteful use of water, and include damage to water-dependent environments. Large dams are built to meet growing demands with effects on land use and occupants of the land. Increased living standards lead to greater demand for water and, since MEDCs are relatively wealthy, they meet the demands because they will get income from supplying water. Longer term environmental or economic consequences are often ignored, however.

Besides these broad issues there are specific issues regarding water supply, especially regarding competition for water. In well-developed areas in which population is growing and water demand is increasing, such as California, there is competition for water between rural areas, primarily for irrigation agriculture, and urban areas which have more political power given their enormous numbers of people. Tourism creates demand for water which is often seasonal, such as in the dry season in Mediterranean areas, and this is in competition with other demands in the area. In LEDCs tourism has the same effect, with water being diverted from local agricultural use to supply hotel complexes and, within agriculture, major water supply schemes are focused on export crops leaving small-scale farmers without water.

El Niño-Southern Oscillation (ENSO)

El Niño is the name given to the change in the normal water circulation in the Pacific Ocean, and Southern Oscillation refers to the associated change in atmospheric circulation in the region. ENSOs disrupt the ocean-atmosphere system in the tropical Pacific with important worldwide weather effects.

The cause of El Niño

Understanding El Niño depends on understanding the normal state of the equatorial Pacific Ocean shown in Figure 8.5.

During El Niño the easterly trade winds weaken and may reverse over part or all of the region as in Figure 8.6.

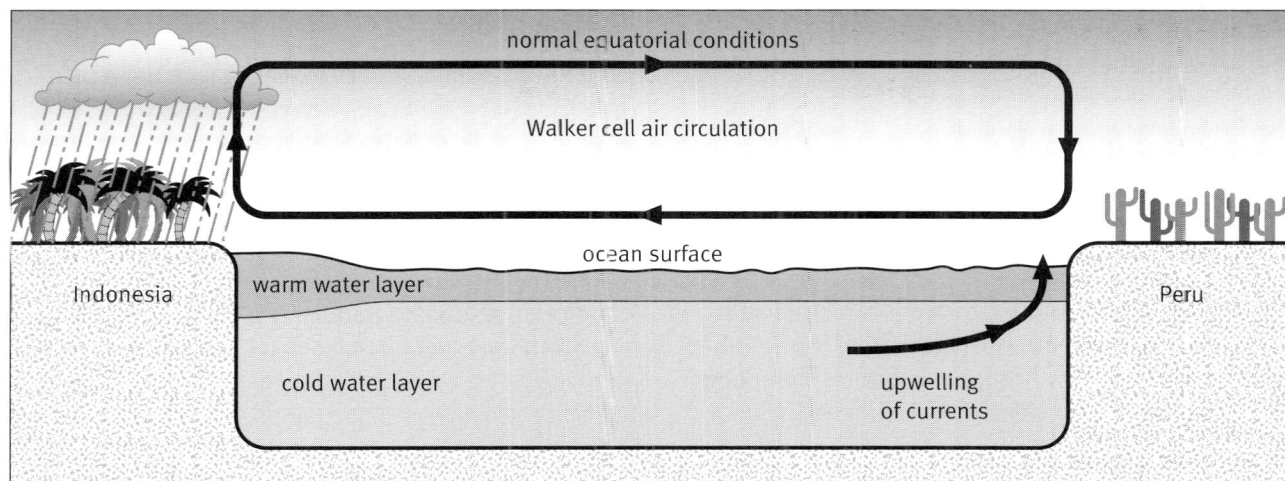

Figure 8.5 Normal equatorial conditions

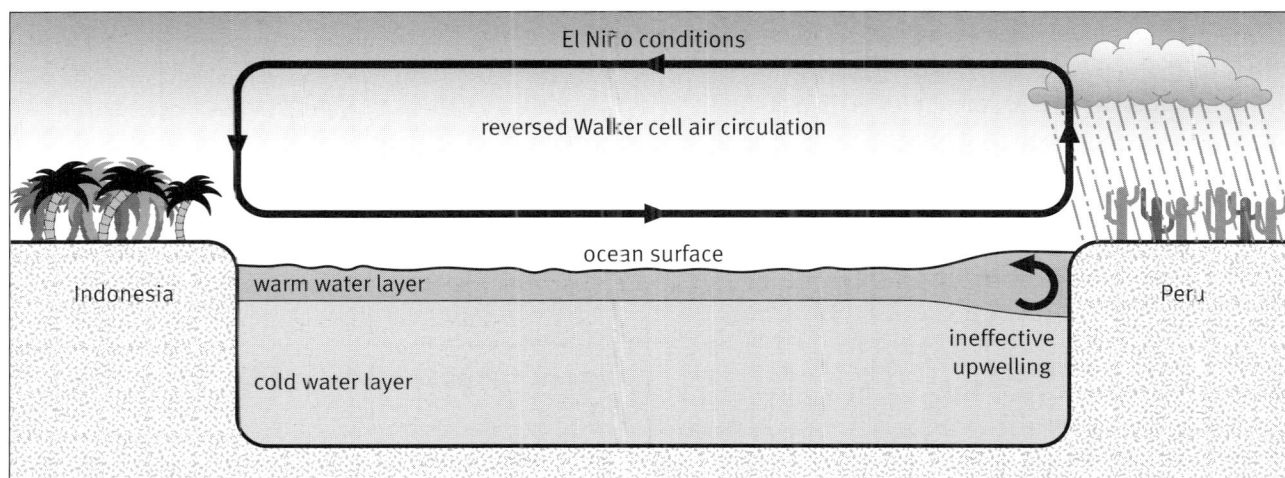

Figure 8.6 El Niño conditions

When does El Niño occur?

El Niño-Southern Oscillation events occur on average every five years but the gap may be anything between two and ten years.

How is El Niño observed?

Observations of conditions in the tropical Pacific are obtained from a network of buoys which measure temperature, currents and winds in the equatorial band. Satellite imagery measuring ocean surface temperatures and showing cloud movements also play their part in the prediction of these short-term (a few months to one year) climate variations.

What is the impact of El Niño?

The eastward displacement of the atmospheric heat source overlaying the warmest water results in large changes in the global atmospheric circulation. In the immediate region, El Niño brings about flooding in Peru and drought and fires in Indonesia and Australia. Peruvians are forewarned as it begins to rain on the

Review task

Make copies of Figures 8.5 and 8.6. Add details from the rest of the section so the diagrams become your revision summary.

Atacama desert, sea level rises by about 30 cm and the temperature of the sea surface rises by 6 to 8°C. Further afield the effects are just as great and El Niño has an impact on every continent with major droughts and major flooding incidents being attributed to it.

Geological hazards

Causes

The major geological hazards are **volcanoes, earthquakes** and **tsunamis**. Earthquakes are more frequent than volcanic eruptions and cause far more deaths. Tsunamis, which are seismic sea waves caused by undersea earthquakes or volcanic eruptions, cause devastation and many deaths in coastal areas.

The background to these hazards lies in the structure of the earth and, in particular, in plate tectonics. The overall structure of the earth is significant because of its effects on the passage of seismic or earthquake waves, and plate tectonics produce both volcanoes and earthquakes.

The earth's structure

Figures 8.7 and 8.8 show the earth's structure and how earthquake waves travel. Recording the differing times of arrival of waves allows monitors to pinpoint the location an earthquake.

Earthquakes are usually the result of a sudden movement at varying depths along the line of a fault within the earth. The most hazardous are those occurring in the shallow zone (0–70 km deep). Earthquake movement is the result of the build-up of stress on the rocks over time, which causes deformation of the rocks, sometimes visible at the surface as offset streams and roads. Ultimately, the stress is too great and the rocks break causing the earthquake. The point of fracture is known as the **focus** and the point at the surface immediately above the focus is the **epicentre** of the earthquake.

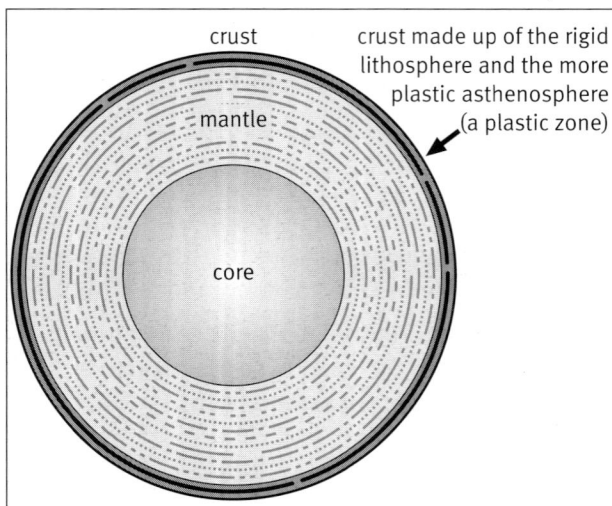

Figure 8.7 The earth's structure

P or **Primary waves** can pass through solids and liquids, are compression waves and travel fastest and therefore arrive at recording stations first.

S or **Secondary waves** can only travel through solid rock, are shearing waves and travel at half the speed of P-waves.

L-waves are surface waves and travel most slowly, arriving at recording stations last. Depending on the type, they have a horizontal movement (Love waves) or a vertical and horizontal movement (Rayleigh waves).

Figure 8.8 Earthquake waves

Plate tectonics

The earth's lithosphere is made is made up of rigid **plates**, seven large and several smaller plates. Some, like the Pacific plate, consist mainly of oceanic lithosphere, while most consist of oceanic and continental lithosphere, like the Eurasian plate. The plates move in relation to each other as a result of deep-seated movement within the earth. Most important processes happen at the plate boundaries of which there are three types (Figure 8.9).

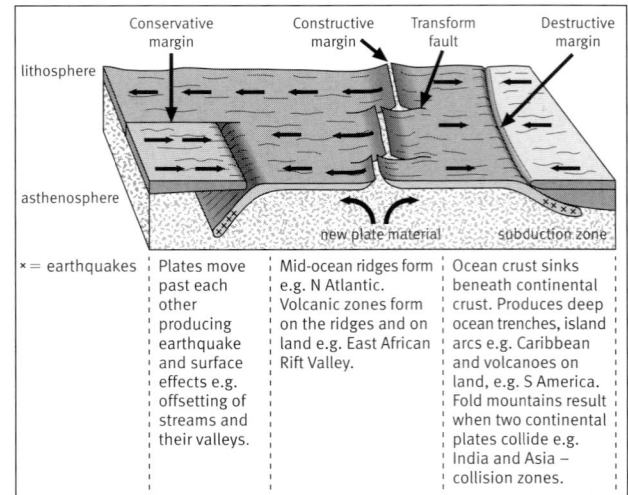

Figure 8.9 Plate boundaries and associated features

Earthquakes

Earthquakes differ from most other hazards in that human constructions play a large part in determining the severity of an event. The Kobe (Japan) earthquake of 1995 illustrates this with damage to 200 000 buildings, the cutting off of water, gas and electricity, fires caused by ruptured gas mains, damage to roads and railways and the port area due to collapse of landfill. The deaths of over 5000 people resulted directly from this damage and the fact that it was a densely built-up urban area with structures inadequately designed to deal with the earthquake hazard.

Earthquakes are described by magnitude and intensity.

Magnitude is a measure of the energy released in an earthquake and based on instrumental observations. It is measured by the Richter scale. This is a logarithmic scale by which each unit increase on the scale denotes a ten-fold increase in magnitude of the seismic waves, so a scale 5.0 earthquake has ten times the magnitude of a scale 4.0 earthquake.

Intensity describes the scale of impact of an earthquake, taking into account casualties, building damage as well as land surface effects, usually measured using the Modified Mercalli (MM) scale which ranges from I (minimal) to XII (major disaster).

Earthquake hazards fall into two categories:

- **Primary hazards** which involve ground shaking, the severity of which depends on the magnitude of the earthquake, distance from the epicentre and the local geology.

- **Secondary hazards** which involve liquefaction of landfill and unconsolidated sediments with a high water content, land slides, rock slides and avalanches in mountainous regions, and tsunamis caused by undersea earthquakes or volcanic eruptions.

Earthquake hazards and vulnerability

Figure 8.10 shows the number of major world cities with a significant risk of earthquakes, 70 per cent of which are at risk of a severe earthquake. Note that the cities of the LEDCs are at greater risk than those of the MEDCs.

Earthquakes cause much greater numbers of deaths in LEDCs than in MEDCs, but earthquakes in the latter cause enormous economic losses. The difference lies in the standards of living of different populations and the measures taken to limit vulnerability to the hazard. The earthquakes in Turkey and Taiwan in September 1999 were of the same intensity and illustrate this point. They also illustrate the need for strict enforcement of building regulations.

Measures taken to reduce the impact of earthquakes by the construction of 'earthquake-proof' buildings have been successful in MEDCs by reducing deaths, but, at the same time, economic losses have risen because they may still be so badly damaged that they have to be demolished, causing, in turn, huge insurance losses. In LEDCs, rising death tolls show the problems where the technological approach used in MEDCs cannot be applied. The causes of increased vulnerability have been separated into:

- **Immediate causes**, such as poor construction standards and rescue services.

- **Underlying causes**, which involve social, economic and political factors, including inequalities and corruption at a local scale and LEDC debt problems at the international scale. Since dependence on the use of MEDC technology to reduce earthquake vulnerability makes LEDCs even more dependent on MEDCs it is thought that more self-sufficient approaches are needed.

Prediction of earthquakes is unlikely to be achieved in the near future although there has been some success based on a wide range of observations of

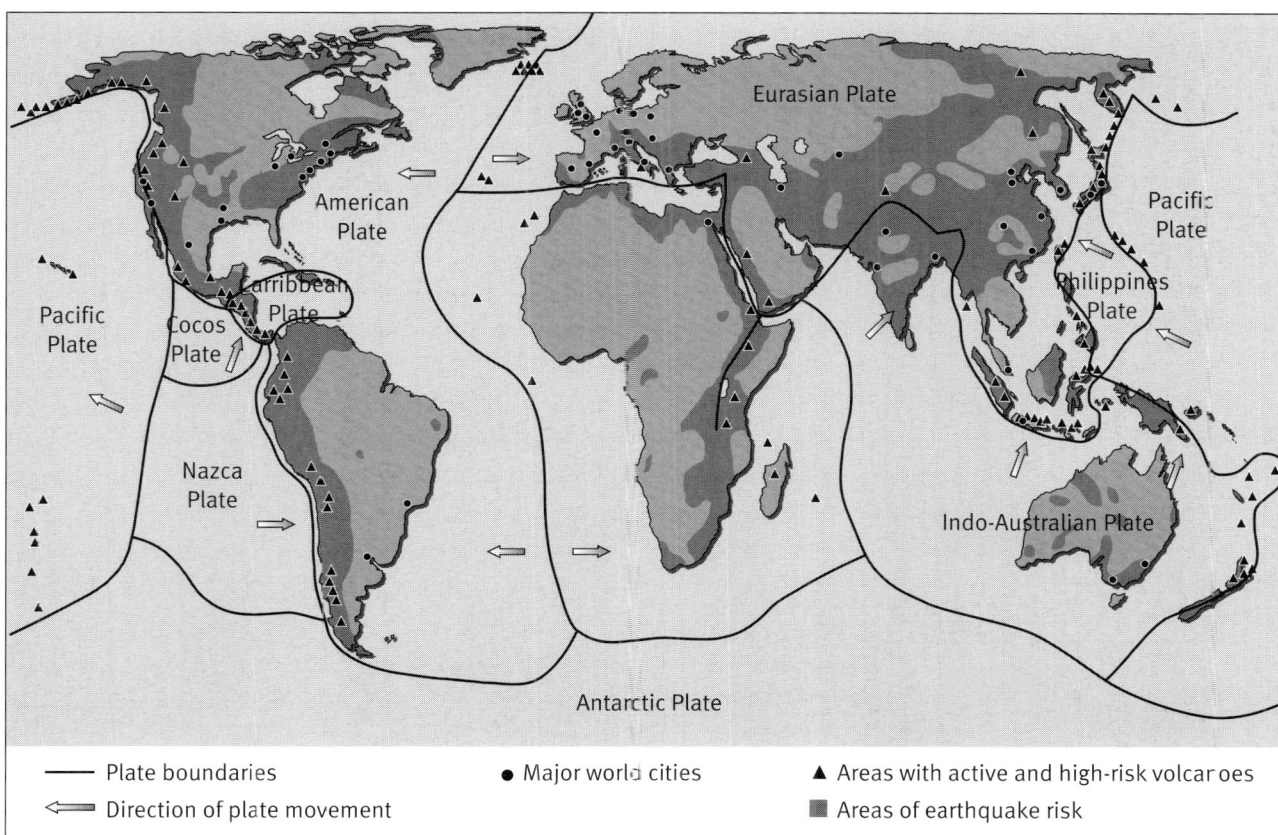

Figure 8.10 Major tectonic plates, volcanoes and recent earthquakes

different phenomena. However, the problem is that warnings may, in themselves, cause panic and associated damage and losses.

An alternative approach is to reduce vulnerability by planning control as well as building regulations. The latter would ensure standards of building to match the likely severity of earthquakes and the former would ensure that, in the long term, the most vulnerable areas would be avoided or built on at a low density.

Volcanic hazards

Volcanic activity is spectacular as is volcanic landscape, but as a hazard it is generally less important than earthquakes because so many volcanoes are in uninhabited or sparsely populated areas or underwater on mid-ocean ridges (see Figures 8.9 and 8.10).

Volcanoes are classified in different ways. As landforms they are shown in Figure 8.11 and as hazards they are shown by types of eruption in Figure 8.12.

> **Review task**
>
> Use your own notes and textbooks to review five types of volcano location. Use the following terms: hot spots, plate margins, rift valleys, mid-oceanic ridges, subduction zones, destructive, constructive, island arcs, fold mountains.

Volcanoes are the result of **extrusive activity** but **intrusive activity** occurs when magma penetrates and solidifies within the earth's crust (Figure 8.13).

As a hazard, volcanic activity depends on the type of eruption, the local population (in number and density) and vulnerable landform elements (such as ice or snow fields).

> **Review task**
>
> Why are earthquakes a much greater hazard than volcanoes?

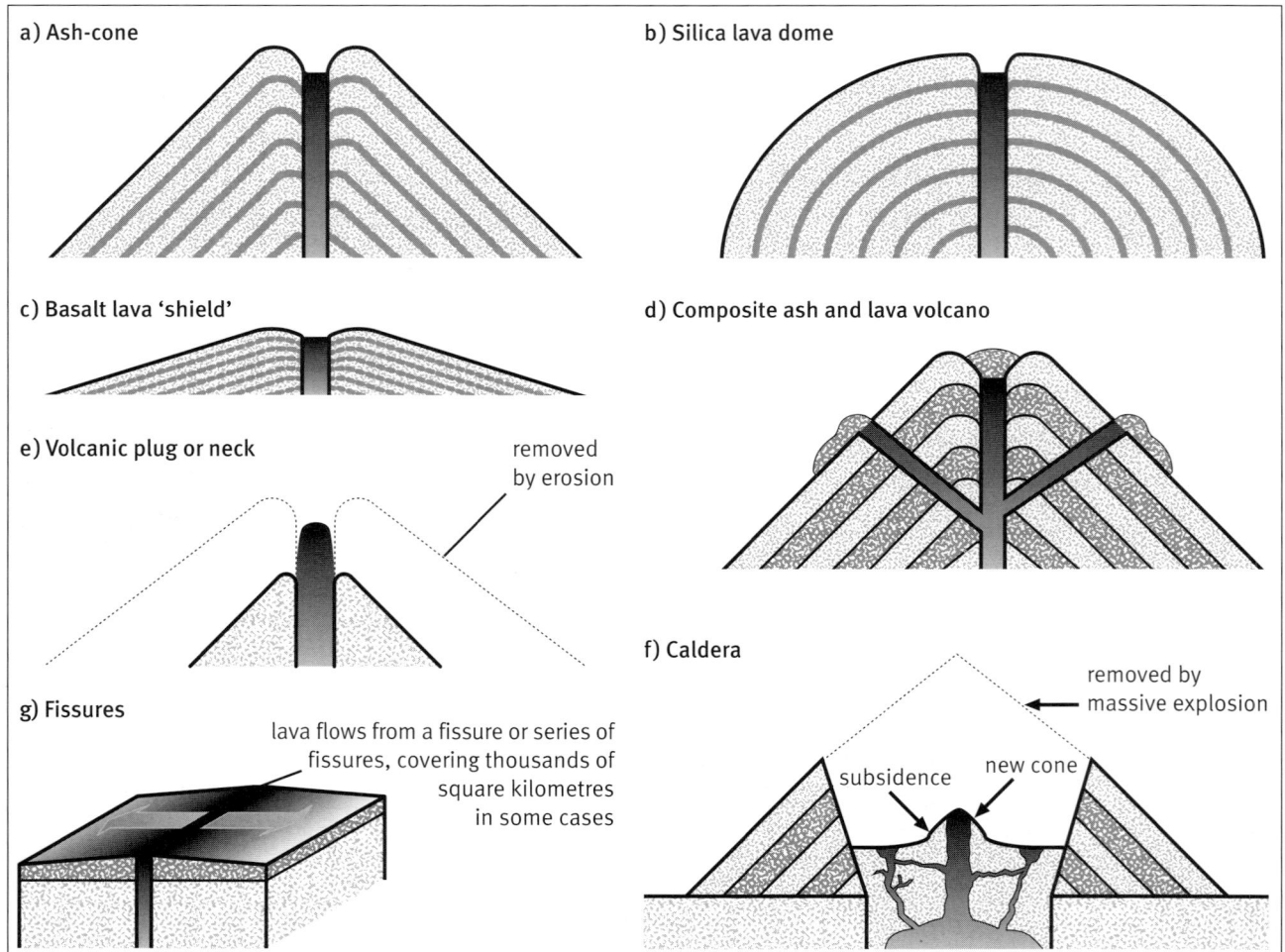

a) Ash-cone

b) Silica lava dome

c) Basalt lava 'shield'

d) Composite ash and lava volcano

e) Volcanic plug or neck — removed by erosion

f) Caldera — removed by massive explosion — subsidence — new cone

g) Fissures — lava flows from a fissure or series of fissures, covering thousands of square kilometres in some cases

Figure 8.11 Volcanic landforms

Hawaiian	Mild eruptions, fluid lavas, no trapped gas to cause violent explosions.
Strombolian	A little more explosive, more or less continuous small eruptions, cone forms of ash as fluid lava flows away.
Vulcanian	Infrequent eruptions may last several months. Explosions throw out large blocks as well as smaller blocks and ash.
Vesuvian	More powerful continuous eruption with gas carrying ash very high in the air, lasting hours and, rarely, days.
Plinian	Most violent, enormous volume of material blasted into the air by huge eruptions or series of eruptions, blowing away much of the volcano and leaving a crater.
Peléean	A 'nuée ardente' or cloud of immensely hot gas and dust shoots out from the centre of the eruption in one or all directions.

Figure 8.12 Volcanic eruption types

Hazards from human activities
Air and water pollution

Pollution means the introduction by human beings, directly or indirectly, of substances or energy into [the environment] resulting in harm to living resources, hazards to human health, hindrance to . . . activities, impairment . . . for use and reduction of amenities.

('Population Resources and the Environment: Critical Challenges', United Nations, London 1991)

Pollution is classified in different ways: what is polluted, how it is polluted and what the pollutant is. For example, a river polluted by nitrates in runoff from a farm might be classed as water pollution, chemical pollution or agricultural pollution. As the water pollution has a direct effect on water quality for drinking purposes and on plant and animal life it is clear that any consideration of pollution involves complex sets of links.

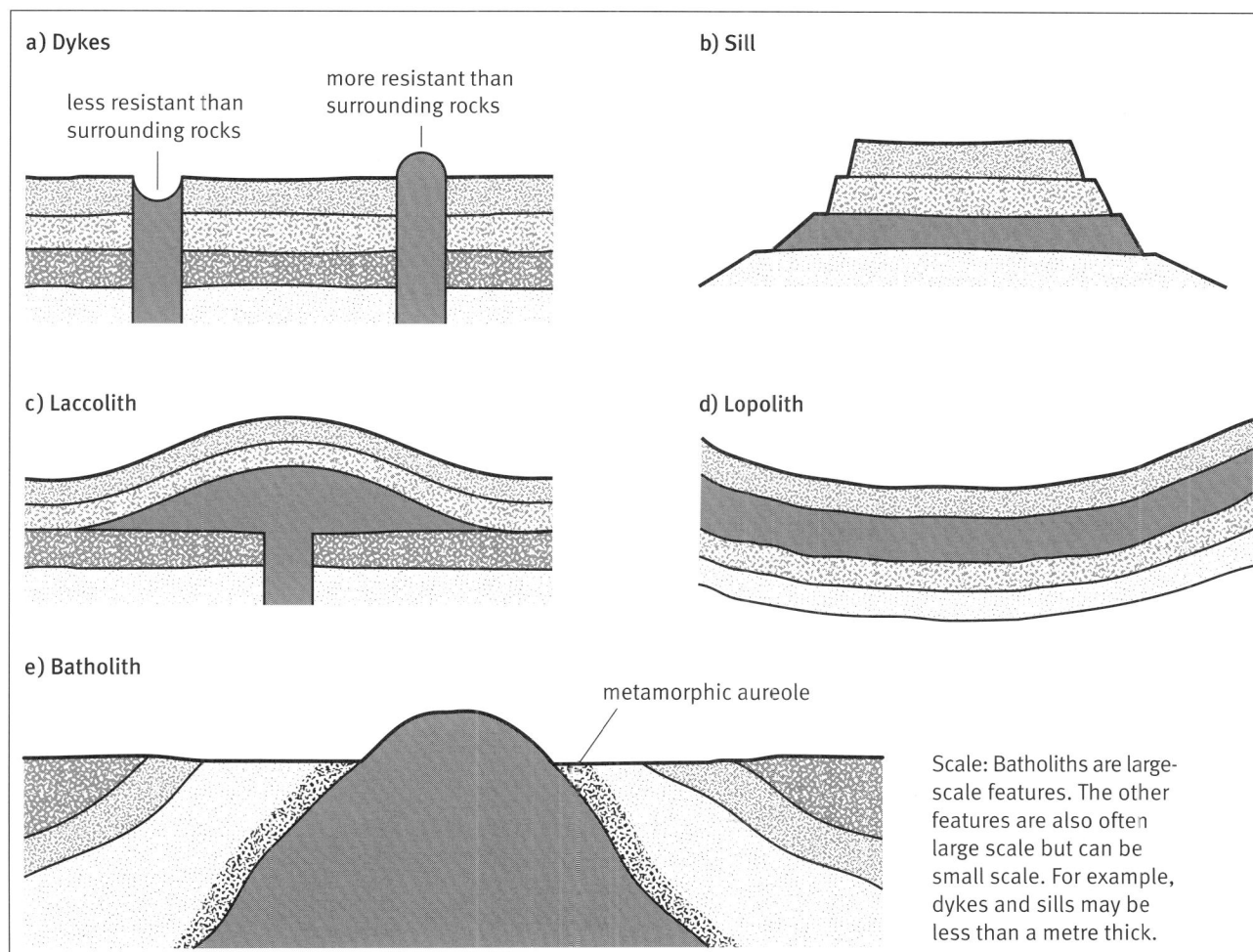

a) Dykes
less resistant than surrounding rocks
more resistant than surrounding rocks

b) Sill

c) Laccolith

d) Lopolith

e) Batholith
metamorphic aureole

Scale: Batholiths are large-scale features. The other features are also often large scale but can be small scale. For example, dykes and sills may be less than a metre thick.

Figure 8.13 Features produced by intrusive activity

EXAM TIP

At the broad scale, the regional and global issues that follow in this chapter are themselves examples of pollution as defined by the UN statement above. This section summarises a range of aspects of pollution. Refer to your own notes and, in particular, *case studies* to ensure you can develop the points.

Agriculture is a major source of **water pollution** (Figure 8.14) in the form of **chemical wastes** and **sediments**.

- Sewage and other oxygen-demanding wastes
- Infectious agents
- Organic chemicals
- Inorganic chemicals and mineral substances
- Sediments
- Radiaoctive substances
- Heat or thermal pollution

Figure 8.14 Types of water pollutants

Nitrate **fertilisers** can bring about eutrophication of fresh water and nitrates can cause health hazards. Eutrophication occurs when waters are enriched by nutrients. This is a natural process but runoff from farmland, sewage discharge and industrial effluents intensify the process with resulting algal blooms and depletion of oxygen in the water with serious effects for plant and animal life.

Pesticides used in farming also cause pollution. Problems arise because of the wide range of chemicals used, the different ways they work, their level of toxicity and the length of time they persist in the environment.

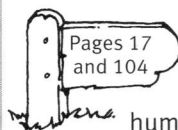

Soil erosion as a result of overcultivation or forest clearance results in one of the most significant human impacts on rivers which is the increase in suspended **sediments**.

Pages 17 and 104

Other sources of pollutants include **mining**, resulting in air and water pollution. In the latter, acid mine-drainage waters may contain toxic metals as well as a range of other elements which often discolour the water and are precipitated as unpleasant deposits.

Major **urban areas**, as you would expect, contribute a huge variety of air and water pollutants from many different sources, of which industrial emissions into the air and water, and sewage are major problems.

Thermal pollution of water is mainly brought about by the release of power-station cooling water into rivers bringing about a rise in temperature. This has a direct impact on plant and animal life, particularly in the immediate vicinity of the outfall.

Air pollution has numerous effects. In urban areas one of these is accelerated weathering of buildings partly due to acidification of rain due to sulphur dioxide emissions. However, for urban areas generally the most apparent form of pollution is that of smoke haze and **photochemical smog**. The former is significant in promoting fog formation as many of the particles are hygroscopic so water vapour condenses on them easily. Smog produced by coal burning has decreased throughout much of the more developed world, but photochemical smog which results from the pollution created by the combustion of oil and petrol by motor vehicles and aircraft as well as the emissions from petrochemical industries has increased. The hydrocarbon pollutants react with sunlight to produce the smog which presents a major health hazard in many cities worldwide.

Acid rain

The problem

Acid rain is the broad term used to describe the increase in acidity of rainfall caused by pollutants. It was first recorded in the 19th century but largely ignored until observations showed threefold increases in rainfall acidity between the 1950s and the 1970s in northern Europe and eastern North

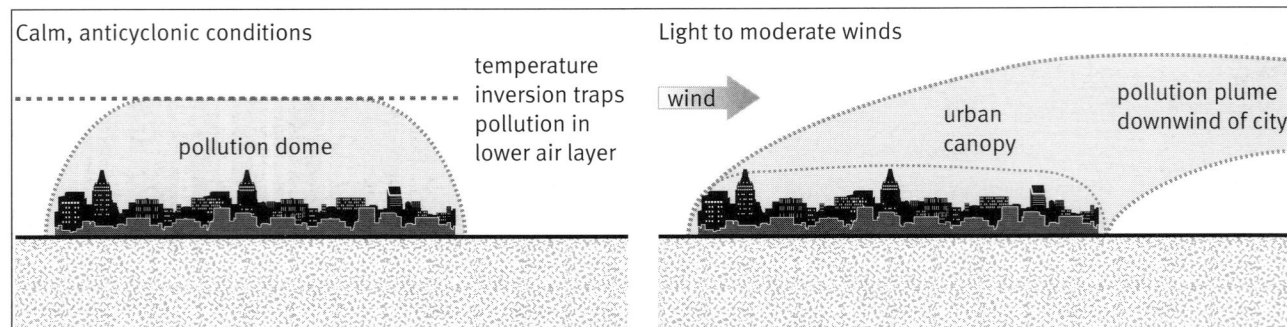

Figure 8.15 Urban pollution

America. The pollutants, sulphur dioxide and nitrogen oxides, come from a variety of sources, most notably power stations and motor vehicles. The high sulphur content of the coal and fuel oil used in power stations and the use of high chimneys easily allows surface winds to transport sulphur dioxide considerable distances across seas and across national boundaries to be deposited in the form of sulphuric acid in the precipitation, although some sulphur dioxide may also be deposited as dry deposition. Nitrogen oxides come mainly from motor vehicles and are generally deposited locally as dry deposition but they also play a part in acid rain formation.

The damage

Acid rain has had an impact on an international scale in North America, Europe and Asia. It has affected vegetation, soils, lakes and water supplies as well as buildings (Figure 8.16).

Affected features	Effects
Vegetation	Forest damage due to changes in the soil affecting water absorption; increased plant stress reduces resistance to disease; loss of drought resistance.
Soils	Acidifying soils affecting agricultural yields and releasing harmful substances like aluminium.
Lakes	Acidifying lakes and declining fish populations through acid 'flush' of spring meltwater and deposition of aluminium washed out of soils.
Water supplies	Acidification and high aluminium levels.
Buildings	Intensified weathering.

Figure 8.16 Effects of acid rain

The complexities

The effect of acid rain varies according to the rock in a particular area. Limestone, for example, will neutralise the effect of acid rain, whereas granite will not, so already acid soils and water will become more acid. The vegetation which will grow on these soils will, in turn, increase acidity levels of surface water. On the other hand, enrichment with fertilisers, for example, can reduce the levels.

It took a long time before it was accepted that pollutant emissions were the cause of acid rain with, in Europe, the UK government, in particular, objecting

to the institution of control measures. This was partly because a direct link between sulphur emissions and environmental damage could not be proved, and partly because of the economic costs of making the technological changes which would reduce emission levels from power stations. Figure 8.17 summarises solutions to the problem.

Technological solutions	Remove sulphur from fuels before burning. Use low sulphur-content fuels. Flue Gas Desulphurisation (requiring large amounts of limestone and producing large quantities of calcium sulphate which then requires disposal).
Political solutions	1979 UN Convention on Long Range Transboundary Air Pollution. 1985 Helsinki Protocol set targets of 30% SO_2 reductions by 1993 (not ratified by UK). 1988 Large Combustion Plants Directive setting reduction targets for 2003 (agreed by UK).
Success	Emissions cut by 50% or more by some European countries but UK managed only 25% cut with imported low-sulphur coal and gas turbines.
Future problem areas	Eastern Europe although much heavy industry has declined. Asia.

Figure 8.17 Solving the acid rain problem

CFCs and the ozone layer

The problem

Chlorofluorocarbons or CFCs have been in use first as refrigerants for over 60 years and then as propellants, solvents and cleaning fluids. In the 1970s their effects on the ozone layer of the stratosphere were identified in the form of declines in total ozone and the appearance of 'holes' in the ozone layer, initially over the Antarctic and more recently over the Arctic and in lower latitudes. These 'holes' appeared in the spring of the appropriate hemisphere.

The cause of the Antarctic hole is thought to be partly due to a change in polar wind circulation but primarily due to the concentration of CFCs in the atmosphere which precipitated the chemical breakdown of the ozone.

The ozone layer functions as a filter for the sun's ultraviolet radiation. UV-A is harmless, UV-B has many damaging effects on humans, animals, plants and some materials but very little reaches the ground under normal ozone layer conditions, and UV-C is fortunately absorbed by ozone as it destroys micro-organisms, nucleic acids and proteins. Depletion of the ozone layer obviously has serious consequences for all life forms.

Consequences of ozone depletion

- **Human health.** The ill-effects of ozone depletion for humans include cataracts, skin cancer and damage to the immune system, weakening resistance to a range of infectious diseases.
- **Plants.** Research suggests that agricultural crops show slower growth and smaller leaves with increased UV radiation with similar effects on forest and grassland productivity. Also significant is the suggestion that changes could upset the balance in natural ecosystems.
- **Aquatic systems.** Ocean life is also at risk and it seems that UV-B radiation is an important limiting factor in marine ecosystems. It is suggested that small aquatic organisms already show damage and that half of ocean fish and all inshore and estuary life are at risk. Plankton, at the very bottom of the marine food chain, are particularly vulnerable, especially in the vicinity of the large Antarctic ozone hole.
- **Climate.** Increased UV-B radiation will make the troposphere more reactive chemically given the many pollutants which influence air quality as well as climate itself. Some of these reactions would be beneficial in reducing methane and the CFC substitutes HCFCs and HFCs. Ozone loss can have effects on global warming, one of which could be the lowering of plankton productivity, thereby reducing the uptake of carbon dioxide by the oceans.

The solutions – international agreements

The process started with unilateral action by the USA before becoming international:

- 1978 USA banned use of CFCs as propellants which caused 75 per cent of worldwide emissions.
- 1979 EU adopts a voluntary agreement limiting CFC production.
- 1981 UNEP drafts an outline convention for the protection of the ozone layer.
- 1985 Vienna Convention signed.
- 1987 Montreal protocol signed.
- 1990 London Amendments with phase-out by 2000.
- 1992 Copenhagen Amendments with phase-out by 1996.
- 1995 Vienna Conference with phase-out of bromides by 2020.

Note that real progress (Montreal) did not take place until after direct evidence of ozone loss (ozone hole, 1985, and its deepening, 1988). Despite progress the largest Antarctic hole ever was seen in the southern spring of 1998 followed by the largest Arctic hole the following northern spring.

Global warming

The **greenhouse effect** is a natural and essential feature of global climate (see Chapter 4), but in the last two decades it has commonly been used in the context of human-induced changes to the atmosphere involving increases in the concentration of **greenhouse gases** – carbon dioxide, methane and CFCs are the most important ones, followed by nitrous oxides. Note that water vapour is an important greenhouse gas which plays an important role by amplifying the effects of other gases.

Global warming is tied up with an **enhanced or anthropogenic greenhouse effect,** whereby an increase in greenhouse gases makes the atmosphere less transparent to long-wave radiation because more of it would be absorbed by the gases. More would be re-radiated back towards the surface and less would be radiated into space creating an imbalance so the earth and the troposphere warm up until the balance is restored.

A doubling of carbon dioxide by itself produces a temperature rise of 1.2°C but **feedbacks** or knock-on effects may double this temperature rise. Note that feedback effects may be positive, as with water vapour, or negative, as with the way clouds reflect solar radiation back to space, illustrating the complexity of the interactions involved. To appreciate the level of complexities look back to Chapter 5, Figure 5.5 showing the nutrient cycle. Carbon in the atmosphere is affected by changes in the features of the diagram.

Human changes to the atmosphere

- The growth of population and industry, overgrazing, forest clearance and desertification have all had effects on the concentration of carbon dioxide in the atmosphere.
- Atmospheric carbon dioxide increased by 25 per cent between 1750 and 1990 and half of that increase was since 1965.
- The main source of the carbon dioxide increase was the burning of fossil fuels.

- The effect of the increased carbon dioxide has been calculated to raise mean air temperature by 0.5°C over 1960s temperatures.

Review task

Use Figure 5.5 and make two lists of changes, one of changes bringing about increases in atmospheric carbon dioxide and one of reductions. Where possible show links between the items you list. In a third list make links to the points above.

What are the predicted effects of global warming?

Predictions are based on **general circulation models** (GCMs), which have become more sophisticated over time incorporating fully coupled atmosphere-ocean systems which develop a picture of climatic changes around the world.

Besides general warming, predictions are for:

- certain areas to become hotter, some drier, some wetter;
- climate to become more variable leading to flooding or drought problems;
- mountain glaciers to continue their generally rapid retreat and for ice caps to begin to melt;
- more and more intense tropical storms;
- sea levels to rise;
- natural vegetation changes to follow climatic changes and for agricultural crops and major production areas to change;
- changes in the incidence of pests and diseases, both those affecting crops and people In the oceans changes in temperature, salinity and availability of nutrients could affect fish stocks, while any major change in ocean currents could trigger dramatic climatic changes. The effects of all these on human activities and populations could well be equally dramatic.

What action can be taken?

Adaptation

Passively, by adjusting to environmental changes as they occur, such as strengthening coastal defences as necessary, is a cheaper possibility for MEDCs as it only involves maintaining systems which already exist.

Actively, by developing strategies that recognise likely developments without following a particular scenario, for example introducing new strains of crops which would be better adapted to extreme conditions.

Review task

Take the list of possible effects of global warming and add to them from your own notes and books. Now reorganise them into groups, such as climate, oceans and agriculture, carrying capacity (Chapter 10) and show them as a Mind Map. Extend each item by considering the consequences of the changes.

Prevention

In 1992 the United Nations Framework Convention on Climate Change was signed at Rio at the UN Conference on Environment and Development. Its aim was:

> To achieve . . . stabilisation of greenhouse gas concentrations in the atmosphere at a level that would prevent dangerous anthropogenic interference with the climate system.

No specific commitments were made at that conference, but in 1995 at Berlin a number of countries decided to reduce their emission levels to below their 1990 levels by 2005. However, no general agreement could be reached. In 1997 at Kyoto the agreement called for a 5.2 per cent aggregate cut in heat-trapping greenhouse gas emissions by industrialised nations below 1990 levels by the period 2008–12. This falls far short of the emission cuts needed in the coming decades – of the order of 50 per cent to 70 per cent according to scientific consensus – to prevent the worst impacts of climate change.

Specific measures

Reduce emissions of greenhouse gases by:

- using alternative energy sources;
- changing from coal to gas;
- limiting deforestation and extending afforestation;
- using pricing and subsidy measures to stimulate changes.

Review task

This is a different kind of task. What are the possible implications of the failure to achieve the emission cuts called for? By the time you read this, more evidence will have been published and more forecasts made. Incorporate them in your answer.

What you need to know:

- **Distribution, density and changes in world, continental and national populations and fertility, mortality and growth rates**

- **Factors affecting and contrasts in fertility, mortality and growth at national and continental scales**

- **The demographic transition model and the significance of changing population structures**

- **Significance and types of migration**

- **Models of migration**

Also, you will need to be able to:

- interpret population statistics and graphs of different types;

- demonstrate your knowledge of ideas and case studies;

- apply your knowledge and understanding of trends, important factors and models to case studies.

Review task

Look at Figure 9.2. Select two countries and compare and contrast their populations.

Population is a topic that covers all scales from global down to local, although most specifications stick to global, continental and national. Here we break up the topic, but a question in the examination could draw on all parts.

Population patterns and trends

World, continental and national patterns of growth

Throughout most of human history the population of the world has grown very slowly. Figure 9.3 summarises this. Notice the acceleration of growth during this century in particular.

Figure 9.3 shows the global situation but actual numbers vary greatly at continental and national scales and so do growth rates. Figure 9.4 provides statistics for the major world regions and for selected countries.

Review task

Using Figure 9.4 make a list of general points from the statistics about:
a) the distribution of population by continent/major world region; b) the differences in growth rates by continent/major world region; c) how the selected countries compare. Summarise the points you have made on a Mind Map, leaving plenty of room around it to add other ideas and examples.

Measure	Expressed as
Total population	Numbers
Population density	People per km^2
Crude birth rate or fertility rate	Number of live births per thousand people in a year, obtained by dividing the total population by the number of live births
Crude death rate or mortality rate	As for crude birth rate
Natural increase	Difference between birth and death rates but stated as percentages
Infant mortality	Deaths up to the age of 1 year, calculated as the number of deaths per thousand live births in that year
Life expectancy at birth	Given in years and generally stated for males and females
Population distribution or urban population	Urban/rural division, given as percentage in urban areas

Figure 9.1 Key statistics

	UK	USA	Brazil	India	Japan	Uganda	Nigeria
Total population (millions)	59.4	272.5	168.0	986.6	126.7	22.8	113.8
Crude birth rate (‰)	12	15	21	28	10	48	43
Crude death rate (‰)	10	9	6	9	7	20	13
Natural increase (%)	0.2	0.6	1.5	1.9	0.3	2.9	3.0
Infant mortality (‰)	5.9	7.0	41	72	3.7	81	73
Life expectation (M/F)	74/80	74/79	63/70	60/61	77/84	41/42	53/55
Urban population (%)	89	75	78	28	79	15	16

Figure 9.2 Key statistics for selected countries (Population Reference Bureau, 1999)

Time	Population
2000 yrs ago	300 million
1750 AD	791 million
1900 AD	1650 million
1950 AD	2521 million
2000 AD	6055 million

Figure 9.3 World population growth

EXAM TIP

Information from Figure 9.3 can, of course, be shown on graphs, so in an examination question be prepared for the resources to be presented in different ways.

Region or country	Population (millions)	Natural increase (%)	Fertility rates (‰)	Mortality rates (‰)
World	5982	1.4	23	9
More developed	1181	0.1	11	10
Less developed	4800	1.7	26	9
Africa	771	2.5	39	14
Asia	3637	1.5	23	8
Europe	728	−0.1	10	11
Oceania	30	1.1	18	7
North America	303	0.6	14	8
Latin America	512	1.8	24	6
Brazil	168	1.5	21	6
Ethiopia	59.7	2.5	46	21
Japan	126.7	0.2	10	7
Germany	82	−0.1	10	10
Philippines	74.7	2.3	29	7

Figure 9.4 Total population, natural increase, fertility and mortality rates for the world, major world regions and selected countries (Population Reference Bureau, 1999)

Annual growth rate (%)	Doubling time (years)
1.0	70
2.0	35
3.0	24
4.0	17

Figure 9.5 Growth rates and doubling time

Rates of change – doubling time

This emphasises the effect of growth rates on population, in that a population's future size is determined by its growth rate as well as its present size (Figure 9.5).

Compare India and China. In 1992 their populations were 880 million (India) and 1.18 billion (China), but by 2050 they are forecast to be 1.68 billion and 1.45 billion respectively. India will have overtaken China because it has a higher growth rate of 2.1 per cent per year compared with China's growth rate of 1.4 per cent, but remember that this assumes growth rates do not change.

Mapping population distribution

Maps of population distribution show clear patterns at whatever scale they are drawn. There are two common methods of showing population distribution, by dot maps and by density shading or choropleth maps. Look in your textbooks for examples. World maps, country maps and maps of small areas within countries all give an apparently clear picture of where people live but you do have to be aware of the limitations of maps.

World population distribution

Remember that this is not a static situation so some regions of the world will become even more important in terms of numbers than they are now. Chapter 10 develops points about the broad factors affecting population distribution.

Review task

Refer to an atlas or textbook map of world population distribution. You should be able to describe the patterns and at least begin to explain them. Some pointers are given in the Mind Map (Figure 9.6) and others in Chapter 10. Add examples from your own notes and course books to the Mind Map to give you a complete summary.

1 Make a copy of Figure 9.6. Develop it by adding lines to incorporate material relating to aspects of world population distribution. Ensure that you include the idea of change.

2 Make another Mind Map for a region of the world you have studied. It can be at any scale from a large country down to a region of the United Kingdom. A good centrepiece is a sketch map of the region.

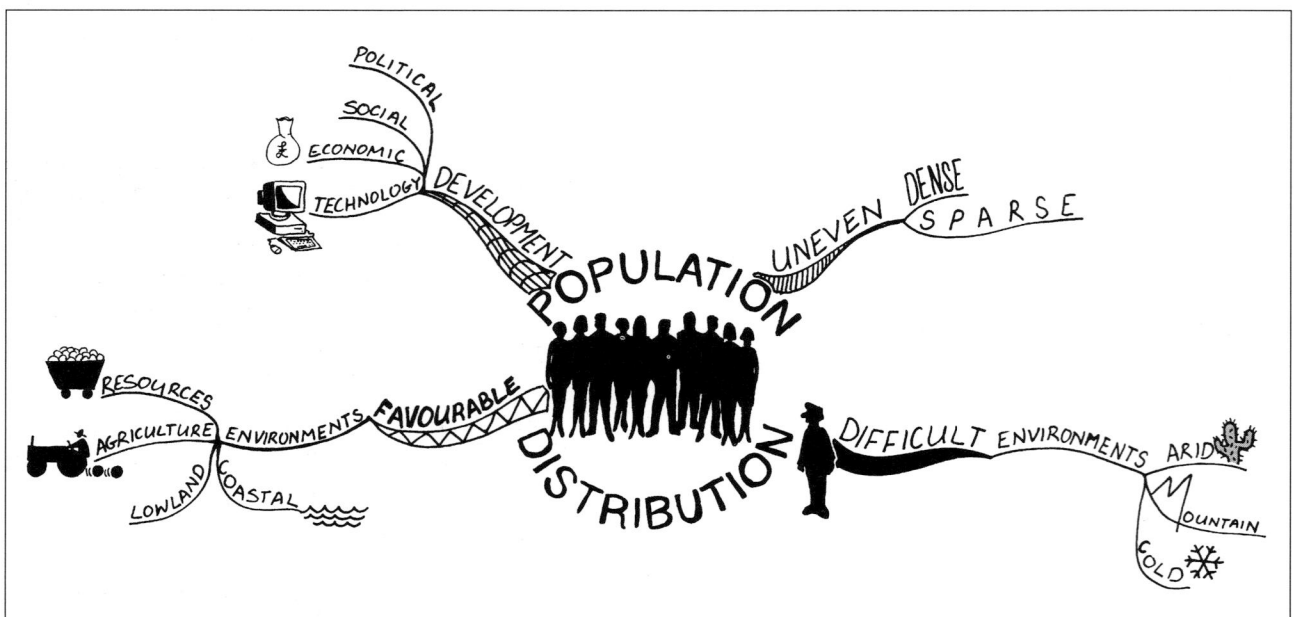

Figure 9.6 Mind Map of world population distribution

Factors of change

Population change

Population change at the world scale is related to birth rates and death rates but at other scales migration, which is studied later in this chapter, must also be considered (Figure 9.7).

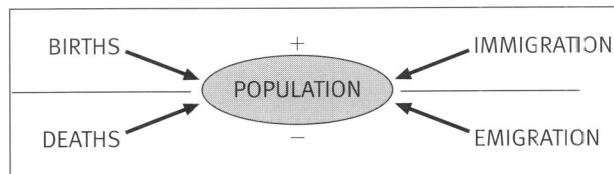

Figure 9.7 Model of population change

Birth or fertility rates

Fertility levels and changes in fertility levels are a result of complex sets of social, economic and political factors. Fertility is the key aspect of population growth in most of the world since levels of mortality or death have already fallen almost everywhere. Patterns of fertility show clear differences between regions of the world and between individual countries (see Figure 9.4).

Factors affecting fertility include:

- Marriage patterns
- Use of contraception
- Level of infantile mortality
- Economic and social value of children
- Economic and social status of women
- Education
- Government policies
- Religion

Death or mortality rates

There are dramatic differences in mortality rates (Figure 9.4) with parts of Africa having markedly higher rates. Note that many of the high mortality rates are due to high infant mortality rates and, as a result, life expectancy rates follow the same pattern.

Factors affecting mortality include:

- Improved nutrition
- Increased manufacturing
- Medical and health care
- Public health developments
- Improved living standards
- Environmental factors

The demographic transition model

Population changes brought about by changes in birth rates and death rates are summed up by this model (Figure 9.8a). It was originally devised and based on the changes observed in countries like the UK and Sweden since the 18th century. The model is used:

- to describe population change, by referring to the phases in the model when making comparisons between countries – this shows the relationship between birth rates, death rates and the growth of the total population over time;

- to predict change – it is difficult to predict the length of time over which changes in birth and death rates occur, because not all countries are following the Western pattern of industrialisation or modernisation and political and cultural factors have importance as well as purely economic factors.

The phases of the demographic transition model are:

1 High stationary phase with high birth and death rates, producing very slow growth.
2 Early expanding phase, with continuing high birth rates but falling death rates, producing growth which is slow at first but increasing as the gap between birth and death rates rises.
3 Late expanding phase, in which death rates are now falling slowly but birth rates fall markedly, so there is still considerable total population growth.
4 Low stationary phase, where birth and death rates even out at low and fluctuating levels producing a relatively static population.
5 A fifth phase has been suggested where some European countries show further falls in birth rates which would, in the longer term, lead to population decline. However, there is no evidence that this will be a long-term change and it could well be part of the fluctuations that would be expected in phase 4.

Review task

Using Figure 9.8b apply the phases of the demographic transition to Sweden and Mexico by pencilling in the phase boundaries for each. What is the startling difference between the two?

EXAM TIP

Case studies are needed to illustrate your examination answers and some specifications specify particular case studies. Make sure you summarise your case studies and pick out the relevant parts to link with the sections here.

Population patterns

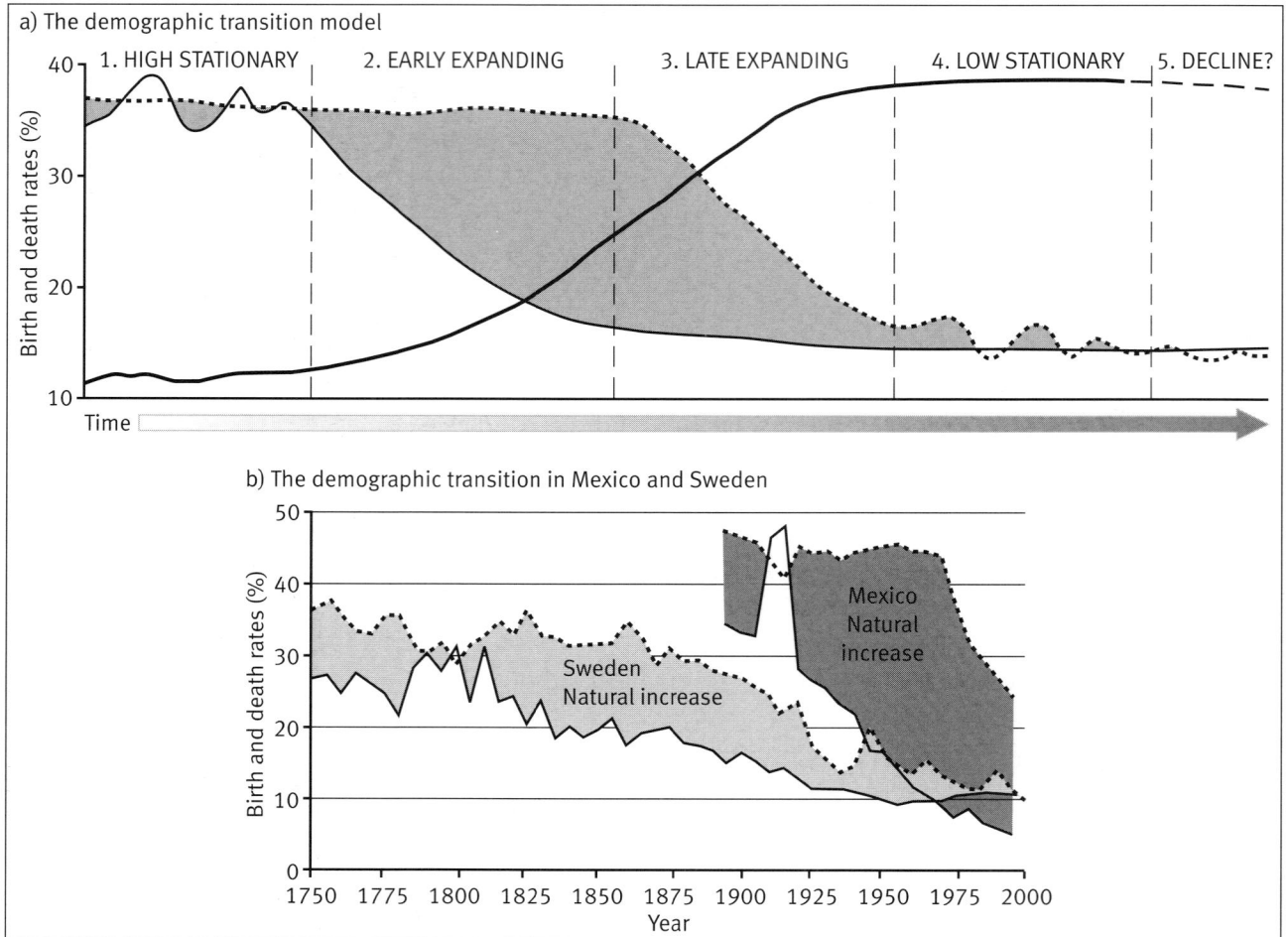

a) The demographic transition model

b) The demographic transition in Mexico and Sweden

Figure 9.8 a) The demographic transition model, b) The demographic transition in Sweden and Mexico

Population structure

Population pyramids or age-sex pyramids are visual means of demonstrating the structure of the population of a country. They are constructed by dividing the population into five-year groups by age and sex and calculating the percentage of the total population in each age-sex group, although sometimes actual numbers are used.

Population structure is important because:

- it affects the rate of population growth, in that if there is a large percentage of the population in the below 15 years age group, there will inevitably be an increase in births as this group passes through child-bearing age;

- it has implications for many social, economic, political and resource issues, such as increased need for expenditure on education where a large proportion is in the younger age groups, or the need to provide for elderly people if the structure shows an ageing population.

The shape of a population pyramid reflects changes in birth and death rates and also migration as well as the impact of major events such as wars or famines.

The contrasting population pyramids in Figure 9.9 summarise differences you can find in population structures at different phases in the demographic transition. Note that a wide base to a pyramid is as much due to a fall in infant mortality as it is to high birth rates.

Review task

Match the descriptions in Figure 9.10 to the sketch models in Figure 9.9a. Then make rough sketches of the two population pyramids in Figure 9.9b and pencil in appropriate labels from Figure 9.10. Where the feature is specific to a particular part of a pyramid use an arrow to show this.

Population pyramids are useful for comparing countries or examining population change within one country. For many purposes a simpler set of statistics is useful – the percentages under 15 years and/or over 65 years. These statistics are useful because they define the parts of the population not usually counted as working or **economically active**, although

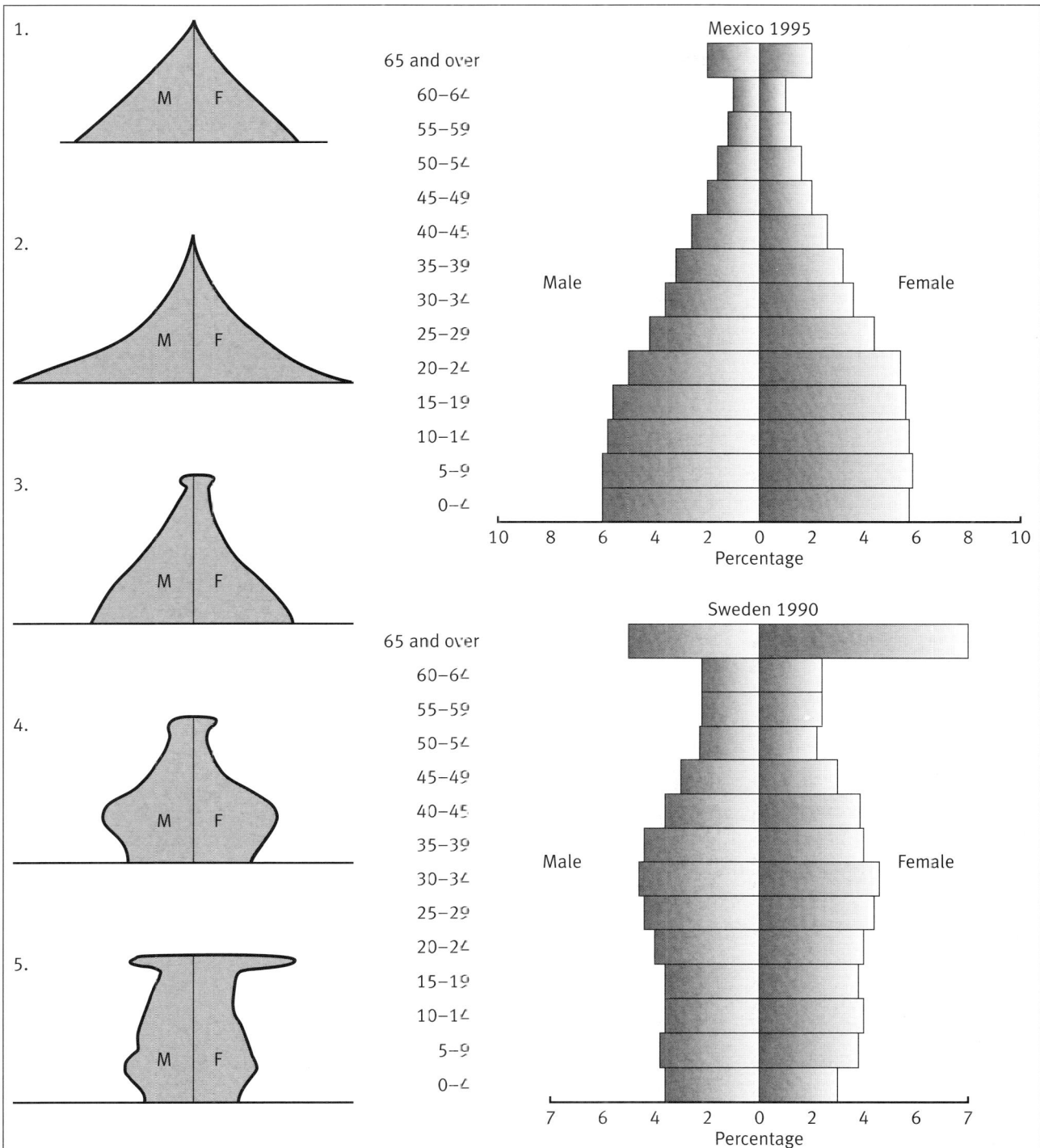

Figure 9.9 a) Population pyramid models, b) Contrasting population pyramids

we know that children have an important economic role in many less developed countries. The under 15 and over 65 population is the economically inactive or **dependent** population.

These figures can be converted into **dependency ratios**. Using actual numbers this is shown as:

$$\frac{\text{Numbers under 15 and over 65}}{\text{Numbers of working age}} \times 100 = \text{dependency ratio}$$

The dependency ratio is the number of people dependent on every 100 working people.

In MEDCs, 19 per cent of the population is below 15 years of age and 14 per cent is over 65 years of age. The corresponding figures for LEDCs are 34 per cent and 5 per cent, giving total dependent populations of 33 per cent and 39 per cent respectively. The dependency ratios are:

$$\text{MEDCs} \quad \frac{33}{67} \times 100 = 49$$

$$\text{LEDCs} \quad \frac{39}{60} \times 100 = 65$$

a) Wide base indicates much of the population is young. High death rates mean later age groups decrease in size. Note the difference between the 0–4 age group and the 5–9 age group. Rapid growth.

b) Birth rates decrease and death rates are low. As the larger groups move through the years there will be a greater and greater proportion of older people. Population will continue rising because even though birth rates have decreased the numbers in child-bearing age groups is much greater. This is known as **population momentum**.

c) Pyramidal shape with wide base and narrow top. Slow growth.

d) The pyramid is almost reversed with more older people and fewer younger people. Birth rates fluctuate so the sizes of age groups vary. Life expectation increases so a large elderly population is evident. Greater proportion of females in the older age groups.

e) Death rates fall and the wide base extends up the age groups. As birth rates stay high there is no narrowing of the youngest group. Rapid growth.

Figure 9.10 Labelling for population pyramids of different shapes

Review task

List the problems associated with a high dependency ratio. Explain how the problems differ according to which end of the age range makes up most of the dependent population.

Migration

Key terms

- **Spatial mobility** – an all-inclusive term covering all movements of people from local daily journeys to work to long-distance permanent international change of residence.

- **Circulation** – short-term movement without permanent or long-term change of residence.

- **In-migrant** – person moving into an area within a country.

- **Out-migrant** – person moving out to another area of the country.

- **Immigrant** – person moving into a country.

- **Emigrant** – person moving out of a country.

- **Migration** – movement from one area to another with a permanent change of residence.

- **Net migration** – the difference between inward and outward migration flows.

Migration can be classified in many different ways. Figure 9.11a is one simple method, but Figure 9.11b has a broader approach and breaks the subject down into issues or questions and begins answering them.

Models of migration

The factors affecting migration are complex, involving economic, behavioural, social and environmental matters. The interaction of these groups of factors means that prediction is difficult, and apparently simple movements involve many conflicting individual forces.

Distance	Time	Origin/destination	Reason
Internal	Permanent	Rural–urban	Forced
International	Temporary	Counter-urbanisation	Voluntary

Figure 9.11a Types of migration

Migrants (Who?)	Large or small numbers	Families or single people	Male or female	Well or poorly educated
Origins and destinations (Where?)	Internal (in-migrants and out-migrants)	International (immigrants and emigrants)	Rural–urban	Counter-urbanisation
Reasons for moving (Why?)	Voluntary or forced	Economic	Social and political	Environmental
Effects (What?)	On place of origin	On the destination	On the migrants	Varied

Figure 9.11b The issues involved in the study of migration

Revise AS and A Level Geography

Models of migration are intended to provide clearer understanding by focusing on key features of migration. They provide a generalised explanation based on principles derived from the study of varied situations, but without the complexities of individual cases.

Push-pull model

The push-pull idea is the simplest form of migration model, although once you start to add different push and pull factors and give them all a different weighting the position can become complex (Figure 9.12).

> Negative factors *push* → *Potential migrant* → Positive factors *pull*

Figure 9.12 A simple push-pull model of migration

Ravenstein's model

This was developed in the 19th century and is based on the rural–urban migration taking place as industrialisation occurred. Ravenstein observed that:

- most migrants travelled a short distance, and set in motion a step-by-step movement creating currents of migration;
- the number of migrants to any particular city decreased as the distance from their place of origin increased;
- migrants travelling a long distance predominantly went to a major city or industrial area;
- each stream of migration was matched by a counter stream;
- migrants were usually adults;
- economic factors were the driving forces behind migration.

Gravity models

These assume that migration movements between two places are in proportion to their population size and that they are inversely related to the distance between them (Figure 9.13). Modified versions of gravity models include other factors besides size and distance.

Intervening opportunities and obstacles model

Obstacles to migration include distance and cost of travel, but they may also be legal, such as political boundaries and limits on migration. Opportunities reflect the fact that migrants will cease to travel long distances if employment opportunities, for example, occur between the place of origin and planned destination (Figure 9.14).

Systems models

These more accurately reflect the complex nature of migration which involves a wide range of large-scale and small-scale processes (Figure 9.15 on page 99).

Rural–urban migration

Migration flows within a country (internal migration) are usually separated from flows between countries (international migration). Rural–urban migration is one major internal flow. It is usual to consider more developed countries separately from less developed countries.

MEDCs

In MEDCs the pattern of population was largely rearranged during the period 1750 to 1950, with a change from mainly rural to mainly urban. Another result was the greater concentration of populations in relatively small areas of high density. The proportion of the population living in urban areas is known as

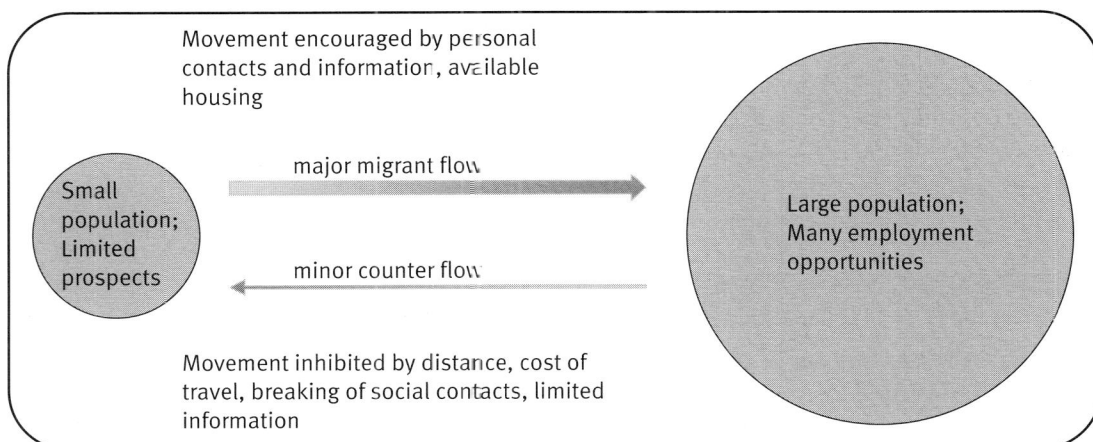

Movement encouraged by personal contacts and information, available housing

Small population; Limited prospects

major migrant flow

minor counter flow

Large population; Many employment opportunities

Movement inhibited by distance, cost of travel, breaking of social contacts, limited information

Figure 9.13 A gravity model of migration

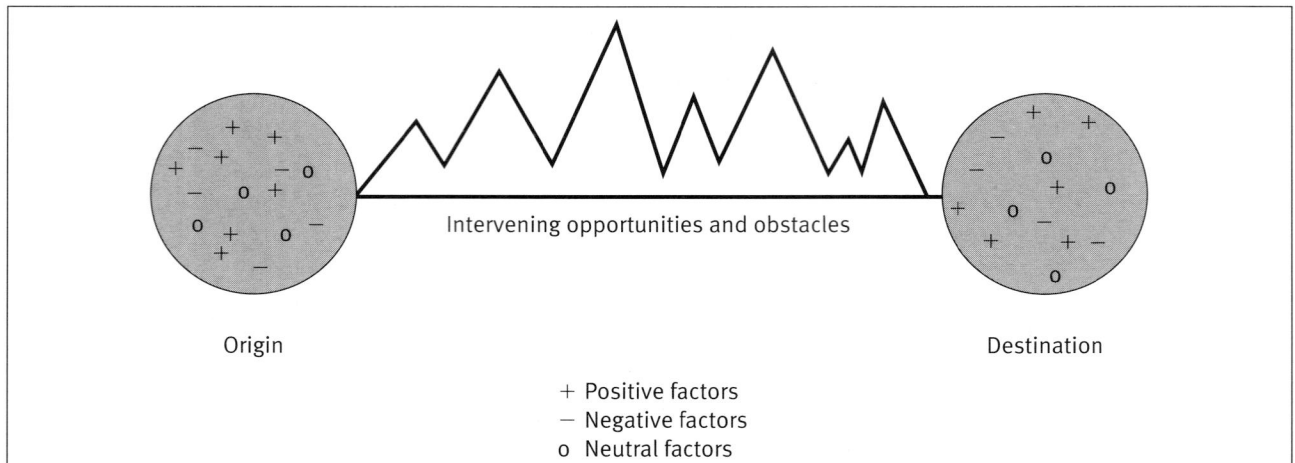

Figure 9.14 An intervening opportunities and obstacles model of migration

urbanisation. To confuse matters the term is sometimes used for what is usually called **urban growth** (see Chapter 11). An increase in the level of urbanisation can be due to:

- rural–urban migration

- greater rate of natural increase in urban areas compared with rural areas.

Over time, the second factor tends to become more important simply because the rural–urban migrants bring about a change to the population structure with a greater proportion being in the family-forming age group. This feature has been observed in developed countries in the past and recently in less developed countries.

In the UK and elsewhere in Europe the Industrial Revolution set rural–urban migration in motion. The overall reason for movement was generally economic but social, political and environmental factors were also significant in different areas. The simple push-pull model is useful here to group together the problems of the rural areas and the attractions of the growing urban/industrial areas. The former include overpopulation, enclosure of common land, poverty, small landholdings and general lack of opportunity. The latter include growing coal industry, iron and steel production and metal-based industries on the coalfields and the growth of manufacturing and trade in other major cities. The movement of people was in line with 'laws of migration' propounded by Ravenstein and over the long term produced the broad population pattern of the first half of the 20th century.

Review task

Recap Ravenstein's laws and apply them to case studies in your course books or your notes, for example population changes in the West Midlands, or the UK as a whole.

Note that the process of migration here was an ongoing one with movement onwards from one destination to another. On top of that were further movements – migration within the urban area, most obviously with the **suburbanisation** of the 20th century. Continued redistribution of population is a feature of rural–urban migration – the process does not end at the urban destination. The continuation of this trend has produced the process of **counter-urbanisation** (see Chapters 11 and 12). Note the difference between counter-urbanisation and **decentralisation**, which includes suburbanisation and the growth of commuter rings around major cities with central functional connections. Counter-urbanisation does not involve such a centralised relationship with connections being much more dispersed.

LEDCs

In LEDCs rural–urban migration is more recent. Since LEDCs make up the greater proportion of the world's population, this translates into greater numbers of migrants so now the majority of the world's urban population lives in the less-developed world.

There are great variations in urbanisation levels. Africa has the lowest levels (30 per cent) followed by Asia (35 per cent) with Latin America having markedly higher levels (73 per cent). There are also, of course, differences between individual countries in all three.

Rural–urban migration in LEDCs differs from that in MEDCs:

- **The importance of the primate city.** The national capital or major city is the main attraction. This is where modernisation and industrialisation is centred and it is the focus of the national transport system. In addition, there is a great bias in the proportion of national resources allocated to it.

Economic conditions: jobs, wages, prices, levels
of commerce and industry
Social welfare and development: education, health

URBAN DWELLER

Adjustment to urban life

Positive
feedback e.g.
encouraging
news

Urban sub-system

Urban control
system

Negative
feedback e.g.
discouraging
news

Migration channels

Rural
sub-system

Rural change
mechanisms

POTENTIAL MIGRANT

Government policies, agricultural practices,
marketing organisation, population movement,
technology, transport, mechanisation

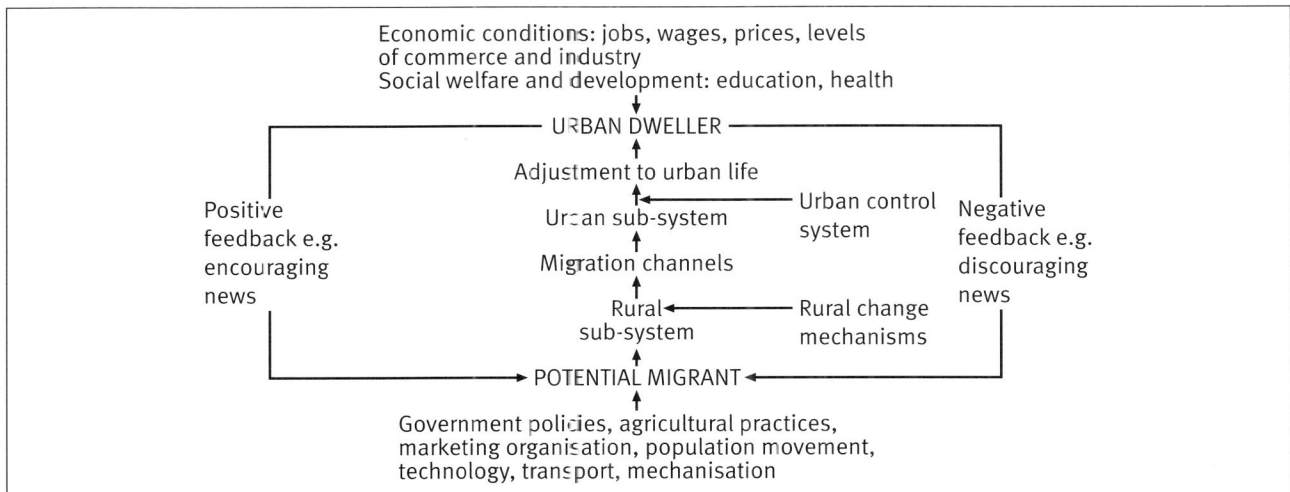

Figure 9.15 A systems model of the rural–urban migration system in Africa

- **Continued links with rural areas.** This varies, with less in Latin America and more in Asia and even more in Africa. The result is a greater tendency to move back and fore and for migrants to feel that the rural area is still home.

- **Urban underemployment.** Since industrialisation and the growth of manufacturing employment has not accompanied urbanisation to the same extent as in the developed world, there has been no major source of employment for new migrants. The result has been the growth of the informal sector and great dependence of limited and temporary part-time work.

Explanations for the rural–urban migration tend to dwell on the push-pull model, but no matter how large the list of push-and-pull factors the one key factor seems to be expectations of migrants. Some migrants do well and so others take the chance that they will do well also and so the flow of migrants keep moving. (See Chapter 12 for the impact of migration on the cities themselves and the way cities have changed.)

International migration

It might seem to be more useful to class migrations as short and long distance rather than internal and international, but national borders are major **barriers** to movement. They result in political controls over movement and they mark social and cultural differences, all of which are obstacles.

> ### EXAM TIP
>
> You can apply most of the migration models to international migrations but the simple push-pull idea is commonly used to summarise factors affecting migration.

> ### Review task
>
> **Rearrange this list into a push list and a pull list. Some may not fit either list. Look back at the models and find the appropriate term or terms to use for new heading(s).**
>   **shortage/availability of land**
>   **distance**
>   **improved transport**
>   **government or commercial inducements**
>   **availability/lack of information**
>   **employment opportunities**
>   **selective immigration controls**
>   **civil and international conflict**
> **Add examples to illustrate each point.**

Categories of migration

Settlement
In modern times there have been the European migrations to New World countries for permanent settlement. These last migrations sent over 60 million people to the Americas in 100 years. The reasons were population pressures in Europe (push) and the availability of land in America (pull), coupled with the improved transport of the 19th century which removed the Atlantic as a major obstacle.

> ### Review task
>
> **Refer to your own notes or course books and select one example of a permanent settlement migration. Summarise the features of the migration in a table or Mind Map, making sure you include changes over time.**

Modern settlement migration is mainly a Third World to Developed World movement involving labour migration to Europe and to North America and Oceania.

Refugees

Forced migrations include refugees and slaves. The latter involves a pull factor, whereas refugee movements are the result of push factors. Major international wars, such as the Second World War, produce very large population movements, but civil wars and wars of independence also create refugees. Ethnic differences and partition of states produce more refugees.

Migrant workers

The pattern of labour migration is very clear in every part of the world where it is prevalent. Many people have applied the core–periphery idea to this pattern and it is clear that there is a strong spatial factor at work. The key difference between this form of migration and permanent settlement is the degree of control, which tries to ensure that stays are temporary and only contracted workers are involved.

The impact of migration

Migration impacts on:

- **The place of origin.** Economically, remittances from family members working abroad provide an important source of income. On the other hand, migration causes a loss of labour, particularly in the young adult age group. Socially, migration produces a skewed age-sex structure with fewer young adults and possibly an imbalance between sexes in different age groups. Traditional social activities suffer and political activity suffers, too. Services of all kinds also suffer.

- **The destination.** Workers from abroad fill gaps in the workforce and at the same time create more need for housing and possibly put pressure on inadequate infrastructures. On the social side there is, at least in the short term, an imbalance to the population structure. Others results are, on the positive side, more cultural mixing, but on the negative side there may be hostility between different cultural groups.

- **Migrants themselves.** These points overlap with those above. On the negative side, migrants often experience hostility, poverty and stress, while on the positive side they gain in income, access to schools and education generally, access to better health care and wider opportunities especially for their children.

Review task

There are three kinds of impact affecting places of origin, destinations and the migrants themselves. They are demographic, social, and economic.

1 Take each one in turn and, using the information in the previous section, write a brief explanation.

2 Make a large Mind Map summarising the whole topic:
 Population distribution and trends
 Natural change
 Population structure
 Migration
Start by brainstorming each sub-section *without referring to it*. This gives you a good idea of how much you know already.
Remember your review plan, so review your Mind Map regularly.

What you need to know:

- **The concepts of overpopulation, underpopulation, optimum population and carrying capacity**
- **The issues and theories involved in the population–resource relationship**
- **The different types of resources and the various ways of dealing with them**
- **The relationship between population growth, the natural environment and available resources**
- **The issues involved in population and resource management**
- **The relationships between population patterns and patterns of social and economic well-being**
- **Distribution and density of population at different scales**
- **Population density, distribution and carrying capacity**

Also, you will need to be able to:

- **describe and comment critically on various theoretical approaches;**
- **demonstrate the relationships involved in population–resource management issues.**

EXAM TIP

This topic covers the relationship between population and resources, but it overlaps with the study of development and, in your own course, the contents of this chapter may come under different topics at either AS or A2 levels.

Population and carrying capacity

This section examines basic concepts to do with imbalances or apparent imbalances in population.

Disparities in population density at world and regional scale (see Chapter 9) are often explained by differences in the availability of resources. Various terms have been used to indicate this and they include: **carrying capacity, overpopulation, underpopulation** and **optimum population**, and, for urban areas, **tolerance capacity**. Remember that the term **resources** includes climate and soils as well as water, minerals and land; in other words providing all the means for sustaining life.

Carrying capacity

Carrying capacity is an ecological idea which has been applied to population geography. It is the maximum number that can be supported by the food-producing system without damaging the productive system itself. If the demands on the system grow beyond its capacity it will collapse with consequent effects on the population it supports. Figure 10.1 illustrates this.

In ecology, the carrying capacity is taken as fixed, but in population geography, of course, this would not be true as technological changes continually increase carrying capacity, so over time it can change.

Review task

Copy and modify Figure 10.1 as follows:
a) **Substitute technological change for column 2 and suggest the kinds of changes that could raise carrying capacity.**
b) **Replace column 3 with new results.**
c) **Add an extra column for longer term effects, good or bad.**
Look over work dealing with solutions to soil erosion problems – sustainable methods of intensifying agriculture, regulation of activities by net size, for instance.

Carrying capacity in population geography includes the idea of a **sustainable population**. This is a population that can be supported without causing degradation to the environment that supports it. Again note how these definitions overlap.

The idea of 'carrying capacity' is most easily applied to local economies based on biological systems, but even here its application is beset with problems due to the complexity of the variables involved. Remember that the carrying capacity (or optimum population) depends not so much on the resource base as on the use of that resource base. Changes in inputs, such as capital investment, can raise carrying capacity, while withdrawal of aid can result in a reduction due to lack of necessary inputs.

Overpopulation

The term **overpopulation** is now usually used in connection with the idea of carrying capacity. In this context it can only apply given constant carrying capacity and a fixed level of technology and standard of living. Under these circumstances the term overpopulation can be applied if a reduction in numbers would produce a rise in the standard of

System	Damage	Result
Ocean	Overfishing	Collapse of fishing activity and economy based on fishing.
Forests	Clear cutting with no replanting/faster than regrowth	Destruction of tree cover and collapse of timber-based local economy. Possible erosion of soils.
Pastoral grazing	Overstocking – growth of animal numbers	Overgrazing, damage to vegetation cover, soil erosion, lack of fodder, collapse of herds and flocks and damage to pastoral economy.
Settled agriculture	Overcultivation, depletion of soil nutrients, damage to soil structure	Decreasing yields, erosion and loss of agricultural land. Collapse of rural economy.

Figure 10.1 The results of overuse on some resource systems

living. Note the use of the term **tolerance capacity** below as a substitute for overpopulation in urban areas. Overpopulation may seem to apply more obviously in areas of high density, but as it is the number of people in relation to the level of resources that is important, sparsely populated areas can be just as easily overpopulated. So it could be applied to a densely crowded agricultural area and to a sparsely populated area used for pastoral grazing.

Underpopulation

Underpopulation applies to areas where an increase in the number of people would produce a rise in overall standard of living by allowing resources to be developed. It is usually used in relation to frontier areas which could support more people (e.g. by mining), although capital from outside the region would probably be required. This highlights the point that in the modern world the concepts of over- and under-population can be valid only if placed in the context of the whole world and its trading and financial systems.

Optimum population

Optimum population can be equated with optimum carrying capacity and again refers to a static situation. Optimum population is the maximum number of people which can be supported at a satisfactory standard of living given a certain level of resource and technology. You can see how the two ideas are similar. Remember that changes in, for example, methods of farming change the optimum population.

These ideas are applied more clearly to rural, agriculturally based or resource-based societies. Urban and industrial areas with their dependence on manufacturing and service industries provide a much more complex situation. The concept of **tolerance capacity** is used here to take account of the stress of greater population numbers.

Cause	Explanation
Cost (economics)	Exploited if competitive compared with other sources.
Technology	New technologies create demands for new resources or new uses for old resources increasing demand and likelihood of exploitation, and vice versa.
Protection	Individual countries may protect their resources by limiting production or may protect an industry which is uncompetitive therefore encouraging production which would not otherwise occur.
Demand	Increases in demand for any reason result in: a) resources being reclassified as reserves; b) exploration being intensified; c) new exploration technologies; d) new exploitation technologies. Under the same circumstances an alternative resource may become viable and be exploited in preference to the original one.
Environment	Environmental impact is a factor in preventing or limiting exploitation but controls vary especially between developed and less-developed countries.

Figure 10.2 What makes a resource a reserve?

Resources

Definitions

- **Natural resources** are natural products that people regard as being useful. The obvious types are energy resources and those used in industrial production. In agriculture there are natural resources like soil, water and regional and local climate. Natural resources important in terms of quality of life include attractive scenery and wild areas.

- **Reserves** are 'currently exploitable resources' as opposed to deposits of minerals which may be too small or too poor in quality or too difficult to extract under existing economic, technological or political conditions. Also, resources that might be exploited in a more-developed part of the world might be left untouched in a less-developed area where there is little demand for them.

- **Renewable resources** can be replaced naturally within a time period which ensures continued supplies and can be used or harvested indefinitely. They are also called 'flow' resources.

- **Non-renewable resources** are finite and cannot be replaced once used. They are also called 'stock' resources.

- **Sustainability** of resources in the long term is the key factor in separating renewable from non-renewable resources, but bear in mind that most renewable resources depend on other resources, like soils, which can be damaged by overuse (see Figure 10.1) and the time period for their recovery is long. Sustainability of even renewable resources depends on successful management of their development. (See also Chapter 16.)

- **Recycling** allows otherwise non-renewable resources to be reused, just as developments in technology allow resources to be used with less waste giving rise to conservation of resources which is possible in many other ways, such as reworking old waste tips for minerals.

- **Substitution** extends the life of resources by using alternatives.

Environmental impacts of resource exploitation and use

Activities that specifically exploit resources often have a more obvious environmental impact than others. Resource exploitation is generally taken to involve those materials that are taken from the earth, like minerals and fossil fuels together with anything else that is moved, like water and forests.

Environmental impact occurs at different **scales**:

- **Local,** where the effect on the environment and people is only in the immediate area. Even then, there is often an indirect effect on a broader scale. For example, quarrying has a very localised immediate effect on landscape, other land uses, local traffic, and noise and dust pollution. In this case the indirect effect may be increased traffic to and from a cement works in the same district and the effects of that works on the environment.

- **Regional,** where the effects from a single activity or a concentration of activities spread well beyond the immediate area to other parts of the region which, of course, often means across national boundaries. An example here could be deforestation resulting in soil erosion in the immediate area and flooding or silting downstream, possibly a great distance away and across international borders.

- **Global,** where the impact directly or indirectly affects all or most parts of the world. The obvious example is global warming to which deforestation contributes but where the major direct source of the current raised carbon dioxide levels is the exploitation and burning of fossil fuels.

Activities, resources and impacts

Broad-scale activities or types of resource that have impacts at a variety of scales include agriculture, water, tropical forests and mining. The exploitation of a resource has very direct environmental effects but the way resources are ultimately used is just as significant for the amount, type and scale of environmental impact. Therefore it is important to bear in mind the impact of manufacturing, transport, tourism and energy.

Agriculture

The major problem is the degradation of agricultural land as a result of overuse. The two types of degradation are soil erosion and desertification.

Soil erosion has been predicted to cause an average decline in food production of 25 per cent from unirrigated croplands between 1985 and 2010.

The direct **processes** involved in soil erosion are water and wind, but the damage done by these processes is made possible or exacerbated by damage done by deforestation, inappropriate agricultural practices, both arable and pastoral, fire, construction and urbanisation. These expose the soil to the effects of water and wind. Water brings about sheet erosion and gullying, and wind takes away the finer particles of the soil. Both water and wind cause further problems in the areas where eroded material is deposited.

The deeper **causes** of soil erosion are due partly to population growth, but also to several factors related to poverty, notably the difficulty for poor farmers in implementing conservation measures. Population pressure obviously induces efforts to increase agricultural yields, which in turn may mean shorter fallow periods, less soil cover for more of the year and greater impoverishment of soil fertility. The result is general degradation and loss of soil by erosion.

Desertification is more severe and in some estimates one-third of the land surface is threatened (45 million km^2) and affecting 850 million people (see Chapter 8).

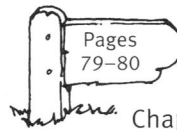
Pages 79–80

The impact of land degradation is greatest on the poorest people in the world (Figure 10.4).

Water resources

Water is a renewable resource but it is often used as a non-renewable resource. Problems include the mismatch between availability of water and population distribution, the shortages of water affecting 2 billion people who live in areas with severe shortages, and the increasing **demand** for water due to rising living standards. Topics of significance are:

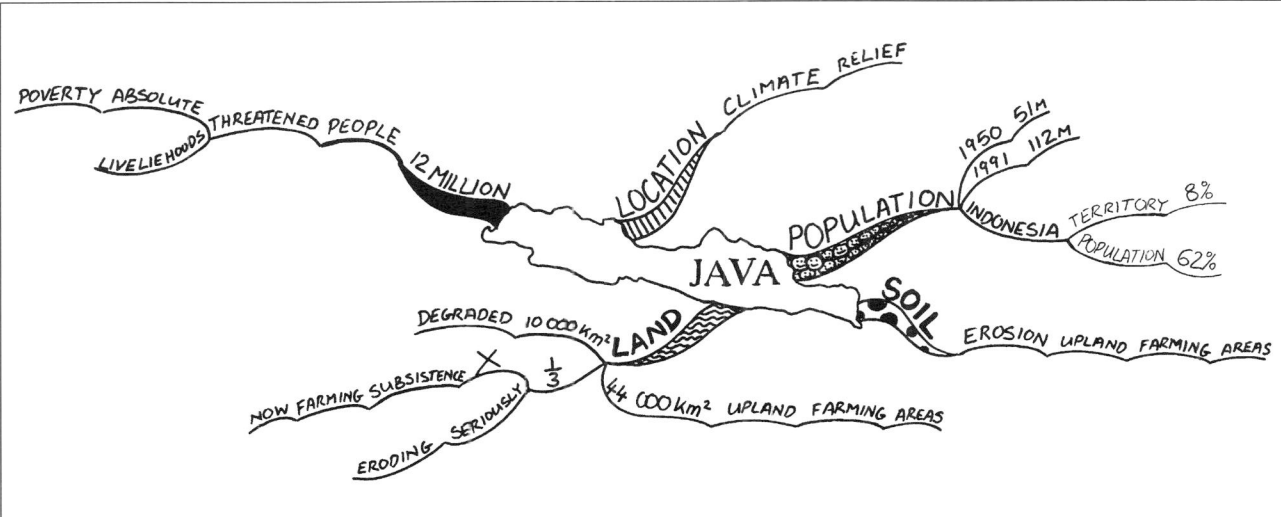

Figure 10.3 Soil erosion and its impact in Java

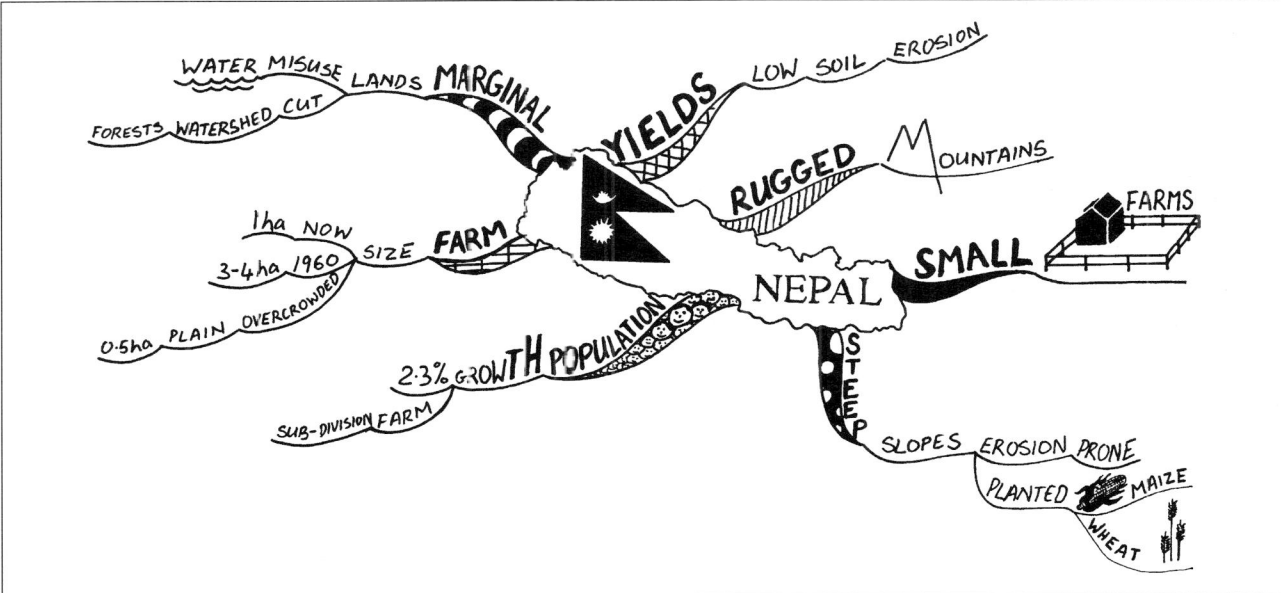

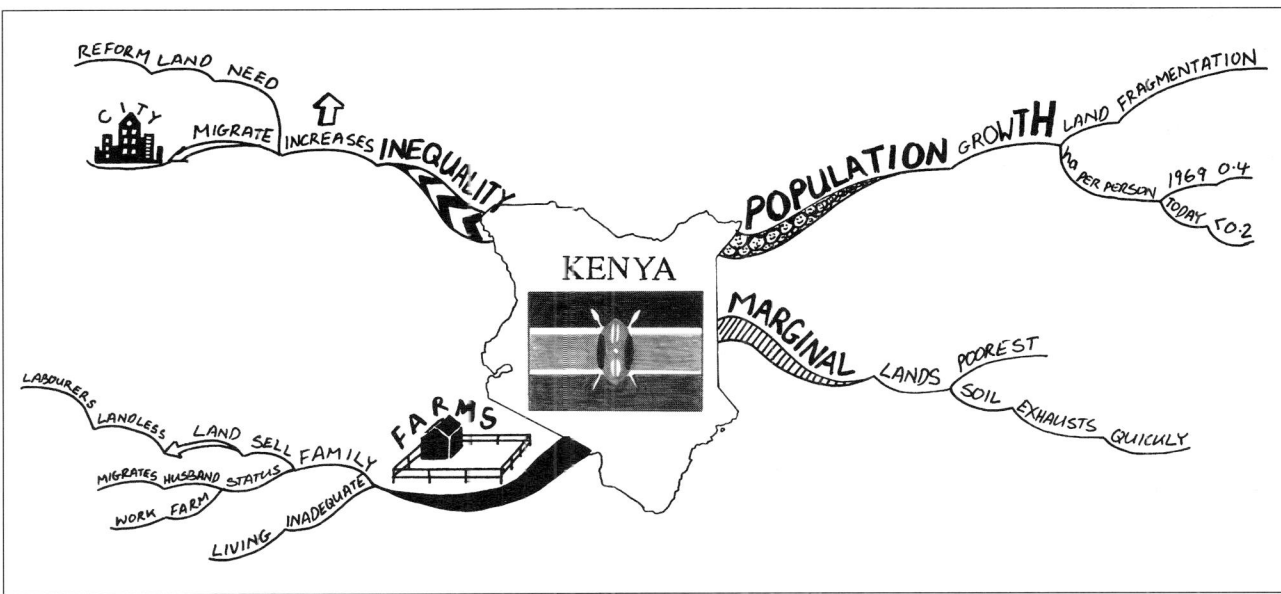

Figure 10.4 Agriculture and land degradation

- **Water and disease** – as a result largely of population growth the numbers of people without access to safe drinking water is increasing rather than decreasing. The use of the stable runoff from rivers in greater proportions has resulted in their contamination and an increased incidence of infectious diseases. Water-related diseases are even more acute in urban communities than in rural, and it has been stated that 'human communities face a growing threat of contaminated-water pandemics' ('Population Resources and the Environment: Critical Challenges', United Nations, London, 1991). Deforestation in upland catchments adds to the problems even in areas with heavy rainfall as in peninsular Malaysia.

- **Water for irrigation** – the increasing demand for water for irrigation (irrigated land accounts for 65 per cent of water use worldwide) has resulted in irrigation-reliant areas having insufficient water at the appropriate times of year. Much more careful management of water is essential if irrigated land production is to be maintained and if irrigation is to be extended.

- **Water stress** – is the state of crisis when water supply falls below 500 m^3 per person per year. This is not necessarily directly related to population growth rates as many areas affected are in arid or semi-arid zones, but population growth does have an effect. Water stress has a knock-on effect on land degradation which in turn produces watershed flow problems and further disruption of supplies.

Tropical forests

Deforestation is a major and increasingly important issue in almost every tropical forest. Globally, the most important **causes of deforestation** are increasing population and the spread of agriculture. In some regions ranching is an important cause of deforestation, while in others it is the use of the forest for firewood. Logging is a major factor in many areas, especially where lack of control and corruption is significant. The combination of many or all of these factors is probably a better description of the real situation throughout much of the tropical forest regions of the world (Figure 10.5).

The results of deforestation of the tropical forests include:

- Soil erosion
- Silting downstream of rivers and reservoirs, irrigation systems and associated farmland
- Rapid runoff and flooding and irregular and less reliable river flow
- Loss of human habitat
- Loss of biodiversity
- Climatic change locally and regionally
- Global climatic changes

Mining

The range of issues affecting mining as an economic activity is wide: methods of extraction and processing and resulting waste as well as possibilities for reclamation; location of mineral deposits and factors affecting their exploitation, particularly constraints and environmental controls. The environmental impact of mining occurs at different scales and at each scale the kind of impact differs:

- Local scale
  - ✱ visual intrusion
  - ✱ subsidence
  - ✱ water pollution
  - ✱ air pollution – dusts and gases
  - ✱ reclamation pits and waste tips

- Regional scale
  - ✱ acidification
  - ✱ metals deposition

- Large scale
  - ✱ air pollution particularly contributing to acid rain

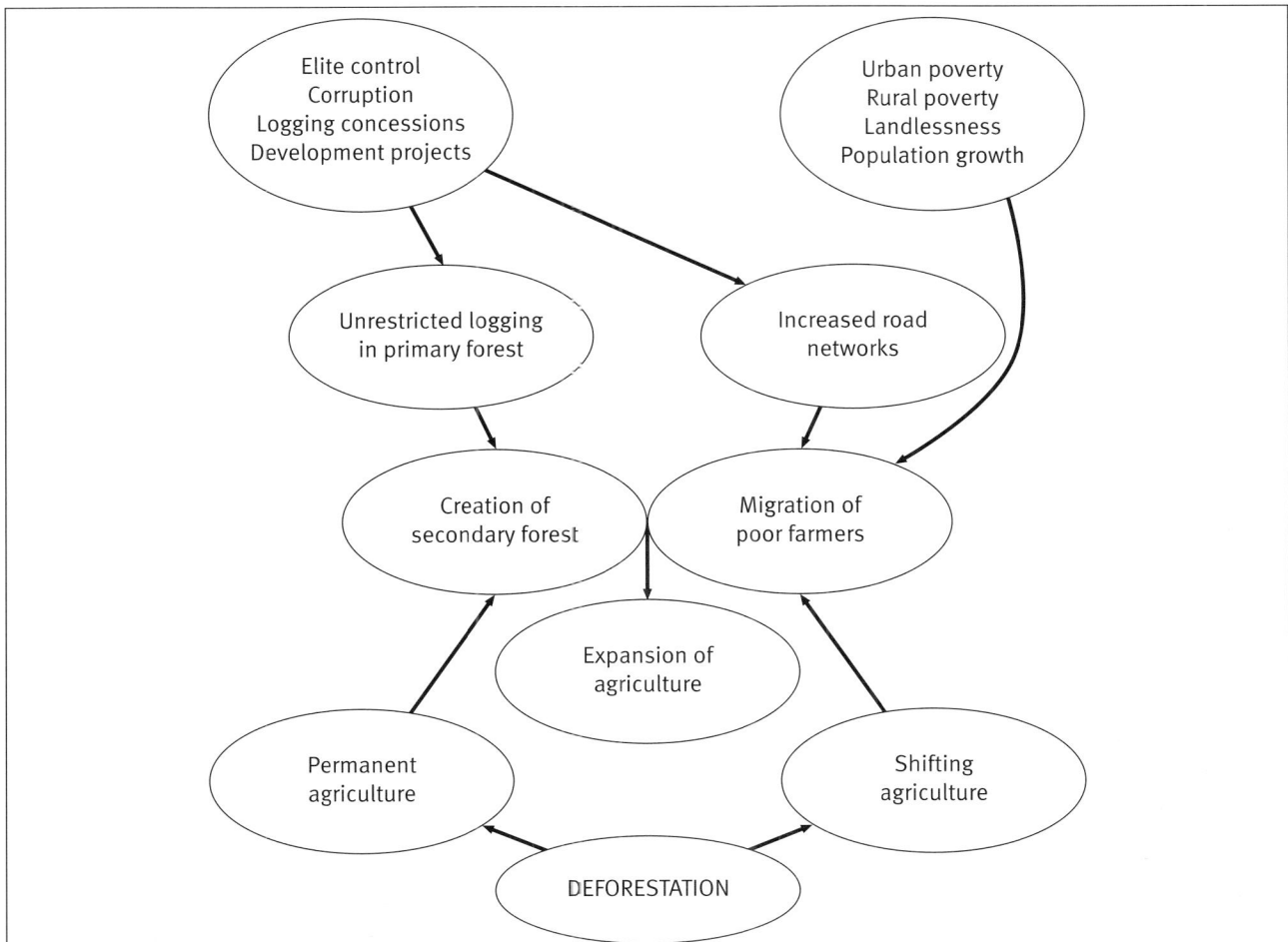

Figure 10.5 Deforestation in the Philippines

Review task

1 **Check your own notes and textbooks for examples of the above. Make a Mind Map based on the above list and add material from your own sources to produce a single visual revision aid.**

2 **Study Figure 10.6 which summarises environmental linkages. Match up the items on it with the classification of environmental impacts above.**

Population and resource imbalances

Population, resources and technology

The idea of imbalance is directly linked to carrying capacity and the other ideas examined earlier in this chapter. The fact that the world's population is distributed very unevenly is the first and key imbalance to note, but that imbalance needs to be related to the resources which support the population. At the most basic level, land is the ultimate resource and, for example, with 9 per cent of the world's land area India and China together have 36 per cent of the world's population and apparently

provide a classic imbalance. However, a population imbalance between these two countries and the rest of world does not necessarily mean that they can be described as overpopulated or that other places are underpopulated. The idea of carrying capacity depends, remember, on the state of technology and not just the available resources. Technology determines how resources are used and the numbers that can be supported and Figure 10.7 shows how the three elements of population, resources and technology are interlinked. Together they all contribute to the quality of life of the population of the area considered.

The relationships or links in Figure 10.7 are dynamic and changes affecting one have a knock-on effect on the others. However, it has certain assumptions built into it:

1 That the internal effects are appreciated, such as the environmental impact of resource development.

2 That the impact of developments elsewhere, such as changes in demand for export products or the prices of imported goods and the general world economy, are also appreciated.

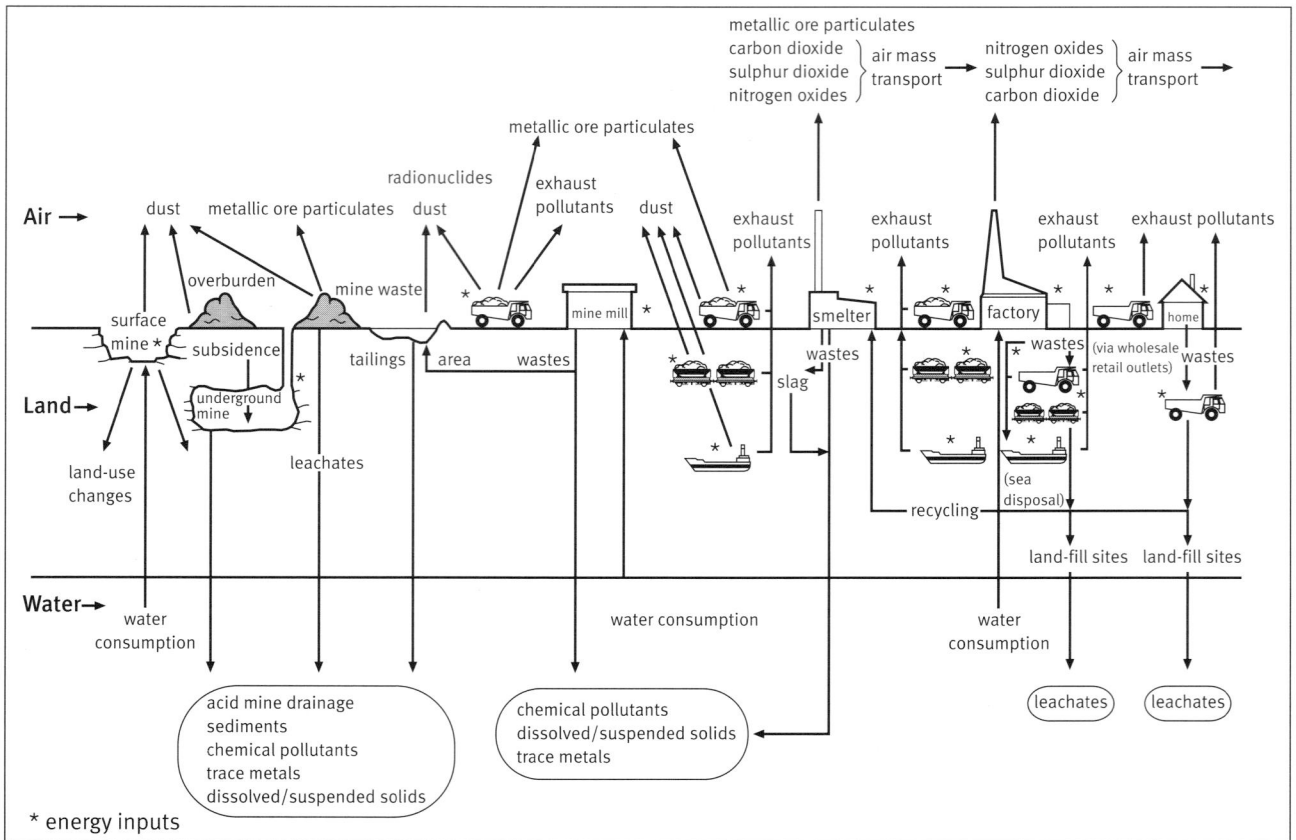

Figure 10.6 Environmental linkages with mineral extraction and use

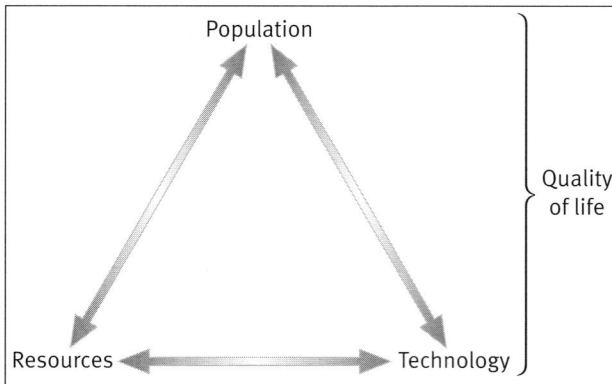

Figure 10.7 The links between population, resources, technology and quality of life

Factors involved in the population–resource relationship

Figure 10.7 provides a useful simplified approach to the population–resource relationship, but there are more factors at work and you probably mentioned some of them in the review tasks. You need to be clear about the full range of factors behind the diagram:

- **Population growth.** A high or low growth or a static population has implications of many sorts, for example, on population structure.

Review task

Make a copy of Figure 10.7 and draw a circle around it. Inside the circle add some examples of internal effects and outside the circle write the external factors and again add a few examples. To do this you may need to refer to your course notes and/or textbooks.

EXAM TIP

The relationship between population and resources is a central issue in geography and it involves all three strands of the subject – physical, human and environmental. The link to quality of life is an obvious one since it is affected by all three. Remember to draw on all aspects of the subject in answering questions on this.

- **Physical factors.** There are some absolute limits to settlement but in general there are areas which are less or more favourable for settlement based on a combination of climate, soils and landforms. Most physical or environmental limits, however, are specific to particular technological circumstances and to cultural perceptions at specific times.

- **Resources** – land, food, energy and raw materials. Changes in the values of these again depend on changing perceptions and technologies (see Figure 10.2). In the past, food and other resources were local but this is no longer so, at least for developed countries, and in less developed countries resources may be utilised to provide exports.

- **Social and economic factors.** Levels of development, levels of technology, trade and the economic and social systems of individual countries are important. However, with globalisation the impact of external factors has grown.

- **Political factors.** Government policies and international organisations are important. Population growth or restriction policies have direct impacts on population but other social and economic policies have significance for imbalances.

Review task

Review the relevant parts of your course notes and textbooks; add examples to each of the points above. Remember that the same examples are probably useful for more than one point. Also, make a note of which other topics they might be useful for.

An overview

A complete view of the links between population and resource issues is shown in Figure 10.8. It describes the situation for one country but you can extend the idea by adding or substituting international organisations in the boxes.

The arrangement of the diagram emphasises the way that the links between population and natural resources are modified and influenced by:

- Technological factors
- Government policies
- Social and economic factors
- Management and use of resources

Note that the links are not all one way as, in most cases, there is a feedback influence which is sometimes direct and sometimes indirect.

Theoretical approaches to population–resource relationships

There are three different views on the relationship between population and resources and the environment and they can be summed up as: pessimistic, optimistic and neutral.

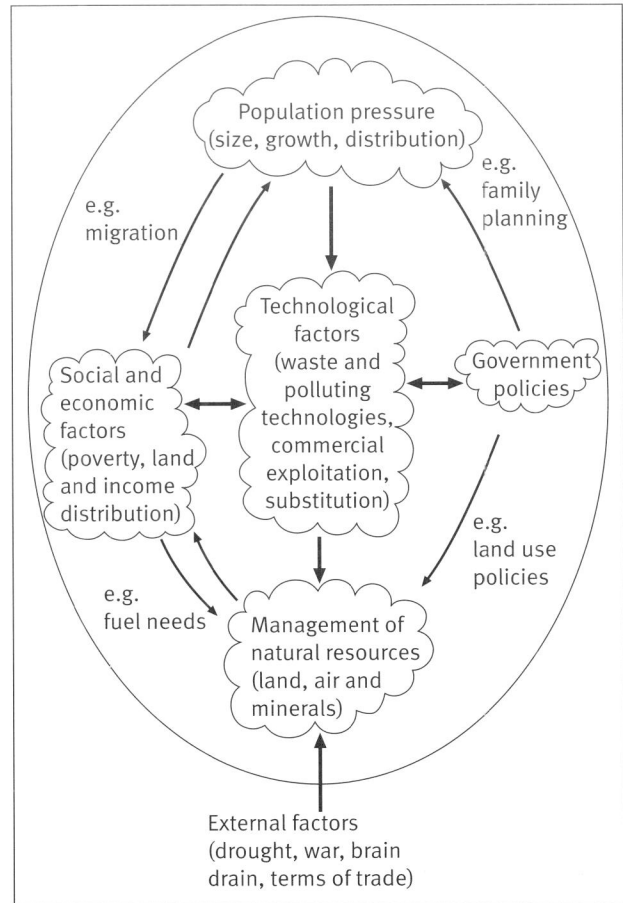

Figure 10.8 Linkages between population and natural resource issues

The Malthusian or pessimistic view

This is based on the relationship between population growth and finite resources. Thomas Malthus put forward the view in 1798 that population would increase faster than food supply and result in famine unless growth was restricted. In modern times this view was revived by **neo-Malthusians** in the 1960s and 1970s in a number of books (e.g. Ehrlich and Ehrlich, *The Population Explosion*; Meadows *et al*, *The Limits to Growth*). This modern approach widens the issue to consider other effects of population growth including environmental problems, exhaustion of mineral resources and loss of biodiversity. The impact of population growth depends on demand or consumption levels and technology and the following equation sums it up:

$$I = P \times A \times T$$

(I = impact, P = population, A = affluence, T = technology)

The pessimists regard population as the most important component and the impacts as being harmful.

The Cornucopian or optimistic view

This argues that population growth is a stimulus to increased food production and the development of new approaches to environmental problems. Boserup was a leading proponent of this. Her examples showed a strong link between population growth, technological change, increased agricultural production and other major social changes in the world. One such example is the way India's food production kept pace with population growth as a result of the Green Revolution. The essential argument is that as population pressure on the land increases so does demand for agricultural products but instead of leading to lower yields and environmental losses it stimulates greater inputs (labour, new technologies) which increase yields and, at the same time, improve the land.

The economist Simon also argues that the problems created by population growth stimulate the development of solutions, so that, in the equation above, population growth sets off technological developments which in turn make the impacts beneficial. As one example, the prices of many metals actually fell between 1980 and 1990, and for Cornucopian reasons – technological developments in exploration extraction and processing, wider exploration and substitution of cheaper materials.

The neutral view

This argues that the situation is not as clear cut as the two groups above make it seem and that population growth is just one factor and is not necessarily harmful or beneficial. Although population growth does clearly increase demand there is a wide range of social, economic and political factors which intervene between population growth and its impact.

The debate between the first two key opposing schools of thought can be summed up in this edited extract from Sarre and Blunden, *An Overcrowded World* (1995), p72:

The solution for the users of resources – Western consumers – is also a disaster for the producers of resources. For example, . . . the world market for copper has collapsed [and] led countries . . . such as Zambia, into debt and economic collapse.

The solution to one resource problem leads to further problems with a different resource . . . new technologies in agriculture meant that per capita world food production increased . . . but . . . malnutrition has not disappeared . . . since it is the distribution of food . . . which determines this [and this in turn] is determined by social, economic and political factors . . . [Then again] intensive farming has brought environmental problems, such as salination. Finally, there are signs that the limits to intensive food production may have been reached.

. . . 'impact' issues tend to get grouped together. However, it is possible that the significance of population varies according to the issue.

Review task

1 **Make a table to list the points for and against on each of the three approaches to the population–resource debate. Refer to material in your course notes and books with relevant case studies.**

2 **Make a Mind Map to summarise the chapter's main ideas. Use the four section headings as a start:**

   - **Population and carrying capacity**
   - **Resources**
   - **Population and resource imbalances**
   - **Theoretical approaches to population–resource relationships**

   Remember to add references to *examples* you have studied in your course.

What you need to know:

- **Urbanisation processes in MEDCs and LEDCs**
- **Suburbanisation and counter-urbanisation**
- **The rank-size rule and urban primacy**
- **Settlements as service centres; central place theory**
- **Models and theories of urban growth and structure**
- **Urban land use zoning in both MEDCs and LEDCs**
- **Residential segregation in cities**

Also, you will need to be able to:

- **interpret settlement patterns from maps;**
- **relate urbanisation processes to case studies on a national scale;**
- **apply models and theories of urban geography to case studies of urban areas;**
- **analyse maps and other data which describe urban zones.**

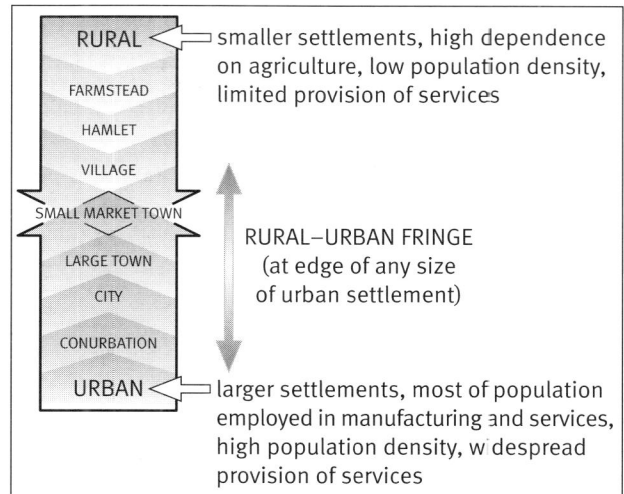

Figure 11.1 The rural–urban continuum

The world's settlements vary greatly, reflecting the economic activities carried out in them, the differences in peoples' culture, wealth and stage of economic development and the contrasting natural environments in which they are located.

Rural and urban

Settlements can be broadly divided into the categories of rural and urban. In 1800 about 97 per cent of the world's population were living in rural areas. By 2000 the proportion was just over 50 per cent. There is considerable global variation in the rural–urban divide. Defining whether settlements are rural or urban can be surprisingly difficult. Different countries may define settlements as urban according to varying criteria of size, employment characteristics or the administrative functions that they may have. In reality there is a **rural–urban continuum** as shown in Figure 11.1, particularly evident in MEDCs.

Sites and situations

The **site** of a settlement is the land on which it was first built. Factors such as water supply, availability of land for farming, availability of building materials, level land, land free from floods and ease of defence influenced the siting of Britain's villages. Usually the sites chosen were minimum-energy sites where all or

most of these conditions were satisfied (as in the village sites in Figure 11.2). Many towns owed their early importance to advantages of site such as bridging points or a natural harbour. More important in explaining the growth and importance of a town than its site is its **situation** – its position in relation to the surrounding area. Thus a port may develop at the site of a natural harbour but will expand only if it has a productive hinterland (trading area).

Rural settlement patterns

The patterns rural settlements make divide between **nucleated settlement** (houses grouped into hamlets and villages) and **dispersed settlement** (single houses and farms scattered over the countryside). Reasons for nucleation or dispersal in an area may be historical, cultural or linked with physical geography. The main reason for nucleation in the UK is the farming system of Anglo Saxon and medieval times when people built their houses together in a village so as to be central to the communally worked open fields. Between the 14th and early 19th centuries, the enclosure of land resulted in the development of individual farms and therefore the dispersal of settlement. This **primary nucleation** and **secondary dispersal** led to the typical settlement pattern of much of lowland Britain (as in the Vale of Pickering, Figure 11.2). In contrast, many upland parts of Britain, for example Wales, are dominated by dispersed settlement. This reflects the old Celtic land holding system (individual ownership and sub-division of a farmer's land among all his sons) and the difficult physical geography of hill areas where only a scattered population can be supported. Within the category of nucleated, distinct patterns emerge, such as the spring-line settlement in Figure 11.2. The **morphology** (form) of villages will vary considerably.

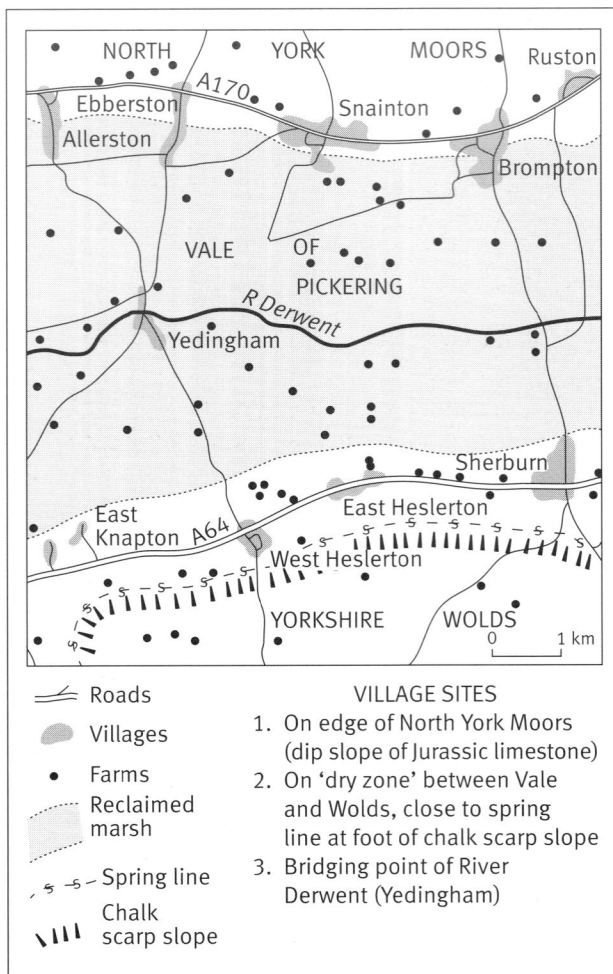

Figure 11.2 Settlement in the Vale of Pickering, Yorkshire

VILLAGE SITES

1. On edge of North York Moors (dip slope of Jurassic limestone)
2. On 'dry zone' between Vale and Wolds, close to spring line at foot of chalk scarp slope
3. Bridging point of River Derwent (Yedingham)

Roads
Villages
Farms
Reclaimed marsh
Spring line
Chalk scarp slope

EXAM TIP

You also need to understand three other aspects of the study of rural settlements: first, the settlement hierarchy (see pages 115–16), second, the impact of counter-urbanisation (considered in Chapter 12) and, third, changes in rural settlements (considered in Chapter 14).

REVIEW TASK

Consider your own case studies of rural settlements. List the factors which have influenced the settlement pattern in general (physical and human) and the sites of individual settlements. Quote examples of settlements, using simple sketch maps, as appropriate.

Urbanisation

In the mid-1980s the world's urban population passed 40 per cent of the total population. In 2000 it was estimated that 50 per cent of the world's population were urban dwellers. **Urbanisation** is the increase in the proportion of the world's, a country's, or a region's population that lives in urban areas. This term must not be confused with **urban growth** which is an increase in the absolute number of people living in urban areas. There are important contrasts between urbanisation in MEDCs and LEDCs:

- Urbanisation in the MEDCs was a consequence of economic development. In most LEDCs economic development has lagged behind urbanisation. Some of the highest rates of urbanisation have occurred in the poorest countries. Only in a few Newly Industrialised Countries (NICs), such as South Korea, can urbanisation be attributed primarily to economic growth.

- Urbanisation in the LEDCs has involved far more people than in the MEDCs and has been more rapid. Between 1950 and 1975 urban growth in Brazil alone exceeded the urban growth of 45 million people experienced in all Europe in the 19th century.

Processes of urbanisation

There have been two key phases in world urbanisation:

1. During the 19th century there was rapid urbanisation in the MEDCs. The demand for labour from industrialisation resulted in rapid urban growth fuelled by migration from the countryside. Imports of food, such as grain from North America, and other resources from colonial empires, supported rapid population growth and urbanisation. Figure 11.3 outlines the patterns and processes of urbanisation in the UK.

2. During the second half of the 20th century rapid urbanisation has taken place in the LEDCs. Migration from rural areas (caused largely by poverty in the countryside and the hope of work and a better life in the city) on a very large scale has been the key process. Rapid rates of natural increase of population are also contribute to the rapid growth of cities, but not directly to urbanisation. Figure 11.4 outlines the key features of urbanisation in India.

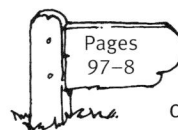

Pages 97–8

Legend

- ● Conurbations
- • Urban areas with population greater than 100 000
- ◯ Densely populated industrial areas (developed on or near coalfields, mainly in 19th century)
- ◯ Other densely populated areas (largely urbanised)

International migration flows
1. from Ireland
2. from Europe
3. from Southern Asia and West Indies (mostly since 1950)

IMPORTS OF FOOD AND INDUSTRIAL RESOURCES

Southern and eastern England; much urban growth 1980–2000

0   100 km

URBANISATION OVER TIME IN UK

1800–1890	Rapid urbanisation linked to industrialisation. Considerable rural–urban migration. Rapid urban growth.
1890–1950	Continued urban growth. Rapid suburbanisation.
1950–2000	Less rapid urban growth of major cities (constrained by Green Belts). Suburbanisation continues. Counter-urbanisation to New Towns and rural areas. Limited re-urbanisation of inner areas.

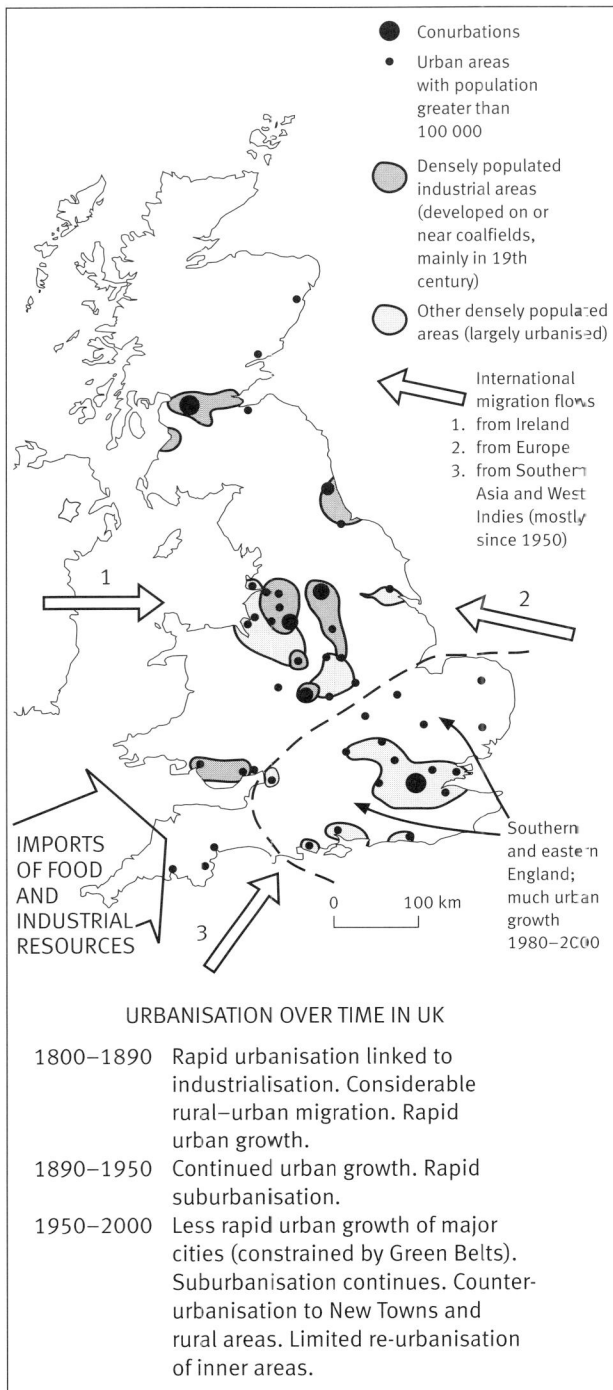

Figure 11.3 Urbanisation in the United Kingdom

Patterns of urbanisation

High levels of urbanisation are found in the MEDCs, but also in some Latin American countries. Levels of urbanisation are generally higher in Latin America than in Asia and Africa. This reflects a long urban tradition in Latin America and, in some countries, relatively high levels of industrialisation. In Asia, although there are many large cities, there are very large, dense rural populations which are not present in Latin America.

Urban growth in the late 20th century

Urban population growth is caused by both urbanisation and natural increase of population. Cities also grow physically as their population increases, so causing problems of urban sprawl.

In MEDCs

From the late 19th century onwards, cities expanded through the process of **suburbanisation**. In the early 20th century urban population growth fuelled this process. Developments in public transport enabled people to commute to work over longer distances and housing became more space consuming. This urban sprawl accelerated with widespread car ownership and increased prosperity. In the UK the pressure of suburbanisation led, after 1945, to policies of **Green Belts** to restrict urban growth and the creation of **New Towns** and **Expanded Towns** to accommodate surplus urban populations beyond the Green Belts.

By the end of the 20th century, many MEDCs were so highly urbanised and had such low rates of natural increase, that there is little potential for continued urban population growth. However, pressures of suburbanisation continue. In the UK this is because:

- Average household size has decreased and therefore more houses are needed for the same number of people.

- People are moving out of the inner cities, wishing to leave behind the problems of such areas and gain the perceived benefits of a suburban lifestyle.

- Increasing average earnings enable people to afford larger houses and to own cars which are used for commuting to work.

- In southern and eastern England many towns *have* experienced population growth, as high-technology industries and the service sector have generated new jobs. People have therefore migrated from other parts of the UK and fuelled the demand for new homes.

Closely linked to suburbanisation has been the process of **counter-urbanisation**. It has been the key process operating in many EMDCs, including Britain, at the end of the 20th century. It involves people moving from cities to rural areas and therefore an increase in the population of villages and small towns. The movement of people has been accompanied by the movement of industry and services. The issues arising from suburbanisation pressures on the rural–urban fringe and counter-urbanisation's impact on rural areas is looked at in Chapter 12.

Pages 126–7

Urban population as % of total

■ Above 1991 national average of 25.7%

■ Below 1991 national average of 25.7%

• Cities with more than 1 million people

Punjab: relatively prosperous, much commercial agriculture

Delhi

Bombay

Southern India: more industrialised than north; coastal areas have much commercial agriculture

Madras

Calcutta

West Bengal: high level of industrial-isation

N

0   1000 km

INDIA'S POPULATION GROWTH

Total population

Note how increase in total population will fuel growth of cities (but *not* rate of urbanisation).

Rural population

Urban population as % of total population

17.3%   18.0%   19.9%   23.3%   25.7%

Urban population

1951   1961   1971   1981   1991

Increasing rate of urbanisation

Slight slowing down of rate of urbanisation (i.e. rise in % of population living in urban areas from 23.3% to 25.7% was less than that from 19.9% to 23.3% in previous decade)

Reasons for slowing down of rate of urbanisation:
1. Many migrants from rural areas move to rural–urban fringe of cities rather than the cities themselves.
2. Government has aided industrial development in rural areas.
3. Increased rural–urban commuting.
4. Less expansion of administrative areas of cities into neighbouring rural areas.

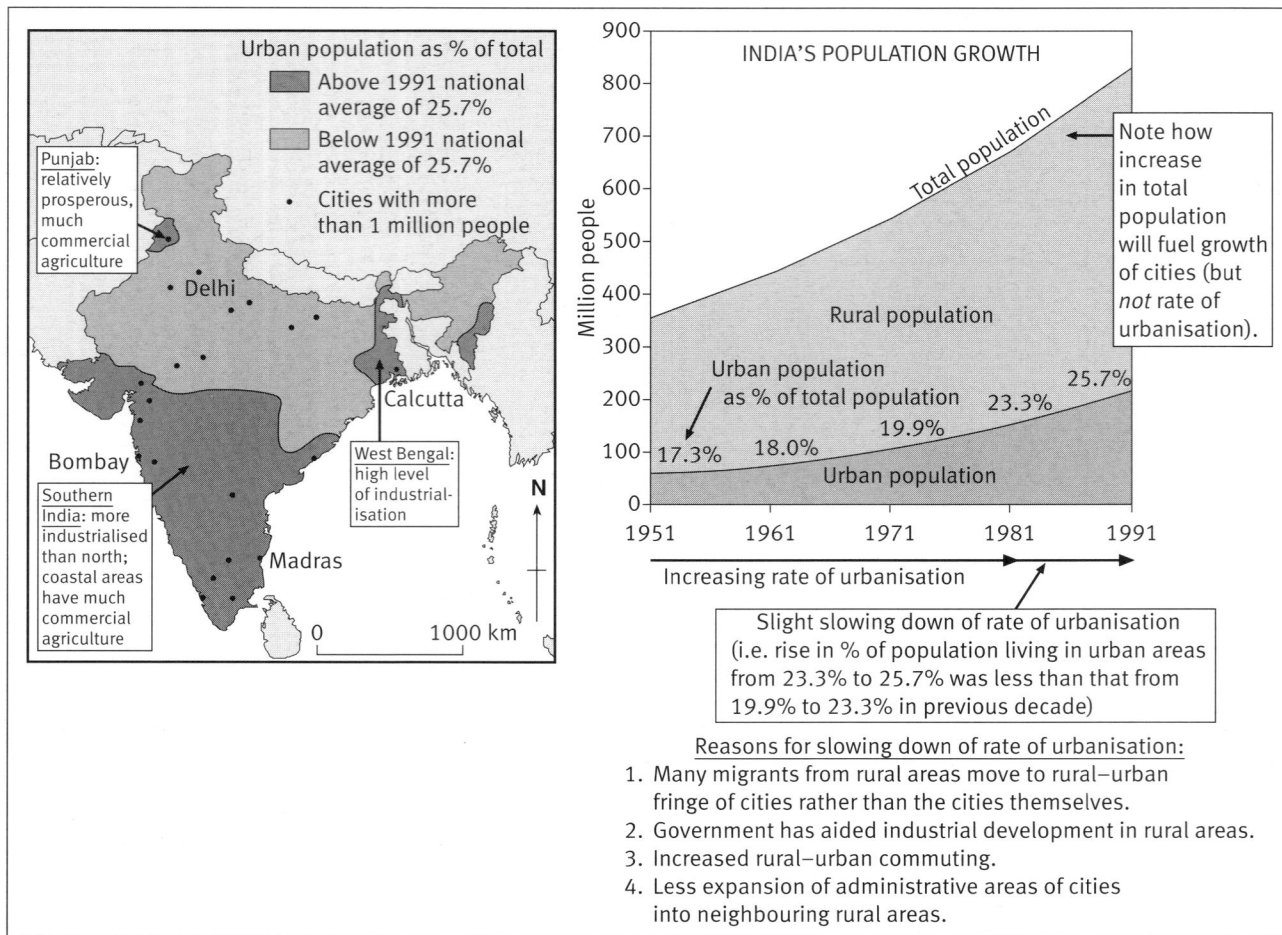

Figure 11.4 Urbanisation in India

In LEDCs

Some of the highest rates of urban population growth are in African countries where the growth is from a base of low urbanisation, there are very high rural–urban migration flows and there are high rates of natural increase in population. Rates of urban growth in other LEDCs are usually higher than in MEDCs because of higher rates of urbanisation and natural increase of population. Chapter 9 looks at the underlying reasons for rural–urban migration in LEDCs and the impact of this flow on rural areas. The problems caused by high rates of urban growth in LEDCs are looked at in Chapter 12.

Pages 97–8
Pages 129–31

Very large cities

Urbanisation and urban growth has been accompanied by the growth of very large cities. In 1950, 70 cities had populations greater than 1 million; by 2000 there were over 150 such cities. An increasing proportion of these cities are in LEDCs. In 1920, 20 per cent of the 24 'millionaire' cities were in LEDCs; by 2000, the proportion was close to 80 per cent. Since the 1960s 'mega-cities' of 10 million

Review task

1 Refer to an atlas or textbook map showing world urbanisation patterns. Check that you are able to describe and explain the global patterns shown.

2 In the UK how do urbanisation processes differ at the present time from the 19th century? (Refer to Figure 11.3.) Make sure that you are able to relate the processes of urbanisation, suburbanisation and counter-urbanisation to a case study area.

3 The following statements describe urbanisation patterns in many LEDCs:
   a) Urbanisation varies regionally within countries, reflecting the location of major cities and contrasting conditions in rural areas.
   b) Population growth fuels urban growth (not urbanisation).
   c) Rates of urbanisation are slowing slightly.
   For either India (Figure 11.4) or another LEDC give evidence and quote examples for these observations.

people or more have emerged; all the world's 10 largest cities now exceed this figure. Although the largest cities in the LEDCs were those that grew most rapidly in the late 20th century, since the mid-1980s the rate of growth of these cities has slowed. In China this has been caused by government policies restricting population growth and rural–urban migration; in India, as Figure 11.4 shows, the causes are more complex. In some LEDCs rapid urbanisation has led to the increasing dominance of a **primate city**, very much larger than the second biggest city.

The rank-size rule and urban primacy

On a national scale the **rank-size rule** describes a numerical relationship between the population size of a country's settlements. The rule states that the size of settlements is inversely proportional to their rank. Thus when cities are ranked, the second-ranked city will have a population half that of the largest city, the third-ranked city will have a population one-third that of the largest city and so on. When plotted on a graph with logarithmic scales, the rank-size relationship will be a straight line. In reality, few countries have city-size distributions that closely follow the rule. A significant variation from the rule is a **primate distribution** in which the largest city is very much larger than the second-ranked city. Primate distributions occur in many LEDCs, but also in some MEDCs, for example, Austria and Denmark. In LEDCs dominance by a primate city often reflects the deliberate concentration of economic activity and political power in one city during colonial rule; this continued after independence. A **binary distribution** occurs where there are two dominant cities, such as Madrid and Barcelona in the case of Spain.

Settlement hierarchies

National city-size distributions may show a stepped pattern, possibly combined with primacy – a **hierarchy** of settlements according to population size. Hierarchies of settlements in a country or region are based on **functions** (or services) provided, besides population size. Figure 11.5 shows the theoretical relationship.

Central place theory

Central place theory aims to describe and explain the relative size and spacing of settlements which act as service centres. It was devised by Walter Christaller in 1933 as a result of his study of settlement patterns in southern Germany. There are several key terms and underlying ideas:

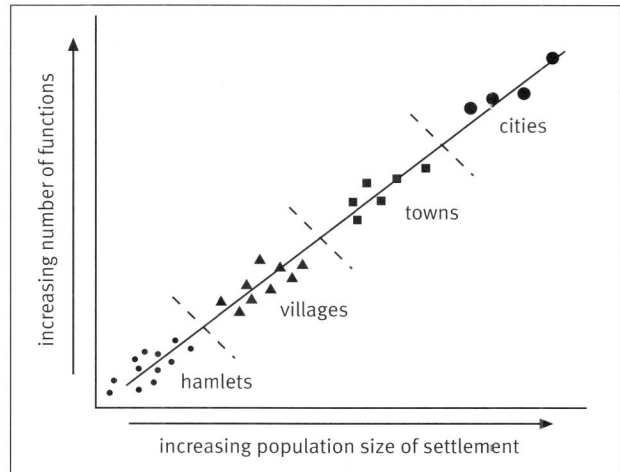

Figure 11.5 Settlement hierarchy based on population size and functions of settlements

- A **central place** is a settlement that acts as a service centre.
- The **range** of a good or service is the maximum distance people will travel to obtain it.
- The **threshold population** of a good or service is the minimum number of people required for it to exist in business.
- **Low-order goods and services** are purchased or used frequently (e.g. **convenience goods** such as bread and newspapers) and have short ranges and low thresholds.
- **High-order goods and services** are purchased or used infrequently (e.g. **comparison goods** such as furniture and electrical appliances) and have long ranges and high thresholds.
- A **sphere of influence** (or market area or hinterland) is the area served by settlement's services.

Central place theory makes two main assumptions. First, settlements occur on an **isotropic plain**, unbounded flat land with transport equally easy in all directions. There are no variations in relief, resources or population density. Second, people show rational behaviour by visiting their nearest central place for goods and services so as to minimise distance travelled.

Christaller stated that people living in the lowest order central places (hamlets) would obtain low-order goods and services from the central places where they lived. They would visit higher order central places (villages and towns) to obtain high-order goods and services. The lowest order central places would have the smallest spheres of influence as the goods and services provided would have low ranges and small thresholds. The higher order central places offer goods and services with high ranges and large thresholds and have larger spheres of influence. (They also provide low-order goods for a smaller sphere of influence). Thus a hierarchy of central places or service centres exists (Figures 11.6 and 11.7). Figure 11.6 shows one of the patterns of hierarchies of central places that Christaller envisaged might occur. Spheres of influence are hexagonal, the most efficient solution for ensuring that the whole area is served without overlapping of neighbouring areas.

The disadvantages of Christaller's theory stem from the facts that isotropic plains do not exist in reality and that people do not behave completely rationally. However, his ideas do act as a basic framework when analysing patterns of service centres. Similarities to the model do exist in reality as the example of the Cambridge area in Figure 11.7 shows.

Determining spheres of influence

Service centres are seldom evenly spaced and they compete against each other for trade. The point between two towns which divides the people who travel to one town from those who travel to a neighbouring town for similar services is known as

the **breaking point**. If several breaking points are established around a town then they can be joined up to show its sphere of influence. The breaking point can be determined theoretically by applying **Reilly's Law of Retail Gravitation**. It states that the degree of interaction between two places is proportional to the products of their populations and inversely proportional to the square of the distance separating them. From this law the **breaking point formula** is derived mathematically. Figure 11.8 states the formula and shows how it is applied.

Spheres of influence may be determined by practical methods: mapping catchment areas of services; delivery areas of major shops; distribution of local newspapers; questionnaires aimed either at shoppers in the towns or the inhabitants of the surrounding areas.

Review task

1 **Criticisms of central place theory.**
   a) **Make lists of ways in which isotropic plains and rational behaviour in terms of shopping and using services do not exist in reality.**
   b) **For the Cambridge region (Figure 11.7) or your own case study area, list ways in which late 20th century changes would affect the hierarchy of service centres.**
   c) **Suggest why the theory is difficult to apply to complex industrial conurbations.**

2 **Advantages of central place theory. How is central place theory useful in explaining the hierarchy of the Cambridge area or your own case study area? Consider the numbers, spacing and arrangement of settlements at various levels in the hierarchy.**

3 **Check that you can apply the breaking point formula to actual examples.**

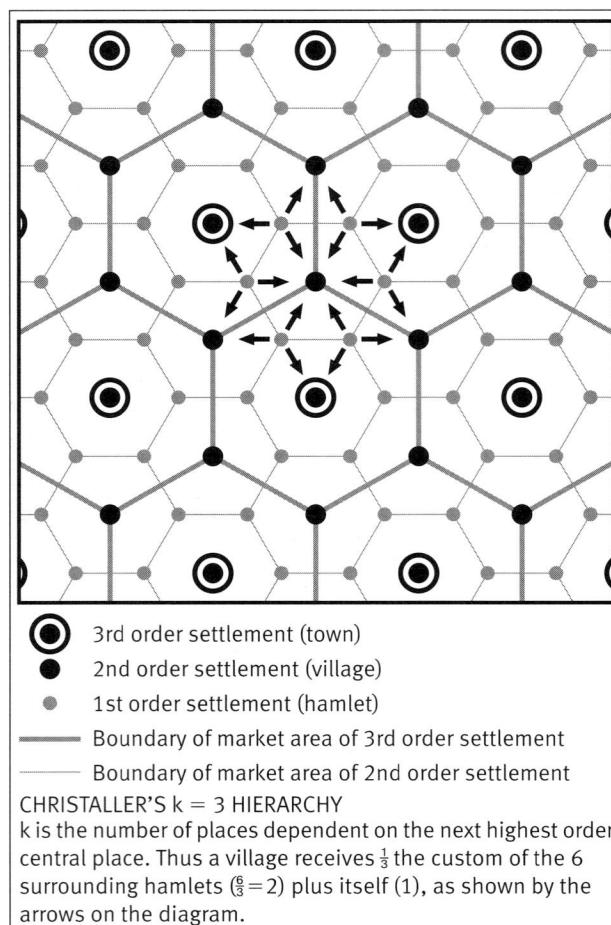

● (⊙) 3rd order settlement (town)
● 2nd order settlement (village)
● 1st order settlement (hamlet)
── Boundary of market area of 3rd order settlement
── Boundary of market area of 2nd order settlement
CHRISTALLER'S k = 3 HIERARCHY
k is the number of places dependent on the next highest order central place. Thus a village receives $\frac{1}{3}$ the custom of the 6 surrounding hamlets ($\frac{6}{3}$ = 2) plus itself (1), as shown by the arrows on the diagram.

Figure 11.6 Christaller's k = 3 system of central places

Urban growth and land use patterns
Models of urban growth and structure

Although each city has its distinctive pattern of functional zones and contrasting residential areas, studies of urban areas have shown similarities in such spatial patterns. As a result models to describe and explain urban growth and structure have been put forward. The model by **Burgess** was based on Chicago in the 1920s. The underlying processes giving rise to the concentric zones shown in Figure 11.9 were held to be those of **social ecology**

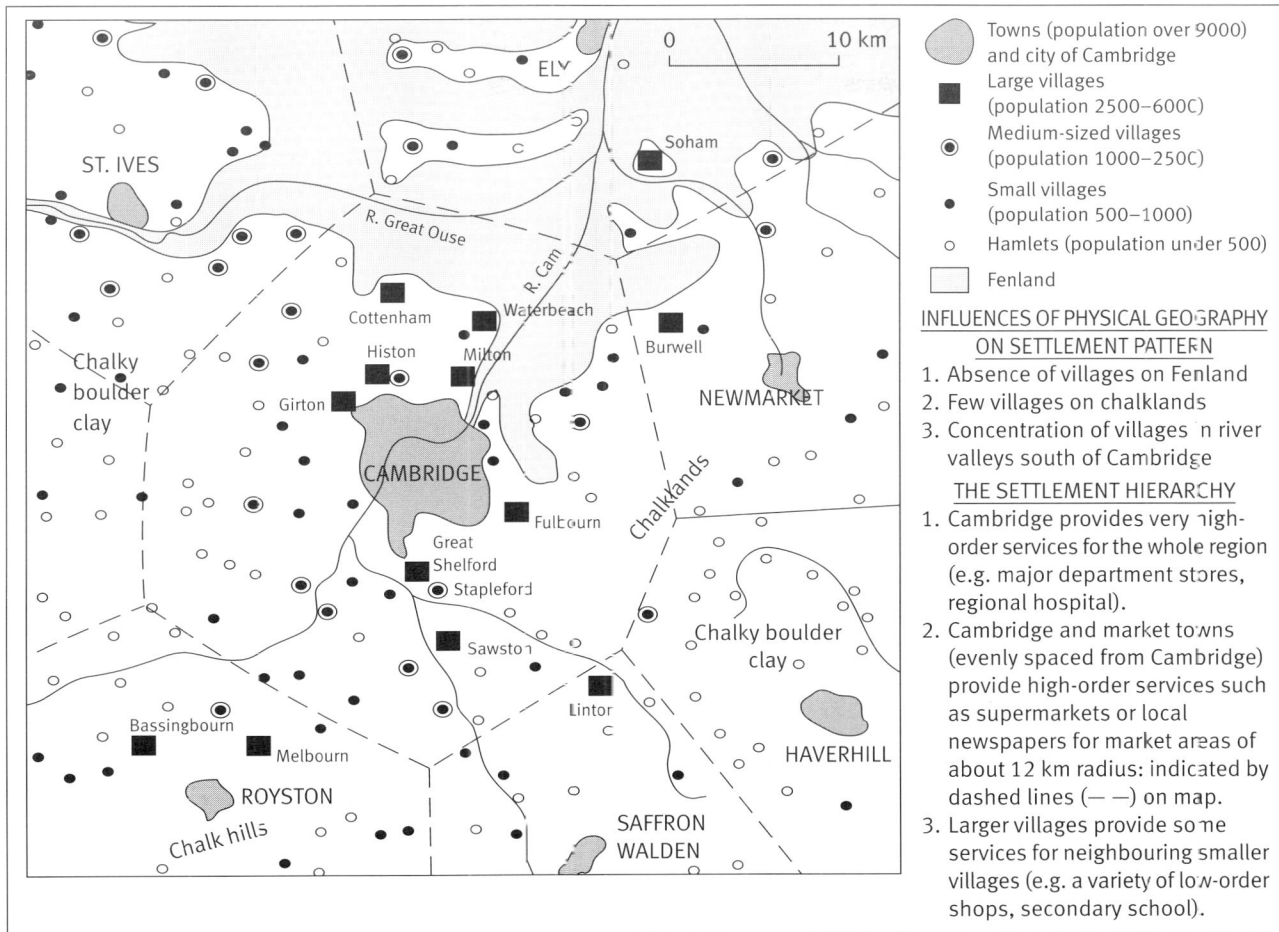

Figure 11.7 The settlement hierarchy in the Cambridge region

Legend:

- Towns (population over 9000) and city of Cambridge
- Large villages (population 2500–6000)
- Medium-sized villages (population 1000–2500)
- Small villages (population 500–1000)
- Hamlets (population under 500)
- Fenland

INFLUENCES OF PHYSICAL GEOGRAPHY ON SETTLEMENT PATTERN
1. Absence of villages on Fenland
2. Few villages on chalklands
3. Concentration of villages in river valleys south of Cambridge

THE SETTLEMENT HIERARCHY
1. Cambridge provides very high-order services for the whole region (e.g. major department stores, regional hospital).
2. Cambridge and market towns (evenly spaced from Cambridge) provide high-order services such as supermarkets or local newspapers for market areas of about 12 km radius: indicated by dashed lines (— —) on map.
3. Larger villages provide some services for neighbouring smaller villages (e.g. a variety of low-order shops, secondary school).

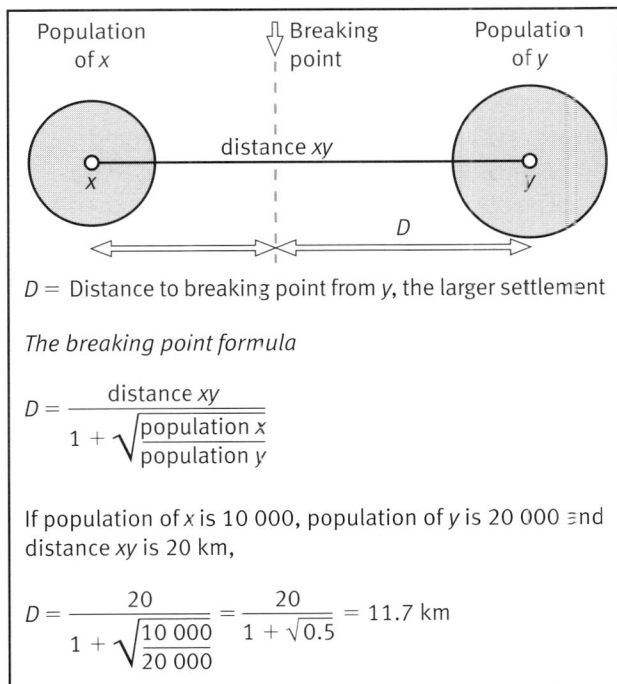

Population of x   Breaking point   Population of y

distance xy

D = Distance to breaking point from y, the larger settlement

The breaking point formula

$$D = \frac{\text{distance } xy}{1 + \sqrt{\dfrac{\text{population } x}{\text{population } y}}}$$

If population of x is 10 000, population of y is 20 000 and distance xy is 20 km,

$$D = \frac{20}{1 + \sqrt{\dfrac{10\ 000}{20\ 000}}} = \frac{20}{1 + \sqrt{0.5}} = 11.7 \text{ km}$$

Figure 11.8 Reilly's breaking point formula

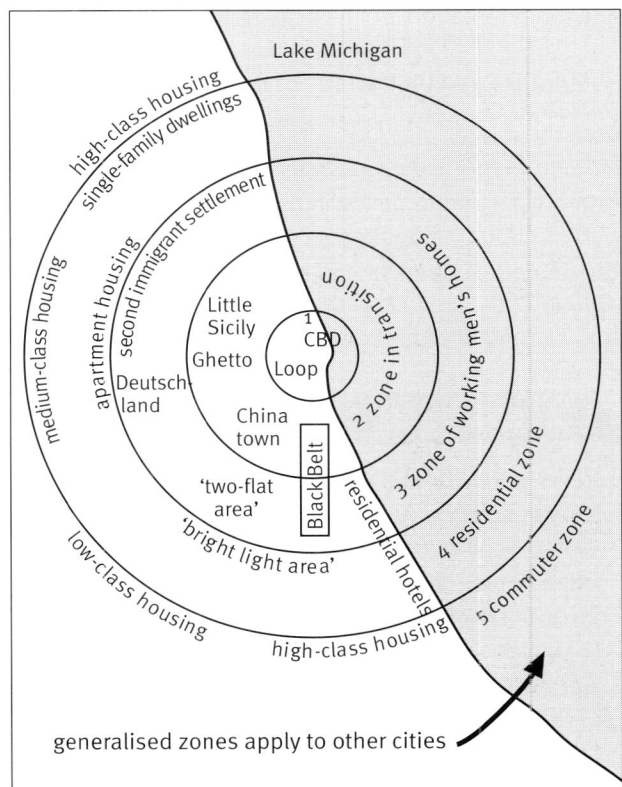

Figure 11.9 Burgess's model of urban structure and growth as applied to Chicago

(ideas based on plant ecology). In the innermost 'zone in transition' **invasion** by newcomers would occur. In the face of such **competition** from newcomers, resulting in their dominance in that zone, existing residents would move outwards to the next zone whose residents would in turn move further outwards (a process of **succession** thus takes place in each zone). Other models include **Hoyt's sector model, the multiple nuclei model of Harris and Ullman,** and **Mann's model** (1965) which combined the ideas of Burgess and Hoyt and applied them to British cities.

Bid-rent theory

The key assumption of **bid-rent theory** is that in a free market, the highest bidder will obtain the use of land. The highest bidder will be able to obtain the greatest profit from a particular site and so can pay the highest rent. At the city centre, competition for land is at its most intense (reflecting the small amount available and its great accessibility) and so retail and office developments can outbid other land uses (Figure 11.10). Bid-rent theory produces a concentric land use pattern similar to Burgess's model. Closely related to bid-rent theory is the **population density curve,** also shown in Figure 11.10. High housing and population densities in the inner city reflect a wish to maximise the use of more expensive land. Both land values and population densities show the phenomenon of **distance decay** from the city centre.

Usefulness and limitations of urban models

Both the models of urban structure and growth and bid-rent theory are very simplified and make rather extreme assumptions (such as Burgess's flat plain with transport being equally easy in every direction). The models of Burgess and Hoyt were devised before widespread car ownership encouraged suburban sprawl, before inner cities were redeveloped and

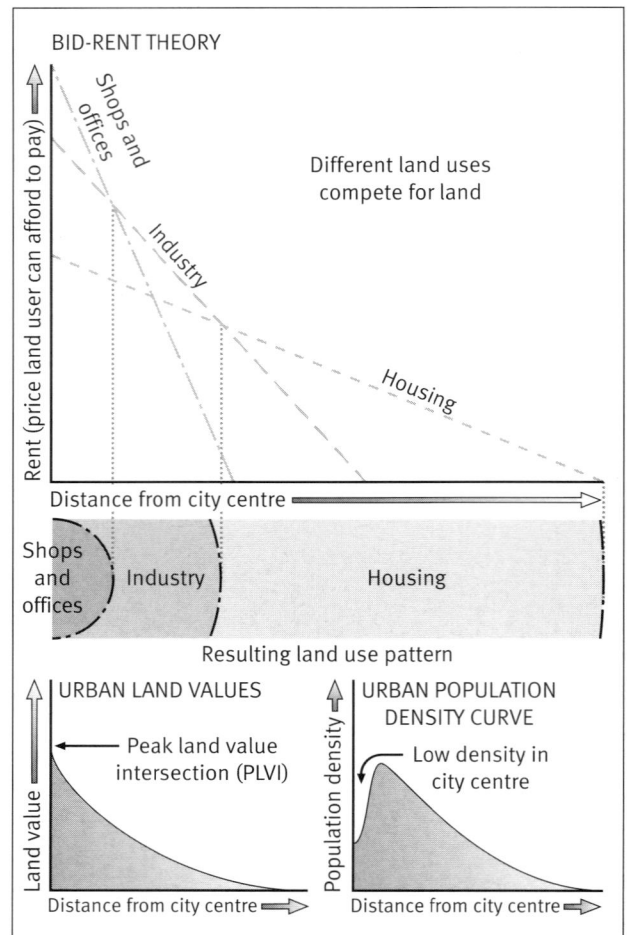

Figure 11.10 Bid-rent theory

before the decentralisation of retailing, offices and industry. However, large areas of American and European cities had developed by the time these models appeared and the models therefore contribute towards explaining much of the cities' structure and growth.

Physical and historical influences

Physical geography influences the original site of a city and the location of urban land uses (e.g. industry on flat valley bottoms). As a city expands, physical geography can provide barriers to development (e.g. ridges of high ground) or can ease development (e.g. an area of flat, well-drained land). It influences transport routes which in turn influence the growth and land use patterns of a city.

The historical context of a city's development will shape its structure; thus a city's CBD which has developed within a medieval street pattern will be very different from one planned in the late 20th century. Changing transport technology over time influences a city's growth pattern:

Review task

Refer to textbooks describing the models of Burgess, Hoyt, Harris and Ullman and Mann. Construct a Mind Map to summarise them as follows: use the names of the models as the initial branches, then for each model split these branches so as to note a) the city or cities they were initially applied to, b) their characteristics, c) their limitations. (It will be best to apply these models to modern-day cities in MEDCs after considering their functional zones in the next section.)

Revise AS and A Level Geography

- Residential development has been shaped in turn by tram routes, railways, bus routes and mass car ownership.
- Industrial location within cities has been influenced in turn by canals, railways, major radial roads and motorways.

Land use zones in cities

The Central Business District (CBD)

Page 148 The CBD is a major concentration of retailing and other service activities. Refer to Chapter 14 for details.

Industrial zones in cities

The location of industrial zones in cities reflects the interplay of many factors such as changing transport technologies over time and the availability of large, low-cost sites for space-consuming industries. Planning policies are important, for example, in locating new business parks and redeveloping derelict sites. Figure 11.11 summarises in map form the location of industrial zones in cities.

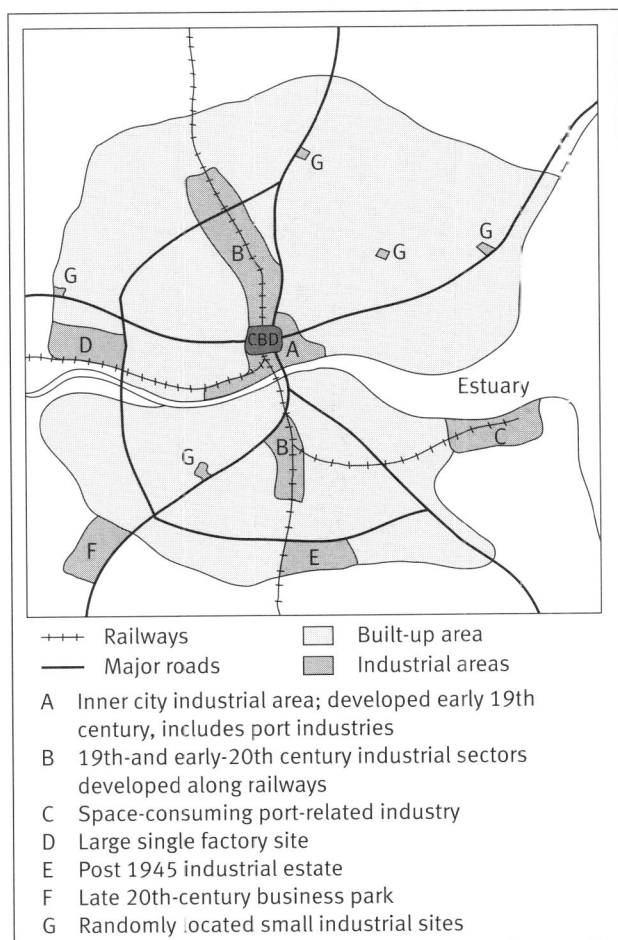

Key	
+++ Railways	☐ Built-up area
— Major roads	▨ Industrial areas

A Inner city industrial area; developed early 19th century, includes port industries
B 19th-and early-20th century industrial sectors developed along railways
C Space-consuming port-related industry
D Large single factory site
E Post 1945 industrial estate
F Late 20th-century business park
G Randomly located small industrial sites

Figure 11.11 Location of industrial areas in a city

Residential zones

Residential zones clearly differ from each other because houses vary in age, type and ownership. This is reflected in the various models of urban structure and growth. Residential zones also differ socially; people with similar socio-economic characteristics tend to cluster together. Reasons for socio-economic segregation are complex and interwoven:

- **Income**: people with high incomes have a greater choice of where to live than those with low incomes.
- **Age and stage in family life cycle**: as people grow older and the nature of their families change, their housing requirements change. The family life cycle model (Figure 11.12) which develops this idea is very generalised and few people follow all the stages shown.
- **Peoples' evaluation or perception of different areas**: people evaluate areas in terms of characteristics including access to shops, access to work, reputation of schools and attractiveness of environment. Attitudes and prejudices towards places and other people play their part in whether an area is seen as desirable or not.
- **Planning policies**: in particular the physical separation of council estates from owner-occupied areas.
- **Operation of housing gatekeepers**: gatekeepers are institutions which act to limit peoples' choice. Thus building societies may 'red-line' areas, drawing boundaries around areas within which they will not grant mortgages. Discrimination towards people on the grounds of, for example, age, occupation or ethnic origin might occur. Access to council housing of different types and in different areas is controlled by local authority housing officers.

Ethnic segregation

In British cities ethnic minorities tend to concentrate in parts of the inner city. Usually these are the same areas of low-cost housing where West Indian and Asian immigrants originally settled in the 1950s, 1960s and 1970s. However, some outward dispersal has taken place and inner cities show some segregation between different ethnic minorities. Similar patterns are found in many European cities, although in France most ethnic minorities live in outer city housing estates. Where conditions of almost total segregation exist, such single-group areas are known as ghettoes, as in areas of US cities populated by African-Americans. Figure 11.13 gives the causes of ethnic segregation in cities.

Settlements: patterns and processes

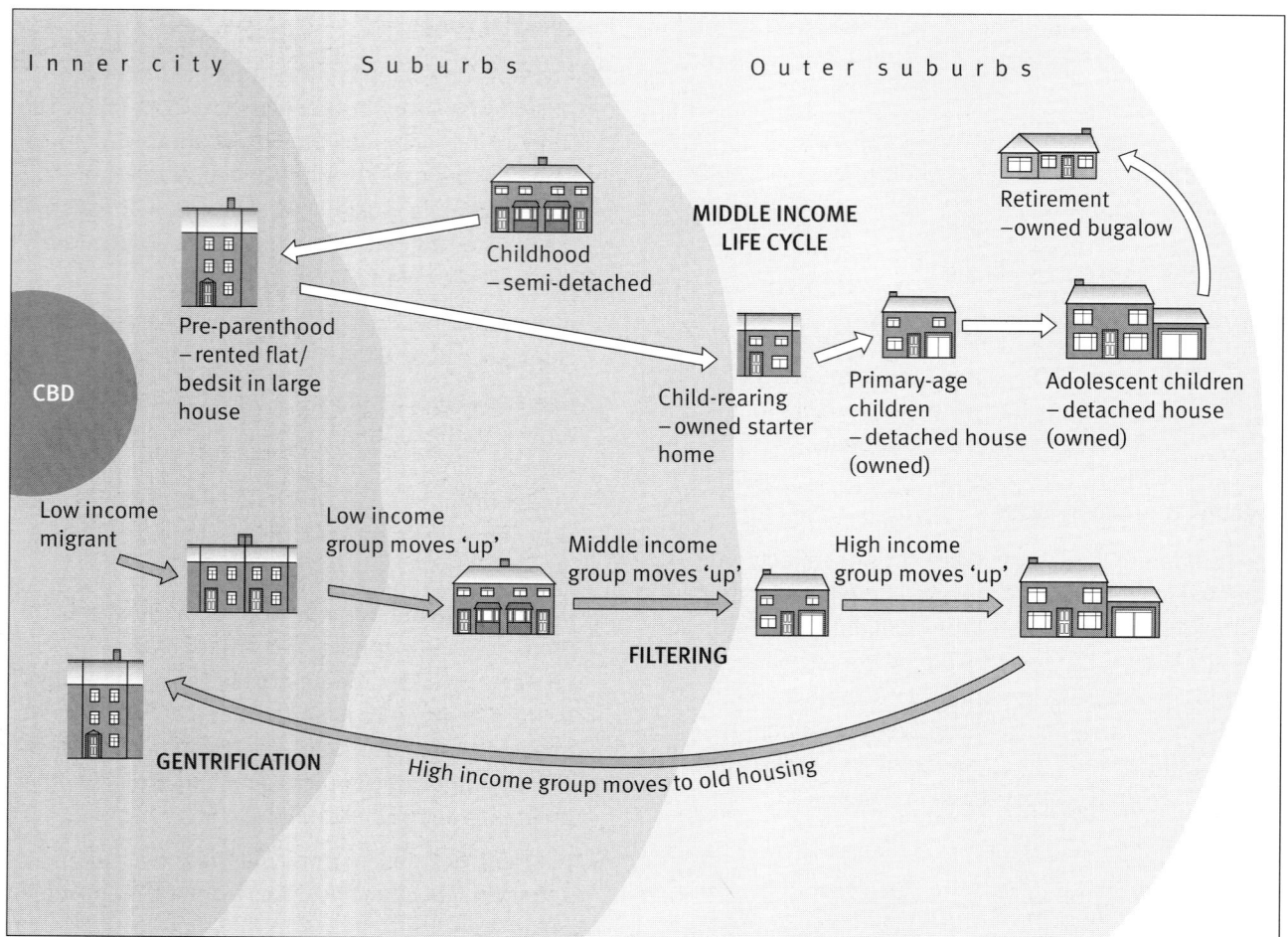

Figure 11.12 Movement in residential areas: the family life cycle model; filtering and gentrification

Internal factors (encouraging ethnic minorities to opt for segregation)	External factors (action taken by majority population which encourages ethnic segregation)
• To reduce feelings of isolation and provide a sense of community and security through own community organisations, places of worship, shops and services.	• Migration of majority population out of areas into which minority population is moving (particularly relevant in the spread of US ghettoes).
• Defensive function: to provide protection from racist abuse and attacks from majority population.	• Ethnic minorities are more likely to earn low wages and be unemployed. Therefore they are forced to areas of cheap housing.
• Attack function: to provide a base for political influence and power; to fight against racial discrimination.	• Discrimination in housing by estate agents, private landlords, government housing agencies, etc.
• To maintain a distinctive cultural identity, through own language, religion and customs. Also, to encourage friendship and marriage within own ethnic group.	• Hostility from majority population, including fear of racially motivated violence.
• Persistence of residential segregation from a time of legalised segregation by ethnic origin (e.g. in cities in South Africa and southern USA).

Figure 11.13 Causes of ethnic segregation

Filtering and gentrification

These are two processes operating in residential areas which lead to changing patterns of social segregation. **Filtering** involves socio-economic groups moving from one area of housing to another, the area they leave behind being occupied by a 'lower' group (Figure 11.12). **Gentrification** reverses this process as high-income people move to inner city housing areas and occupy renovated houses. These are attracted by the closeness of the city centre and houses with 'character'. An upwards spiral of an area becoming more fashionable, with rising house prices, is created. Closely related to gentrification is the building of new housing and the conversion of industrial buildings into flats, both aimed at high-income people, in inner city areas such as London Docklands. This is sometimes referred to as **re-urbanisation**.

Spatial pattern of residential zones

The operation and interaction of all the processes described so far give rise to patterns of residential zones more complex than that described in the Burgess model. Figure 11.14 illustrates this in the case of Birmingham.

Review task

You should have available a case study of an urban area in the UK (or another MEDC), including a map of its functional zones. Apply the content of the relevant sections of this chapter to your case study as follows:

1   **Physical and historical influences.**

2   **The application of the major models of urban structure and growth.**

3   **The location of industrial zones.**

4   **Residential zones: socio-economic and ethnic segregation in residential areas (patterns, reasons); gentrification and re-urbanisation in inner city areas (if these processes are present).**

You should be able to construct a sketch map of the patterns of urban zones similar to that of Birmingham in Figure 11.14. As in Figure 11.14 you need to be able to locate and describe examples of relevant districts. (Knowledge of aspects of other cities you have studied will help to supplement your main case study.)

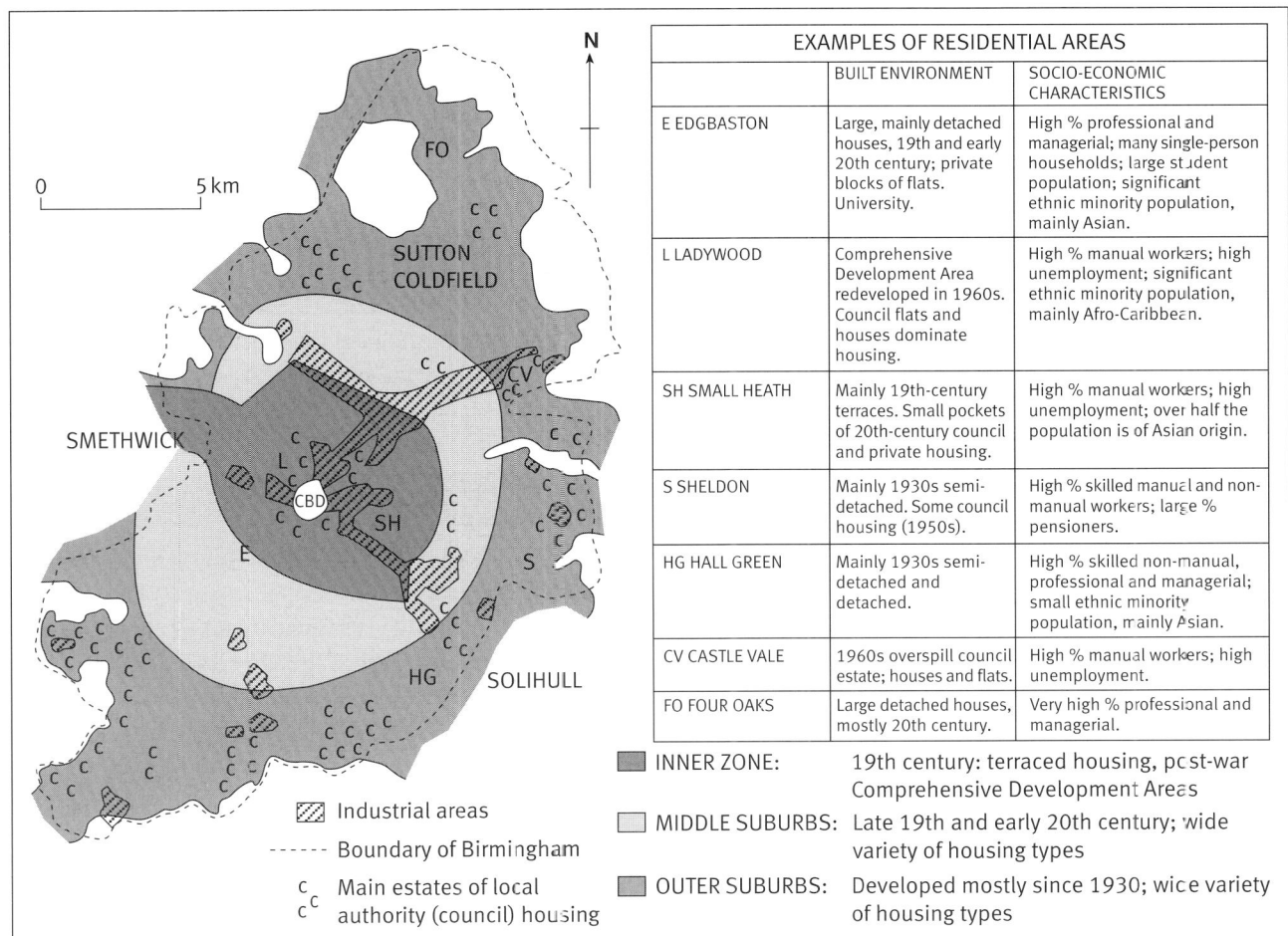

EXAMPLES OF RESIDENTIAL AREAS		
	BUILT ENVIRONMENT	SOCIO-ECONOMIC CHARACTERISTICS
E EDGBASTON	Large, mainly detached houses, 19th and early 20th century; private blocks of flats. University.	High % professional and managerial; many single-person households; large student population; significant ethnic minority population, mainly Asian.
L LADYWOOD	Comprehensive Development Area redeveloped in 1960s. Council flats and houses dominate housing.	High % manual workers; high unemployment; significant ethnic minority population, mainly Afro-Caribbean.
SH SMALL HEATH	Mainly 19th-century terraces. Small pockets of 20th-century council and private housing.	High % manual workers; high unemployment; over half the population is of Asian origin.
S SHELDON	Mainly 1930s semi-detached. Some council housing (1950s).	High % skilled manual and non-manual workers; large % pensioners.
HG HALL GREEN	Mainly 1930s semi-detached and detached.	High % skilled non-manual, professional and managerial; small ethnic minority population, mainly Asian.
CV CASTLE VALE	1960s overspill council estate; houses and flats.	High % manual workers; high unemployment.
FO FOUR OAKS	Large detached houses, mostly 20th century.	Very high % professional and managerial.

INNER ZONE: 19th century: terraced housing, post-war Comprehensive Development Areas

MIDDLE SUBURBS: Late 19th and early 20th century; wide variety of housing types

OUTER SUBURBS: Developed mostly since 1930; wide variety of housing types

Figure 11.14 Land use zones in Birmingham

Urban land use patterns in LEDCs

Differences between cities in LEDCs and those in MEDCs are largely a reflection of three factors:

1. The rapid urban growth in the late 20th century of cities in LEDCs.

2. Contrasting historical and cultural contexts.

3. The existence in cities in LEDCs of large numbers of people living in poverty.

Although the cities in the LEDCs have much in common with each other, considerable differences also exist, resulting from contrasting histories, cultures, political systems and levels of economic development.

Land use models

The differences mentioned above mean that the models devised in the context of American and European cities are not very relevant to cities in the LEDCs and that a variety of models for cities in LEDCs exist. In parts of India and north Africa with a long history of urban development, colonial urban development often took place in a location adjacent to the old existing city; more recent urban development then surrounded these cores (Figure 11.15a, Clarke's model). In Latin America most cities developed from a colonial core (initially modelled on Spanish cities) with the rich living close to the centre, as in pre-industrial Europe. The rapid urban growth of the late 20th century involved considerable in-migration, increased overcrowding of inner city slums, the growth of squatter settlements (mainly on the periphery) and the movement outwards of the rich elite, usually in a distinct sector (Figure 11.15b).

Other features of land use zoning in cities in LEDCs 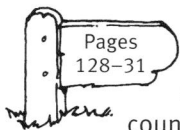 include industrial areas in distinct sectors or scattered throughout the city and ethnic segregation in some countries' cities.

Pages 128–31

The problems and issues related to the rapid growth of cities in LEDCs are considered in Chapter 12.

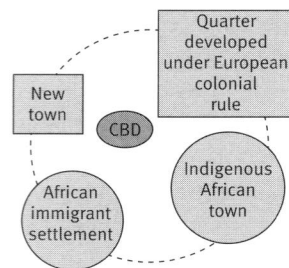

a) CLARKE'S MODEL OF AN AFRICAN CITY

- Quarter developed under European colonial rule
- New town
- CBD
- Indigenous African town
- African immigrant settlement

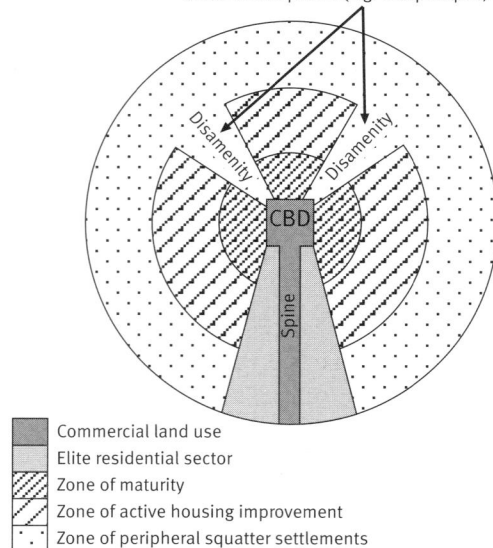

b) LAND USE MODEL OF A LATIN AMERICAN CITY

Sectors where negative factors restrict conventional urban development (e.g. steep slopes, flood plain)

- Disamenity
- Disamenity
- CBD
- Spine

- Commercial land use
- Elite residential sector
- Zone of maturity
- Zone of active housing improvement
- Zone of peripheral squatter settlements

Figure 11.15 Land use models of cities in LEDCs

> ### EXAM TIP
>
> **You are likely to encounter a variety of models of land use models for cities in LEDCs in textbooks and examination papers.**

> ### Review task
>
> **As with cities in MEDCs, it is important to have knowledge of a major case study and other relevant examples. Draw a simple sketch map of your case study city in the centre of a large sheet of paper and around it add detailed labels to describe and explain its structure: influence of physical factors, location of commercial and industrial zones, different types of residential zones, etc. Does an urban structure model (such as those in Figure 11.15 or others) help to describe and explain your city? List the relevant points of comparison.**

What you need to know:

- **The nature of inner city problems in MEDCs**
- **Government policy towards inner city issues in the UK**
- **Other urban issues in MEDCs, including pressures on the urban fringe**
- **The impact of counter-urbanisation**
- **Issues of rapid urban growth in LEDCs, in particular housing**

Also, you will need to be able to:

- **apply the various issues in MEDCs and LEDCs to case studies;**
- **analyse urban problems and evaluate different policies towards these problems.**

Inner cities in MEDCs

Although inner city areas are perceived as 'problem areas', it is important to remember that deprived areas also exist elsewhere, especially in edge-of-city council estates, and that prosperous areas exist within the inner city, either long-standing or gentrified.

The nature of the inner city problem

- **Declining and changing population** caused by:
  * slum clearance (1945–75) and the linked movement of people to edge-of-city estates and New Towns;
  * attraction of the suburban lifestyle and home ownership;
  * declining job opportunities in the inner city; the outward movement of people was selective; the more skilled and enterprising were most likely to move out; the immigration of people from the Commonwealth partly replaced those who left the inner city between the 1950s and early 1970s.
- **Declining employment opportunities** caused by:
  * loss of manufacturing jobs; the inner cities have had a disproportionate share of declining, traditional industries; more successful enterprises left so as to avoid ageing buildings and poor infrastructure;
  * higher-skilled commuters from outside the inner city increasingly filling what jobs were available, especially in the service sector;
  * loss of other jobs, for example in the docks of port cities.

- **Poor built environment** characterised by:
  * 19th-century housing with problems of physical decay, poor basic amenities and overcrowding;
  * housing built in post-war redevelopment schemes which is deteriorating physically and provides difficult living environments, for example tower blocks and deck-access flats.
- **Poverty and deprivation.** With about 7 per cent of the British population, the inner cities have 20 per cent of households identified as suffering from housing stress, 14 per cent of unskilled workers, about 10 times the national proportion of people qualifying for Income Support (i.e. living below the official poverty line), twice the national average of single-parent families, unemployment rates of 50 per cent above the national average and a higher incidence of illness. Although the needs of people in the inner city are particularly great, provision of and access to public services is poor.
- **'Collective despair'.** The decline and deprivation described above interacts to produce feelings of dismay and no hope among inner city residents. Social unrest and racial tension can develop. The existence of 'collective despair' forms a breeding ground for crime, involvement in drugs and vandalism.

Explanations and solutions

Describing the nature of inner city problems helps to identify some causes, but not necessarily underlying root causes. The partly completed Mind Map in Figure 12.1 shows the main attempts which have been made to offer explanations of the root causes of inner city problems. They are important because they form the basis of the various government policies which have aimed to provide solutions.

Government policy towards inner cities

Since 1945 there have been a huge variety of policies, programmes and initiatives towards the inner cities, reflecting changing circumstances (such as the deterioration of housing built in the early post-war urban redevelopment schemes or the economic recessions of the late 1970s and early 1990s) and the changing attitudes towards causes, as shown in Figure 12.1. It is therefore important to focus on the following:

- The main phases of government policy.
- Key, relatively recent programmes, such as Urban Development Corporations (UDCs) and City Challenge.
- Case studies of major programmes

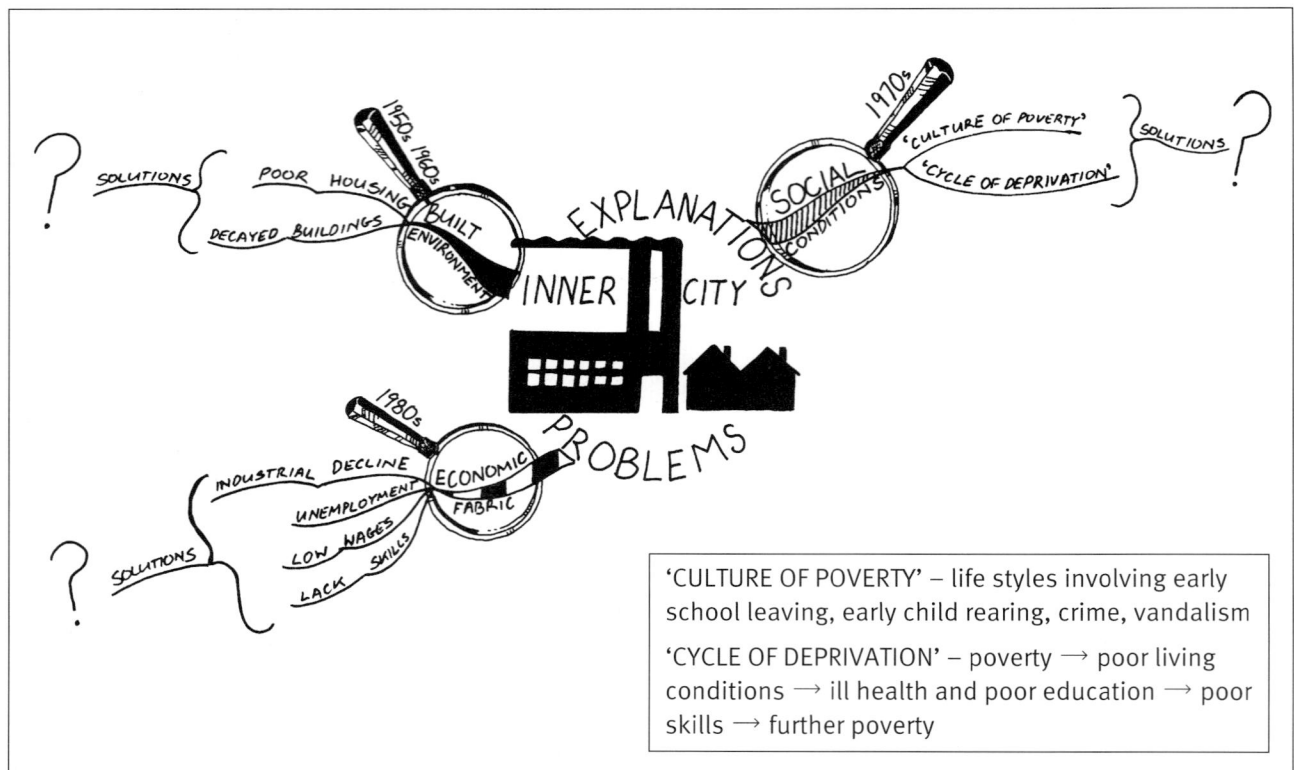

'CULTURE OF POVERTY' – life styles involving early school leaving, early child rearing, crime, vandalism

'CYCLE OF DEPRIVATION' – poverty → poor living conditions → ill health and poor education → poor skills → further poverty

Figure 12.1 Explanations of inner city problems

The main phases of government policy

1. 1946–67

The programme of **Comprehensive Development Areas (CDAs)** involved the clearance of 19th-century slum housing and the construction of new council-owned housing, mostly in the form of flats. Redevelopment was only part of a wider urban policy of decentralisation involving the building of edge-of-city estates and New and Expanded Towns to take the **overspill** population from the inner cities. Problems of this policy included the considerable cost, major errors in design and construction of housing and the break-up of established communities.

2. 1967–77

The key policies were **rehabilitation of housing** and the development of projects aiming at **improving social and economic welfare. General Improvement Areas (GIAs)** and later **Housing Action Areas (HAAs)** were declared to improve or rehabilitate housing rather than to replace it. Grants were made available to improve housing and the surrounding environment. In 1968 the **Urban Programme** was set up to tackle areas 'with multiple deprivation and where there had been large-scale immigration'. Many small-scale projects were introduced to tackle social problems, targeting for example immigrant communities and education. Problems of these policies included insufficient funding, the GIA/HAA policy helping

gentrifiers in some cities and the limited impact of many small-scale social welfare projects.

3. 1977–90

Economic regeneration was the key policy with environmental, social and housing projects receiving less emphasis. The government's Urban Programme was expanded to fund projects involving **partnerships** between central and local government; the aim was economic, social and environmental regeneration of large inner city areas. With the change from Labour to Conservative government in 1979 there was a change of emphasis: even more importance given to economic development, more control by central government and more involvement by the private sector. During the 1980s 26 **Enterprise Zones (EZs)** were set up with a 10-year lifespan, to encourage growth of industry and services in locations of industrial decline (not all were in inner cities). Most important were **Urban Development Corporations (UDCs)**. The government funds and empowers an UDC to buy land and improve the infrastructure in an area with serious economic decline and much derelict land. The plan is then to attract private companies to develop service industries, manufacturing and housing. The first two UDCs were set up in 1981 in London Docklands and Merseyside. Nine more of varying size were set up in the late 1980s. Some local authorities went into partnership with development companies to establish programmes on a similar scale, for example, Salford Quays and Birmingham Heartlands (which became the twelfth and last UDC in1992). Many successes have been claimed for the UDCs but many criticisms have also been made (Figure 12.2).

4. 1990 onwards
Policy has moved back to more involvement by local authorities (although in partnership with central government and the private sector) and towards a less fragmentary approach – the **Urban Regeneration Agency** (1998) replaces the involvement of many government departments. The **City Grant** programme replaces previously separate grant schemes to reclaim derelict land and buildings for development

by the private sector. In 1991 **City Challenge** was launched in which local authorities competed for the funding of major inner city regeneration schemes. An example of a successful bid was Hulme City Challenge, Manchester (Figure 12.3).

Hulme City Challenge proved to be a largely successful scheme. Also, in the Manchester area, the developments at Salford Quays and in the area of Central Manchester UDC have resulted in new economic activity (mainly in the service sector) and new housing attracting people to live in the inner city. However, only a short distance away, 19th-century terraces in Salford are being abandoned and unemployment levels remain high throughout inner Manchester and Salford.

Achievements

- Streamlined approach to redevelopment and regeneration through extensive powers given to UDCs.

- Much new housing (e.g. London Docklands Development Corporation (LDDC), 22 000 units between 1991 and 1996).

- Creation of jobs (e.g. LDDC, over 30 000 jobs).

- Major 'flagship' projects (e.g. Canary Wharf, London) encourage further investment.

- Improvements to infrastructure (e.g. Docklands Light Railway and Jubilee Line Extension).

- Removal of derelict land and buildings.

- New cultural, leisure and tourist developments (e.g. Bridgewater Concert Hall, Manchester; London Docklands Sports Centre; Albert Dock, Liverpool).

Criticisms

- New housing not affordable by the local population.

- Many new jobs required specific skills not possessed by the local population. This resulted in commuting from elsewhere and unemployment remaining high.

- Some UDCs created relatively few jobs, e.g. Liverpool.

- Heightened social and economic contrasts between relatively poor local population and wealthy newcomers.

- Criticisms of inadequate involvement by and consultation with local population.

- Early 1990s: cutbacks in government funding to UDCs. Also, recession led to reduced private sector investment.

Figure 12.2 Urban Development Corporations: achievements and criticisms

Hulme before City Challenge

- **Hulme in the 1950s:** a large area of decaying 19th-century terraced housing. Also, old factories and the shopping centre of Stretford Road.

- **1965–72:** 60 ha cleared. Old housing replaced by deck-access, system-built flats, including 1000 units in 4 giant 7-storey crescents, 'inspired by the Georgian crescents of Bath'.

- **1975:** Manchester City Council admitted that the Hulme redevelopment had failed and all families with children were to be moved out. Problems included cockroach and rat infestation, ineffective expensive electric heating, major construction faults allowing damp penetration and noise.

- **1980s and early 1990s:** continued deterioration. Major problems of crime and drugs. In 1991 Michael Heseltine, Environment Secretary, described Hulme as 'one of the most notorious examples of inner-city deprivation'.

1991: Hulme City Challenge announced

- Government funding of £37.5 million spread over 5 years (1992–7).

- Demolition of 2830 deck-access flats, including the 4 crescents. 800 'walk-up' flats improved and 950 brick houses retained (all council owned).

- 1000 new houses and flats built by housing associations and a further 1000 by the private sector for sale or rent. Also, environmental improvements, expansion of local businesses and new shopping developments.

- Partnership between central government, Manchester City Council, housing associations, local community organisations and the private sector.

Figure 12.3 Hulme City Challenge, Manchester

The widening gap between disadvantaged and affluent groups in cities

In the UK in the 1980s and 1990s there was increased socio-economic segregation in cities; a widening gap between rich and poor. Contributory factors have been:

• A general trend for the least well-off to get poorer and the rich to get richer. Levels of state benefits tended to stagnate as did earnings in occupations that were low paid. On the other hand, high earners tended to have higher rates of increase in earnings and benefited from reduced taxation.

• Many inner city areas have increasingly become districts where the unemployed and low earners have been 'left behind'.

• A similar trend has occurred in some edge-of-city housing estates. The least attractive estates, with deteriorating older housing or high-rise developments, have become areas with high rates of poverty, unemployment and social problems. In some cities, for example Glasgow and Edinburgh, such estates are the main areas of deprivation.

Problems of urban growth in MEDCs

Although urbanisation (i.e. an increasing proportion of the population living in cities) has ceased in most MEDCs, physical urban growth through suburbanisation continues to be a major planning problem.

Page 113

(Figure 11.3 in Chapter 11 describes the processes of urbanisation and urban growth in the UK.) The pressures of urban growth are concentrated on the urban fringe.

Pressures on the urban fringe

Rural areas close to cities have always felt under pressure. Besides facing the threat of urban sprawl, they are used for purposes linked with cities such as reservoirs, sewage farms, waste disposal sites, airports and golf courses; land uses which take up too much space or cause too much of a nuisance to be located within cities. The rapid expansion of Britain's cities through suburbanisation in the early 20th century led, after the Second World War, to the policy of Green Belts to restrict urban growth. The redevelopment of densely populated inner city areas after 1945 led to even more pressure for housing on the edge of cities. Besides being housed on edge-of-city estates the overspill population of London and other large cities was accommodated in New Towns and Expanded Towns. The rapid growth of population in the South East in the post-war period also led to much housing being built in other towns beyond London's Green Belt. The planning response was to

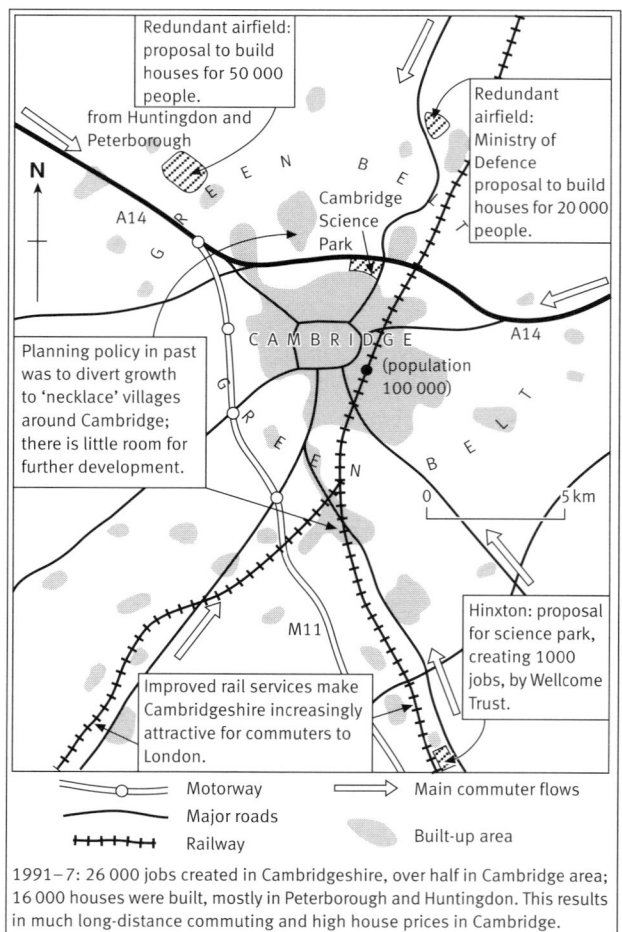

1991–7: 26 000 jobs created in Cambridgeshire, over half in Cambridge area; 16 000 houses were built, mostly in Peterborough and Huntingdon. This results in much long-distance commuting and high house prices in Cambridge.

Figure 12.4 Pressures for urban growth, Cambridge

extend the Green Belt. In the last quarter of the 20th century pressures on London's Green Belt increased further from developers of housing, industry, offices, shopping centres and leisure facilities. Pressures for development have been particularly great to the west of London near Heathrow Airport and at locations where the M25 crosses radial routes from London.

Pressures for development on the urban fringe have been great around other cities as well; Figure 12.4 looks at the example of Cambridge where problems have been particularly acute and Figure 14.7 in Chapter 14 shows examples on the southern fringe of Solihull near Birmingham. The central issue concerning such urban fringe zones is should the Green Belt be preserved at all costs, even if some of the land is partly urbanised, or should the national interest and the need to put new economic life into a region be the priority?

Page 151

Other problems of urban growth in MEDCs

The physical growth of many cities in MEDCs combined with the decay of inner city areas creates other problems including road congestion, air pollution, public transport provision, waste disposal problems and water supply problems. Figure 12.5 lists these problems and gives details of some of the planning responses to them.

The government-sponsored report of the Urban Task Force, produced in 1999, attempted to offer solutions for urban areas as a whole. Recommendations included:

- The creation of Urban Priority Areas, mainly in inner city areas, where developers would have a wide range of incentives for housing and commercial developments.

- Undeveloped greenfield sites should not be released while alternative brownfield land is available.

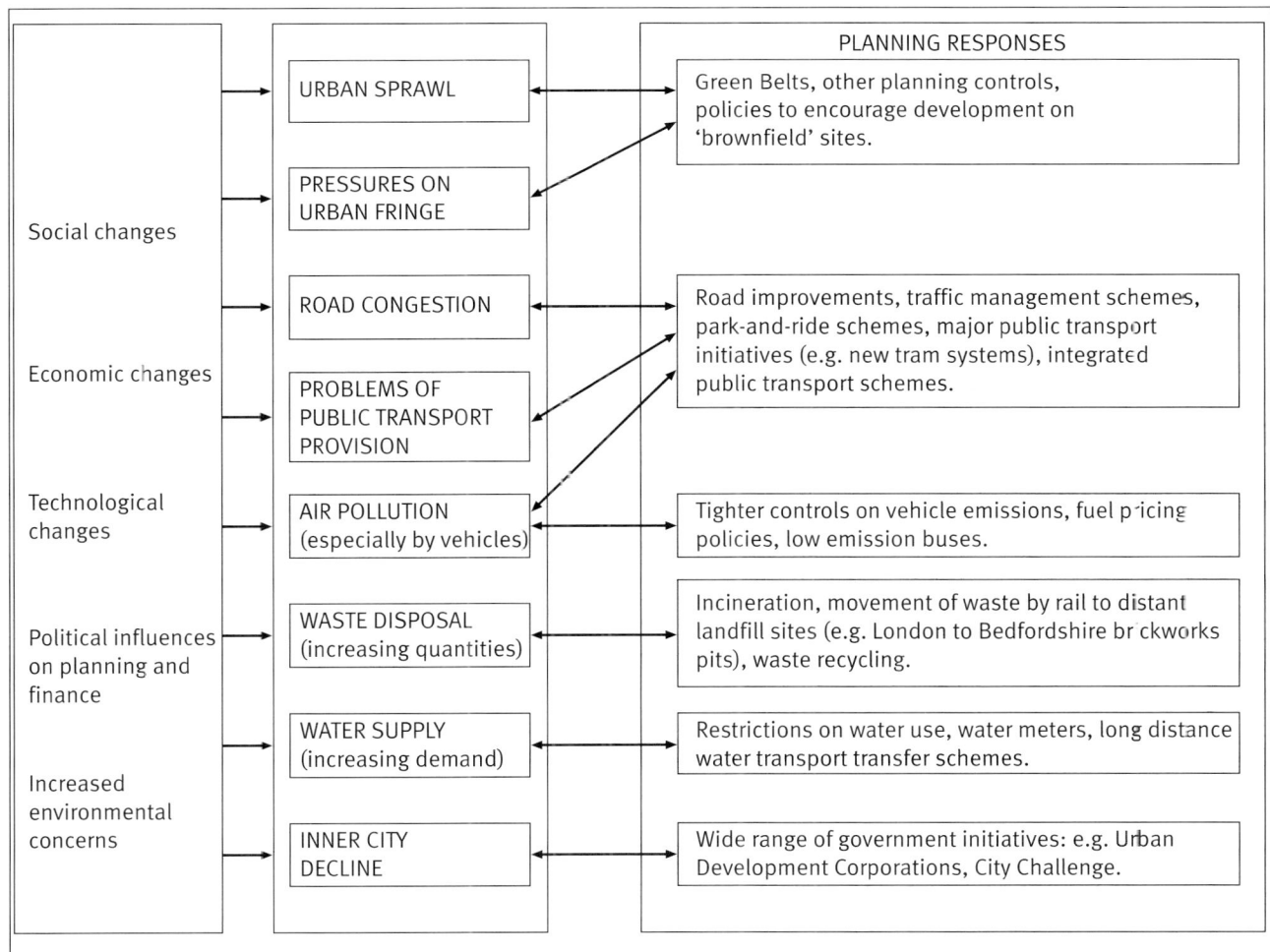

		PLANNING RESPONSES
Social changes	URBAN SPRAWL	Green Belts, other planning controls, policies to encourage development on 'brownfield' sites.
	PRESSURES ON URBAN FRINGE	
Economic changes	ROAD CONGESTION	Road improvements, traffic management schemes, park-and-ride schemes, major public transport initiatives (e.g. new tram systems), integrated public transport schemes.
	PROBLEMS OF PUBLIC TRANSPORT PROVISION	
Technological changes	AIR POLLUTION (especially by vehicles)	Tighter controls on vehicle emissions, fuel pricing policies, low emission buses.
Political influences on planning and finance	WASTE DISPOSAL (increasing quantities)	Incineration, movement of waste by rail to distant landfill sites (e.g. London to Bedfordshire brickworks pits), waste recycling.
	WATER SUPPLY (increasing demand)	Restrictions on water use, water meters, long distance water transport transfer schemes.
Increased environmental concerns	INNER CITY DECLINE	Wide range of government initiatives: e.g. Urban Development Corporations, City Challenge.

Figure 12.5 Problems of urban growth in MEDCs

- Measures to bring 1.3 million derelict empty homes, mainly in the north, back into use and restore derelict land.

- An environmental impact fee charged to developers of greenfield sites.

- Tax incentives, such as lower council tax, for inner city residents.

- Measures to tackle air pollution.

- Transport spending to be increasingly focused on projects to benefit pedestrians, cyclists and public transport users.

However, there seem to be no clear solutions for a major underlying problem of the UK's urban areas: a continuing loss of manufacturing jobs in many parts of the north and economic growth, involving job creation, in the South East.

Counter-urbanisation and the countryside

Counter-urbanisation in many MEDCs, including Britain, has involved people (and economic activity) moving from cities to rural areas and therefore an increase in the population of villages and small towns. Greatest growth has been in areas within commuting distance of major cities, with good road links and expanding employment in industry and services, such as East Anglia (Figure 12.6). Reasons for counter-urbanisation are:

- An increase in car ownership, combined with improved roads.

- Developments in information technology allow individuals and businesses to operate successfully away from cities.

- Dissatisfaction with urban lifestyles leads to a wish to escape from urban problems.

- Increasing migration for retirement.

- Industry and commerce have become increasingly 'footloose' in their location.

- Attractions of the rural environment and the perceived benefits of 'country living'.

The impacts of counter-urbanisation include:

- Villages closer to urban areas become commuter or suburbanised villages.

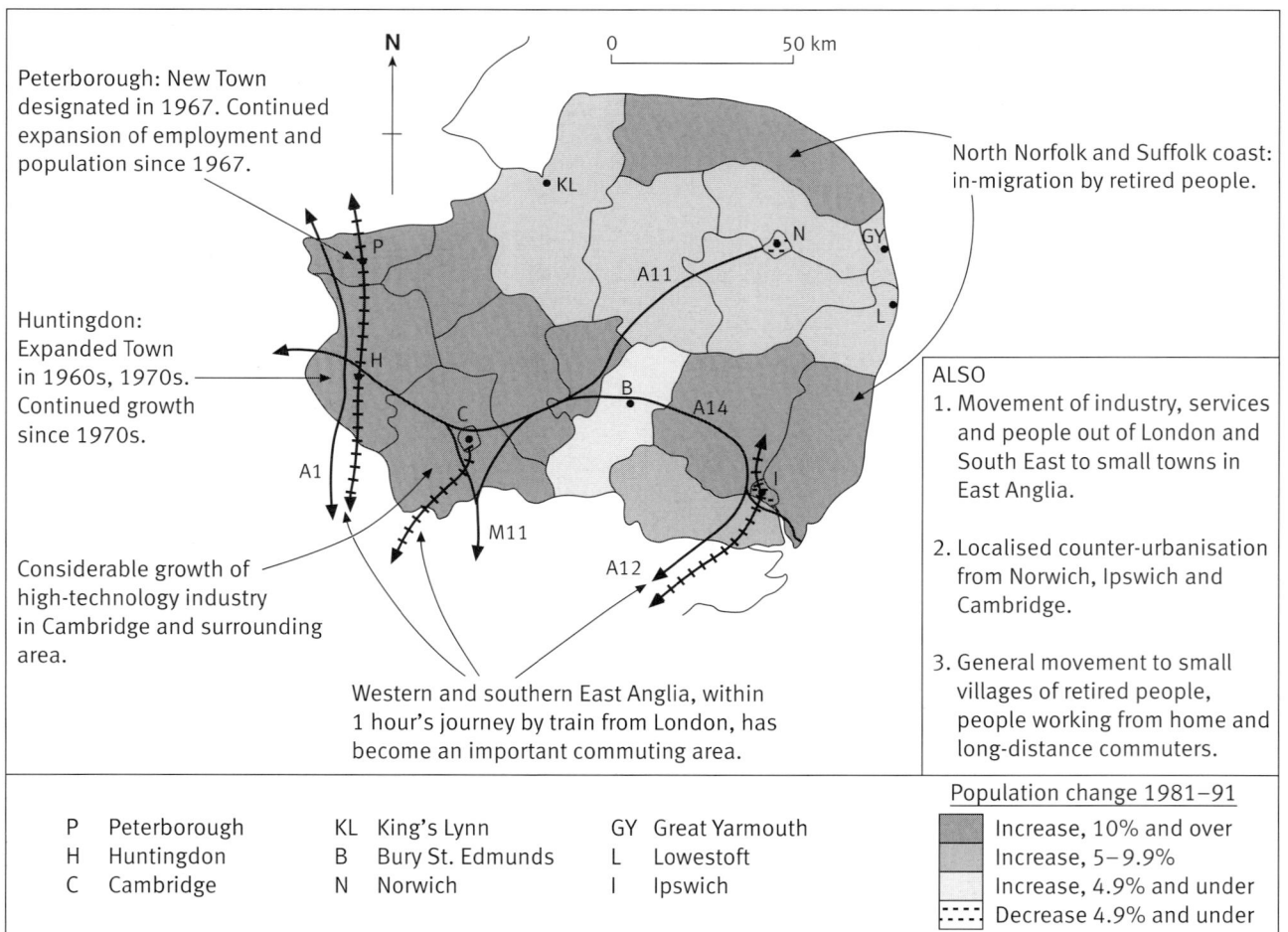

Peterborough: New Town designated in 1967. Continued expansion of employment and population since 1967.

North Norfolk and Suffolk coast: in-migration by retired people.

Huntingdon: Expanded Town in 1960s, 1970s. Continued growth since 1970s.

Considerable growth of high-technology industry in Cambridge and surrounding area.

Western and southern East Anglia, within 1 hour's journey by train from London, has become an important commuting area.

ALSO
1. Movement of industry, services and people out of London and South East to small towns in East Anglia.

2. Localised counter-urbanisation from Norwich, Ipswich and Cambridge.

3. General movement to small villages of retired people, people working from home and long-distance commuters.

P	Peterborough	KL	King's Lynn	GY	Great Yarmouth
H	Huntingdon	B	Bury St. Edmunds	L	Lowestoft
C	Cambridge	N	Norwich	I	Ipswich

Population change 1981–91
- Increase, 10% and over
- Increase, 5–9.9%
- Increase, 4.9% and under
- Decrease 4.9% and under

Figure 12.6 East Anglia: counter-urbanisation

- The morphology or form of villages and small towns changes with the addition of housing estates and new industries.

- Increased commuting by car increases road congestion and air pollution.

- The social structure of rural areas changes with the influx of mainly middle-class people with an urban background. Conflict may occur over rural noises, smells and traditions (such as hunting).

- The influx of prosperous retired people to areas of attractive scenery may force up house prices to the disadvantage of younger local people who wish to remain in the area in which they have grown up.

- The impact on rural services is complex. Considerable expansion of villages may result in services such as shops and primary schools expanding; on the other hand, commuters tend to use services in nearby towns and so service provision may decline in some areas. (See Chapter 14 on changing service provision in rural areas.)

Page 151

Review task

1 **To analyse the pressures on the urban fringe and other problems of urban growth of an urban area you have studied, draw a sketch map in the centre of a large sheet of paper, labelling problems and planning responses. If your area is a large city, you may find it easier to focus on one side or sector of the city.**

2 **Consider the various causes and impacts of counter-urbanisation. Again, apply them to an area you have studied. A Mind Map would be a useful way of doing this.**

Issues in cities in LEDCs

Cities in LEDCs suffer from many problems caused by their rapid growth and the poverty of many of their inhabitants. Figure 12.7 summarises them and shows how they interact with each other.

Housing for the poor in cities in LEDCs

The majority of people in most cities in LEDCs cannot afford or find good quality basic housing. This is especially true of recent migrants to the cities. The poor therefore find accommodation in inner city slums or in squatter settlements, usually on the periphery of cities.

Inner city slums are permanent buildings where housing conditions have worsened through age,

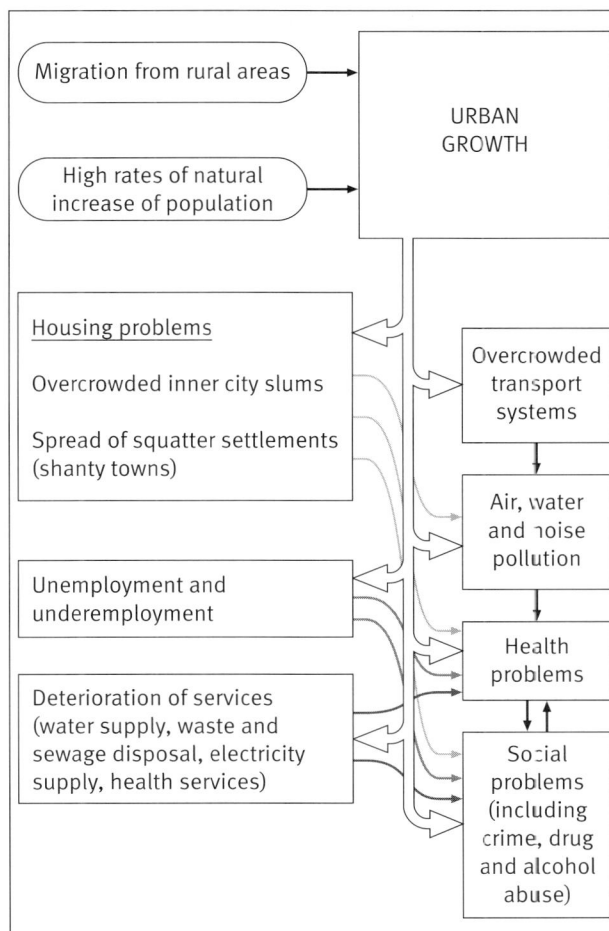

Figure 12.7 Urban problems in LEDCs

neglect, and sub-division. In some cities, for example Cairo, they are the dominant housing for the poor. In Cairo the pressure of people has resulted in the building of rooftop squatter shelters.

Squatter settlements are so called because they are built illegally and so their occupants have no security of tenure. They are also known as **spontaneous settlements** or **shanty towns** and are referred to by various local names, such as *favelas* in Brazil or *bustees* in India. Houses are built from any materials available and there is a lack of basic services such as water supply and electricity. Shanty towns are often on land unsuitable for other development, such as on steep hillsides or poorly drained land and may occupy such locations closer to city centres. In time, some squatter settlements are upgraded through the actions of the residents themselves or the city authorities.

Turner's model

Based on his studies of Latin American cities, J F C Turner developed a model to explain the housing

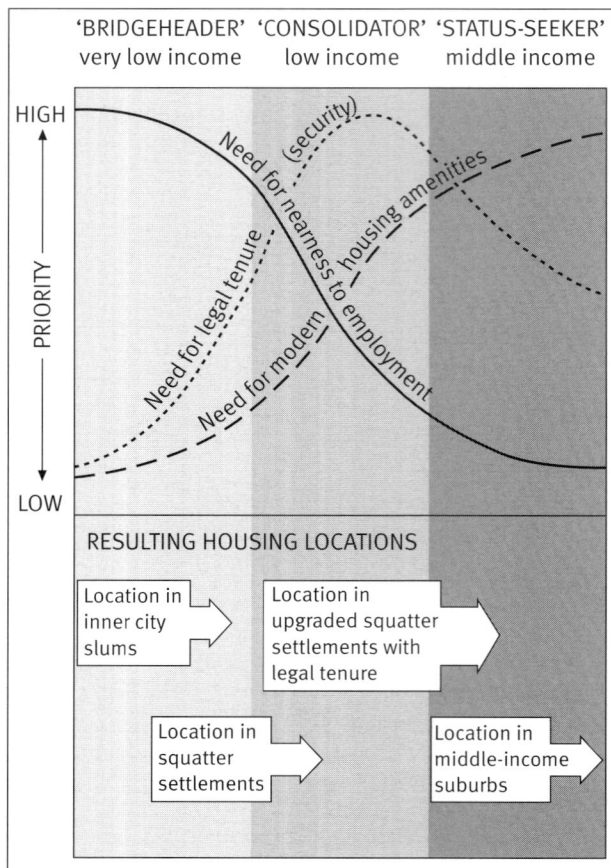

Figure 12.8 Turner's model of housing priorities for low-income families in LEDCs

locations of low-income people (Figure 12.8). Turner argued that because of their poverty, people had to choose between the usual ambitions for priorities of families (job proximity, security of tenure, modern amenities). Thus the recent arrivals to the city (**bridgeheaders**) locate close to the employment opportunities of the central city and later follow the sequence shown in Figure 12.8.

Government responses towards housing the poor

Governments in LEDCs face a housing problem which is enormous in size and continues to grow, yet funds available for 'solutions' are usually very limited. Governments have responded in a number of ways:

- **Demolition** of squatter settlements and the **eviction** of their inhabitants. Such action might 'solve' a locality's problem, but the displaced people are likely to squat elsewhere.

- **Public housing schemes,** including high-rise complexes, usually on the urban periphery, and **new or satellite towns**. Such schemes are very costly, the poor are often unable to afford the

rents charged and schemes are often located well away from jobs. Success has been achieved with this approach in relatively prosperous countries such as Singapore.

- **Urban renewal** of some slum areas through demolition followed by rebuilding is seldom practical as it involves a temporary loss of permanent housing and is very costly. Inner city slums are sometimes cleared by development companies, but for commercial land use or high-income housing.

- **Aided self-help schemes** paying to use the labour of the squatters to upgrade their homes and environment. The city government helps by supplying materials and training and by providing basic services. Such schemes reflect a changed attitude towards squatters; they are seen as people whose energies should be used rather than a nuisance to be evicted. Also, they are not displaced. Such an approach reflects or reinforces Turner's model which sees the bridgeheaders becoming consolidators and then status seekers.

- **Site and service schemes** involve guiding future development of low-cost housing. Prepared lots with basic services and security of tenure are provided for people to build their own houses.

Figure 12.9 shows how various approaches have been attempted in Rio de Janeiro, Brazil. Whatever approaches to housing are taken, underlying problems remain. Large numbers of people remain poor, living together in distinct areas, in effect ghettoes of poverty. There is usually a lack of involvement by people in their own area, even in self-help schemes. In particular, housing and social problems cannot be solved if large numbers of people remain unemployed or have low incomes.

Employment problems

The rate of urban population growth in LEDCs is usually much greater than the creation of jobs. Most rural–urban migrants lack education and are unskilled. The result is much unemployment, under-employment and the growth of the **informal sector** of employment. The informal sector covers all activities that are not registered or taxed; it includes jobs in small-scale manufacturing, many services, transport and building. Also included are many illegal activities such as theft and drug pushing. The informal sector may be capitalised upon in self-help housing schemes or exploited by entrepreneurs in the formal sector. Where demand for jobs in the formal sector greatly exceeds supply, employers can exploit the situation by paying low wages. For the employees, the

Built-up area
(population 9.5 million)

—·—· Boundary between
main urban zones

Main areas of shanty towns (*favelas*)
are in suburbs and outer urban area,
although the largest (Rocinha,
housing 80 000 people) is in the
central area. 20% of Rio's population
live in *favelas*.
Since 1990 policy has been to widen
streets, lay water pipes, provide
electricity, improve sewage disposal
and replace shacks with basic
houses. Schemes aim to consult local
people and employ them in building.
Such schemes have helped only
about 15% of the *favelas*.

Large city authority housing schemes
(*conjuntas habitacionais*) built before
1975, mainly in outer urban area,
house 1 million people. They were
built to rehouse *favela* dwellers from
the centre, but are of poor quality
and deteriorating.

Barra ca Tijuca,
affluent new
suburb housing
100 000 people.

Centre includes main commercial areas,
port, upper- and middle-class housing
(some on land cleared of *favelas*) and
favelas (on steep slopes).

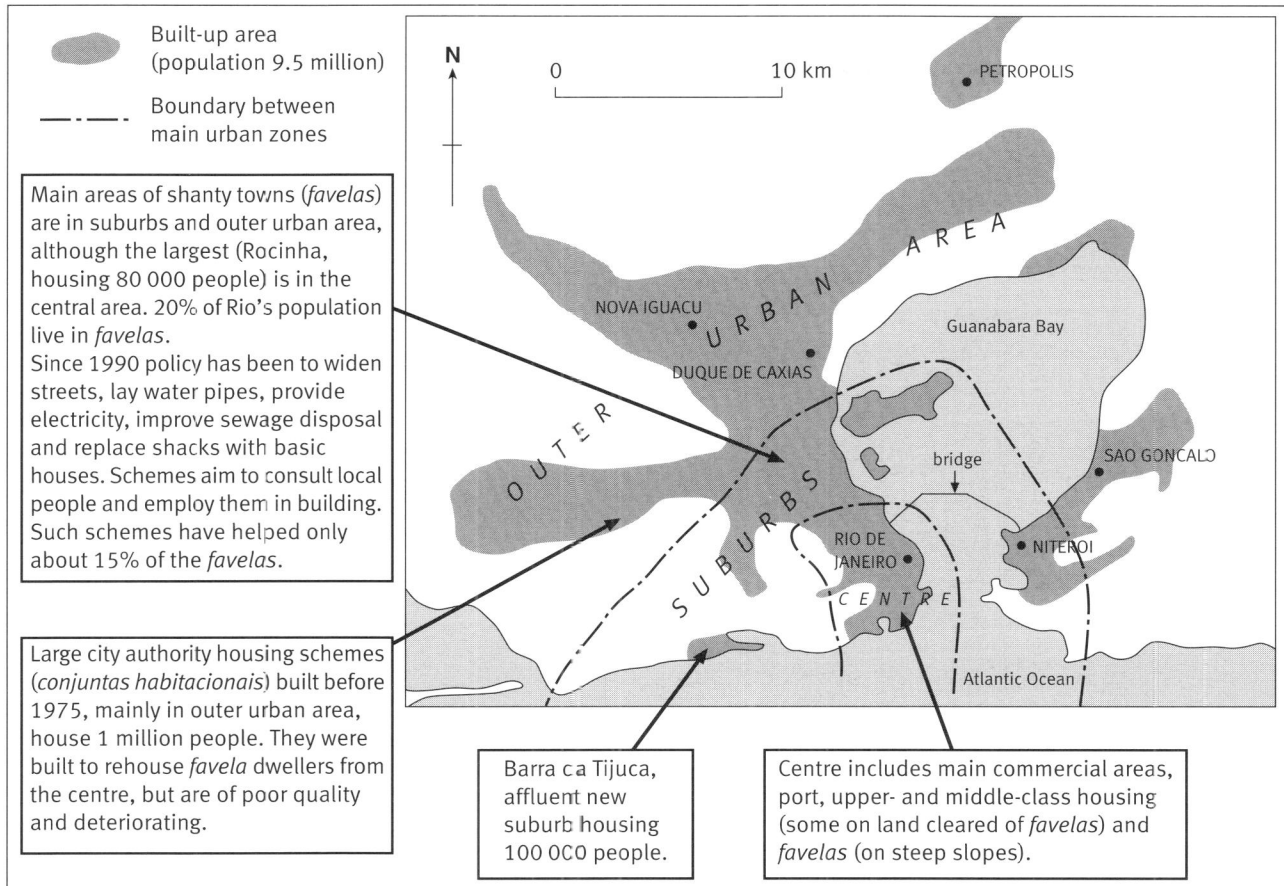

Figure 12.9 Rio de Janeiro, Brazil

low wages can still represent a major improvement
over opportunities in the countryside.

Environmental problems

Rapid urban growth causes environmental problems
of sewage and waste disposal. For example, in Mexico
City there are no legal landfill sites and few recycling
plants. Illegal dumping of waste, much of it
hazardous, has led to fires, pollution of underground
water and health problems.

In LEDCs air pollution by factories and vehicles tends
to be far less regulated than in MEDCs. Vehicles are
often poorly maintained and road systems in large
cities are inadequate for the amount of traffic. In very
large cities air pollution is therefore a major problem.
Mexico City with a population of about 18 million has
3.5 million vehicles and thousands of factories
producing about 12 000 tonnes of noxious gases
each day. Smog is a major problem and the city has
the world's highest concentration of ozone.

Review task

1 **Make sure that you are familiar with a case
    study of a large city in an LEDC.**
    a) **List separately its problems of housing,
        employment and the environment. Add
        arrows between items in the lists to show
        interrelationships between problems.**
    b) **What types of housing scheme have been
        attempted to solve problems? To what
        extent have they been successful? Is
        Turner's model relevant to your city? (If
        so, explain how.)**

2 **Construct a Mind Map to show the various
    responses of governments in LEDCs to urban
    housing problems. The five main approaches
    can be represented by the initial branches.
    Include problems, advantages and examples
    of the various approaches.**

What you need to know:

- **Classification of main types of economic activity**

- **The sector model of changing employment structure**

- **Changes involving industrialisation, de-industrialisation and re-industrialisation**

- **Industrial location factors including the influence of government and government agencies**

- **Changing patterns of industrial activity at global, regional and national levels**

- **Issues of problem regions – growing, declining and marginal**

- **The role of transnational corporations**

- **Consequences of manufacturing growth and decline – economic, social and environmental**

Also, you will need to be able to:

- **apply your understanding to new situations;**

- **describe, explain, and give examples to illustrate, factors affecting industrial location and changing trends.**

Economic activities

Economic activities can be conveniently divided into three sectors: **primary** (farming, fishing, forestry and mining), **secondary** (manufacturing and construction) and **tertiary** (services) and, in MEDCs, it is now usual to identify a fourth category called the **quaternary** or information sector which encompasses research, development and information handling activities.

Sector model of changing employment structure

This links the classification of economic activity to economic development. It suggests a series of stages through which economies pass, with increasing percentages employed in each of the three sectors (see Figure 14.1 for sectoral change in the second half of the twentieth century). Figure 13.1 shows, over a longer period of time, the full sequence of changes of the **Clark-Fisher sector model**, as it is also known, from pre-industrial to fully developed

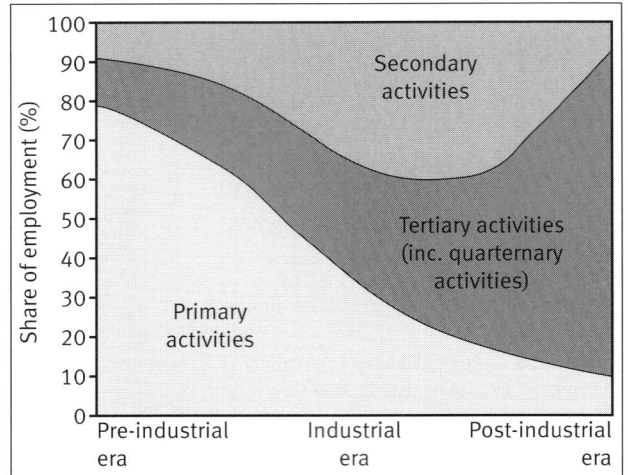

Figure 13.1 The Clark-Fisher sector model

'mature' or post-industrial economy. It highlights not only the relative changes in primary, secondary and tertiary employment but also the growth of the quaternary sector. The main feature of the model, as far as the bulk of this chapter is concerned, is to do with the decline of manufacturing or de-industrialisation.

De-industrialisation, post-industrial society and re-industrialisation

De-industrialisation is linked to tertiarisation (see Chapter 14) and is the phase of development that much of the more developed world appears to have experienced or is experiencing. It involves the decline of manufacturing in absolute terms as well as the relative decline caused by the rise of service activities. In Europe de-industrialisation involved considerable manufacturing decline, particularly of employment. The losses were mostly in the old established 'smokestack' industries, which were mainly located on the old industrial regions based in the coalfields.

Page 144

This change has given rise to the term **post-industrial society** in which the tertiary and quaternary sectors appear to be the main focus of growth. Note, however, that although employment has shifted from manufacturing to services this is not true in terms of output, partly as a result of increased productivity per worker. In fact, in many industries wealth production and levels of export earnings are expanding. In addition, a good proportion of employment in services depends directly or indirectly on demand from manufacturing industry. Figure 13.2 summarises the features of de-industrialisation and the post-industrial society.

- A relative decline in manufacturing's share of employment.
- An absolute decline in manufacturing industry employment.
- An excess of manufacturing imports over exports.
- A contraction or almost total elimination of some industries.
- Increase in service share of employment in relative and absolute terms.
- Increased reliance on service activities to maintain a balance of payments.

Figure 13.2 Features of de-industrialisation and the post-industrial society

Re-industrialisation is the term given to the process of:

- new investment and growth in new technologies such as bio-technology;
- the adoption of new computer-based technologies by existing industries;
- small-firm and new-firm resurgence;
- a surge of new, inward multinational investment.

Accompanying these new developments has been a spatial reorganisation of industry with rural and otherwise 'attractive' environments providing the locations rather than old established industrial areas. Where the latter have attracted new industries it has often been accompanied by government aid. It has been suggested that re-industrialisation represents the most recent of a number of waves of development or innovation (Figure 13.3).

Location factors

A range of factors has been identified in the past as influencing the location of manufacturing activities.

Review task

Make a list of points linking together the sectoral model of employment, de-industrialisation, post-industrial society and re-industrialisation. Pencil the points in the appropriate places on Figure 14.1 or on a large copy of it.

The relative importance of factors varied according to the type of industry and they changed over time.

Review task

As you read through the following sections on different location factors, check your own notes and use Figure 13.4 to record the names and locations of examples of case studies. You could also make a large copy of the table and summarise the main features of each case study.

Raw materials

Early industrialisation was primarily influenced by the location of raw materials, and industries grew up in locations where the raw materials were located. This was reflected in the pattern of industry which developed in Britain in the 19th century. A **raw material location** minimised the cost of assembling the raw materials and, especially where processing or manufacture brought about a great weight loss, it also reduced the costs of distribution. As industries developed which used a greater range of raw materials and a variety of components from different sources this dependence on a raw material location

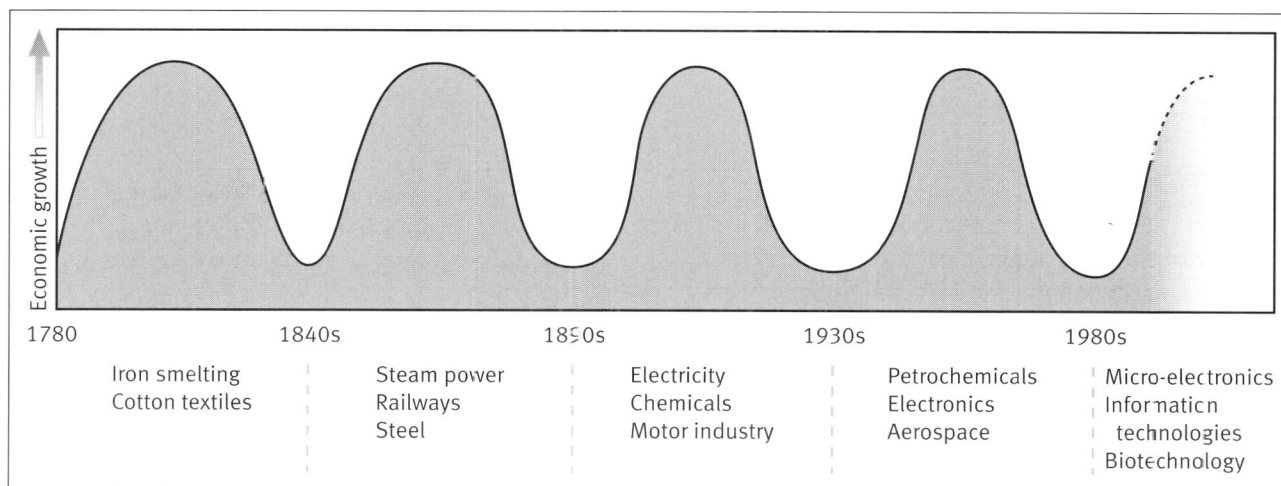

1780	1840s	1890s	1930s	1980s
Iron smelting Cotton textiles	Steam power Railways Steel	Electricity Chemicals Motor industry	Petrochemicals Electronics Aerospace	Micro-electronics Information technologies Biotechnology

Figure 13.3 Waves of development

Location factor	Case studies
Raw materials	
Market	
Labour requirements	
Availability of capital	
Importance of transport	
Need for enegery	
Land	
Influence of government and government agencies	
External economies of scale (agglomeration economies)	
Environment	

Figure 13.4 Location factors and case studies

waned. Port locations where raw materials are processed at a **break-of-bulk location** represent an intermediate location between the source of the raw materials and the market and given the number of raw material processing industries which depend on transported raw materials this is an important type of location for some industries, like iron and steel, oil refining and grain milling.

Market

Industries that draw on a large range of raw materials or components are regarded as **footloose** (in that no particular location is more favourable than any other) or **market-oriented** or market-based, where locating in the major market brings an advantage. Here distribution costs are minimised, although where a whole country or an international market is served the situation is rather more complex. One major market may be the ideal location with products transported elsewhere, since the size of different markets is important as well as their location.

Transport

Transport is an element of industrial costs at all stages from the assembly of raw materials to the final movement to the point of sale. Elements which have to be considered are the **fixed costs** (docks, etc.) and the **variable costs** (equipment and fuel, etc.), which are themselves complex because so much depends on the nature of the product and the form in which it is transported, pricing policies and the mode of transport. Developments in transport technology bring about changes in these relationships. Changes in

these factors can make dramatic differences to costs and change the significance of particular locations.

Labour

Labour is important in terms of its cost, quantity and quality. Industries that are large-scale employers generally need to locate in areas of relatively high population, simply to have a large enough **quantity** of potential workers. This can be interpreted in many ways. For instance, an industry that wants to employ a large number of relatively low-waged women might well choose a location that has very little employment opportunities for women, irrespective of the fact that the area might be thinly populated. **Cost** of labour is a major factor in the relocation of industry in relation to global restructuring where much lower-grade work has been relocated in LEDCs where wage costs are much lower. **Quality** takes account of the skills and knowledge levels required. For example, assembly work in the electronics industry is relatively low level in skills and training requirements, whereas research and development requires very high levels of education and training. Hence knowledge-based activities within a company may have a different location from the low-skill activities, to the extent of being in a different region, country or even continent (especially in high-technology industry).

Capital

This is a central requirement especially as industrial activity has become more capital intensive. The two main types of capital are **physical**, which includes land, buildings and machinery and is largely immobile, and **money**, or working capital, which has to be raised, usually by borrowing, and is more mobile. The fixed nature of physical capital is relevant in the context of **industrial inertia** (see page 136). The mobility of capital is very significant in the context of transnationals (see page 141), since they are able to direct their capital investment to wherever in the world meets company plans. Government policies (see page 135) attempt to redirect capital investment by giving incentives to companies to invest in particular areas. **Venture capital**, which is the provision of capital required to start up a new enterprise, is variable in its availability, making it easier to start up in some locations than others. For example, one major centre in the USA is the San Francisco/Silicon Valley area.

Energy

The role of energy as a factor in early industrial location is covered below (see Patterns of industry,

Revise AS and A Level Geography

page 136). As new forms of energy were developed and as transport technology improved, the relationship between energy location and industrial location has weakened in all but a few industries where energy requirements make up a significant proportion of overall costs. The common example given of one of these is the aluminium smelting industry and a source of cheap energy is a locational attraction. In developed countries, at least, electricity from a national power grid is readily available in all locations so for most industries energy is a relatively neutral factor in location.

Land

A few industries have a very specific requirement of a large area of generally level land, perhaps adjacent to a port, but these apart, the issue is a complex one. A number of factors combine in different ways to make some locations preferable to others. The cost and availability of land combine and become important when a company wants to expand. An established site is likely to be in an urban area and adjacent land on which to redevelop and expand is likely to be unavailable and/or expensive. This may be one reason for the relocation of much industry to **greenfield** sites in urban fringe areas or in rural areas. An additional point is that the redevelopment of **brown land** (land previously used for industry) sometimes involves considerable reclamation costs.

Influence of government and government agencies

One result of the changes in industrial location during the 20th century has been the decline of some industrial areas and the growth of others. Governments have followed policies which have been intended to counter some of the resulting difficulties by developing policies to deal with the problems of growth and decline. There have been two sets of policies which incorporated remedial measures to deal with:

- **Declining areas.** Here policies have been aimed at providing incentives for companies to expand or relocate in areas of declining industry (or problem areas) to bring employment to them.

- **Growth areas.** In these areas policies have aimed to limit growth to reduce problems of congestion and pressures on land and other resources.

The combination had the intention of limiting or reversing the economic decline of 'old' industrial areas and in the UK, over a 60-year period, the detailed policies changed (Figure 13.5) with varying areas of the country receiving different levels of assistance or controls. As well as giving incentives to

UK companies to relocate, great efforts were also made to attract foreign companies to whom a variety of financial incentives were offered. This has been in competition with the governments of other countries offering similar incentives.

External economies of scale or agglomeration economies

This has been given as the explanation for the development of industrial areas which are based on closely related activities. For example, the metal-working industries of the Black Country area of the West Midlands were closely linked in that there was a skills base overlapping between different industries and sets of backward and forward linkages between small and large companies. The motor industry is the most important example with many small companies supplying components to large car plants. The skills base emphasises another aspect of the overall industrial infrastructure. In high-technology industry the same skills element is a key factor in the agglomeration evident there (see page 135).

Environment

The location of high-technology industry in pleasant environments is given as one example of a trend to locate where it is easier to attract workers. This applies particularly to highly skilled workers (see high-technology industry, page 138).

> ### Review task
>
> **Remember to review your case study chart regularly.**

Concentrations of industry

Industry tends to concentrate in particular areas. Historically, these concentrations responded to changing circumstances, such as new energy sources or new methods of production. Present patterns of industry show the same tendency, whether around medium-sized towns, like Swindon, away from major conurbations or along major routeways, like the M4 axis.

The growth of concentrations of industry has been explained by the **cumulative causation** process, developed by Myrdal. Essentially, it is a cumulative process which involves a **multiplier effect** so that the initial development attracts suppliers (a backward linkage) and users (a forward linkage) (Figure 13.6). Other similar firms establishing in the same area are likely to develop horizontal linkages. All the extra

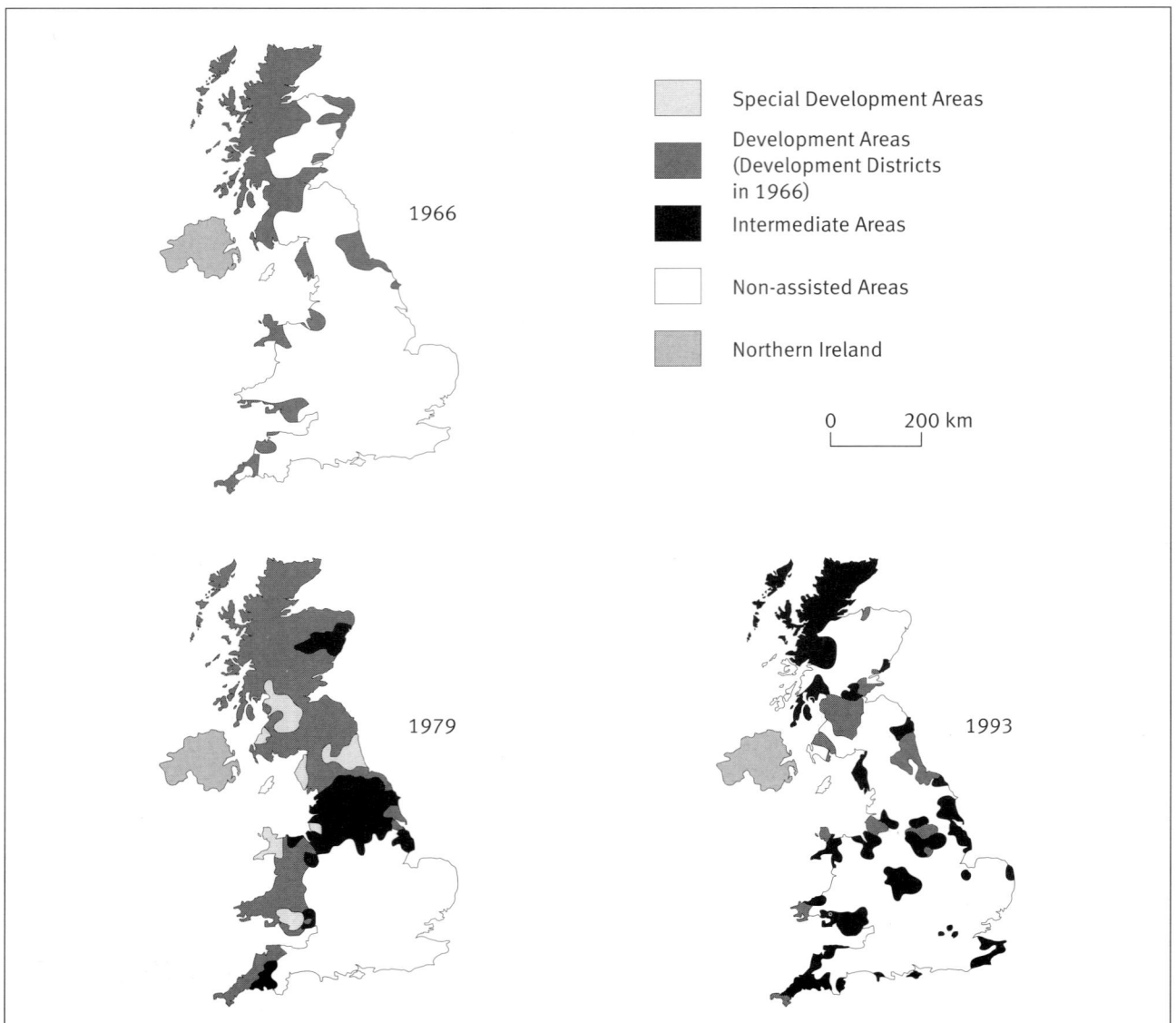

Figure 13.5 Assisted areas in the UK at different times

activity provides the market for other activities, such as specialist services, and firms seeking workers in a growing area. The overall clustering is known as **agglomeration** and provides benefits such as improved infrastructure and external benefits of scale in access to related industries and services.

Agglomeration is not entirely advantageous and problems may arise because of the great concentration of development causing dispersion from the core of the region to the periphery or even to another region. This would result in **diseconomies of scale**, where growing congestion and associated problems raise costs of manufacturing, and would also cause social and environmental problems. Such diseconomies often lead to dispersion, but where this does not happen **industrial inertia** is the term used to describe the situation where a firm remains in an area long after the original advantages have disappeared.

This is usually explained by the reluctance to write off investment of a large amount of capital or by the existence of a large skilled labour pool.

Patterns of industry

The importance of the location factors examined earlier varies according to the type of industry. To understand the pattern of old established industrial regions in the UK (Figure 13.7) and other developed countries one further factor, history, must be added. It is impossible to explain the growth of the coalfield-based industrial regions without considering the circumstances at the beginning of the Industrial Revolution and the factors which were key in determining industrial location.

Coal was the source of energy, iron ore was the key raw material, and other raw materials, such as

Revise AS and A Level Geography

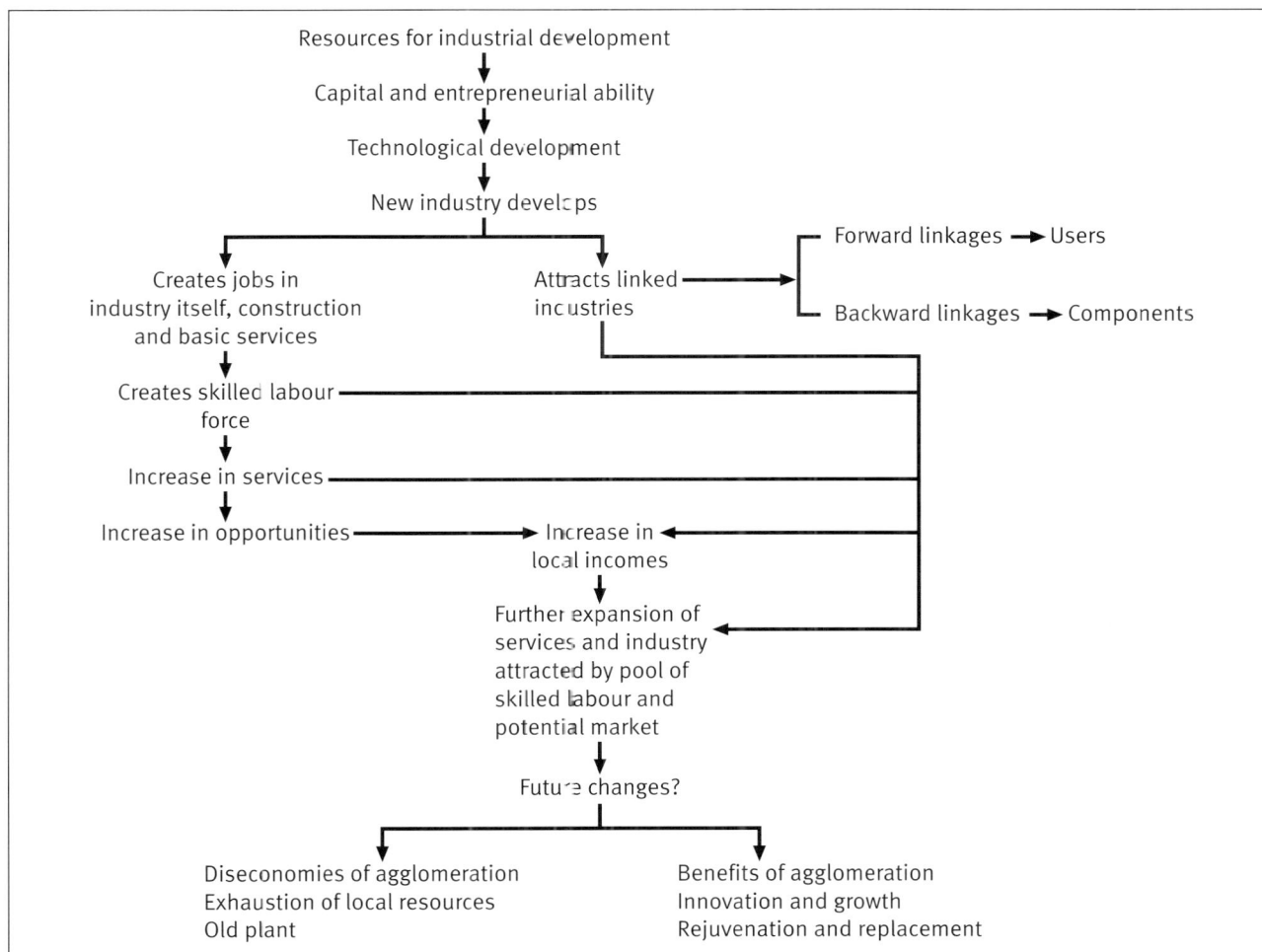

Figure 13.6 Cumulative causation and the multiplier effect

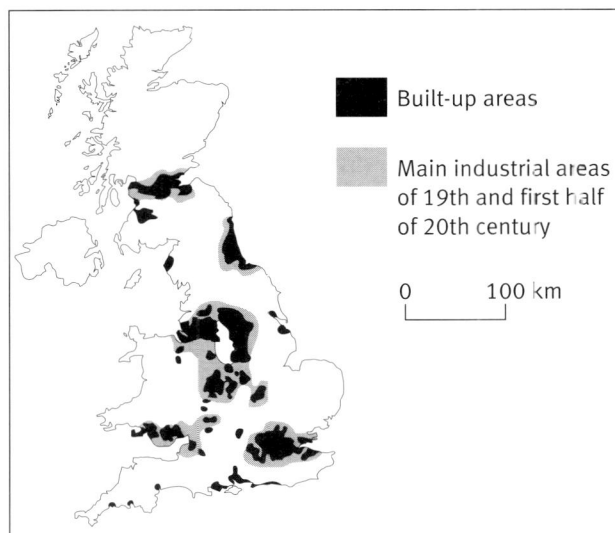

Figure 13.7 Old industrial regions in the UK

limestone, were usually found in close proximity. Production technology meant that there was a large weight loss in manufacture, and relatively high transport costs meant that the most economic location was on the coalfield. This was accentuated by the fact that most iron ores were low grade resulting in large weight losses. The pattern of heavy industry was therefore established and other industries using iron and steel, such as shipbuilding, also located on coastal sites on or adjacent to coalfields.

Over time, changes took place. Developments in production technology meant that small plants were replaced by larger plants, relocating within regions as coastal locations became preferable to facilitate the use of imported iron ores. However, the essential pattern was one of coal mining and heavy industry, with ports dependent on the export of coal and the other products of the region. Urban agglomerations developed which were very much dependent on this limited range of activities dominated mainly by large companies.

The decline of these industries was due partly to exhaustion of resources, partly to higher production costs. Combined with competition from areas with lower production costs and the lack of diversification, these areas suffered from high unemployment, and from the 1930s onwards government programmes were introduced, intent on attracting inward investment by other industries to widen the industrial and employment base.

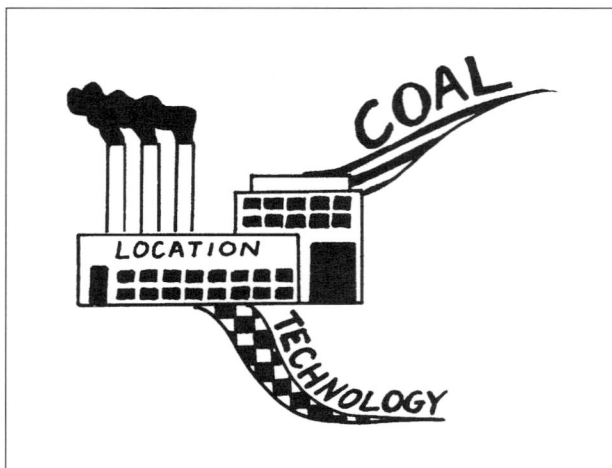

Figure 13.8 Mind Map for review task

The changes involved the use of government funds and the establishment of industrial estates to which light industries were attracted by grants, creating more employment opportunities for women, the decentralisation of some government activities from south-east England, the improvement of transport and communications and, most recently, the attracting of foreign investment in the form of companies from the USA, Japan and other European countries. This last phase can certainly be placed in the re-industrialising category, but, as with this trend generally, the majority of new industry is highly localised within any particular region.

Changing patterns of industrial activity

Changing patterns in high-technology industry

High-technology industries produce new and advanced products by rapid technological change and expenditure on scientific research and development. They have a vital role in the changing British industrial scene and have four key features:

1 Rapid technological changes produce short, and shortening, product life cycles.
2 There is a rapid growth in market demand for the new products.
3 Scientific research and development is vitally important to success in highly competitive environments.
4 There has been a great deal of restructuring of businesses with growth of foreign multinationals and new, small home-based firms.

Even in such a modern field there have been changes in the patterns of activity. London and south-east England had the greatest concentration of high-technology industry up to the 1980s, but there have since been two kinds of regional shifts, both in location and kind of development:

1 To the outer parts of southern England, with East Anglia being the fastest growing, followed by the south-west, where the growth has involved large-firm head office and research establishments and small locally based companies.
2 To Wales and, to a lesser extent, Scotland, both being 'assisted' regions, and here development has been in the form of branch factories which have head offices and main research establishments elsewhere.

As well as this regional shift, there has been a sub-regional shift with a very clear urban–rural move from big cities to smaller towns and rural areas throughout the country. The factors which have been identified as being important to explain these trends are:

- Availability of highly qualified workers. These are concentrated in the South East but there has been a mass movement of people out of big cities to small towns and rural areas. This movement has to explain above-average growth in other rural areas, such as North Wales and North Yorkshire besides those in southern England.

- Environmental influence has drawn people out of the big cities.

Revise AS and A Level Geography

- Concentrations of scientific research attract development, with the Cambridge region being Britain's leading high-technology growth centre.

- Good communications as high-technology firms generally deal with international markets and need access both to London and its airports.

- Movement to rural or semi-rural settings avoids the constraints of factory sites in congested urban areas.

- Government regional policy financial assistance is the major factor in Wales and Scotland where the main developments have been of multinational companies seeking a European base.

> ### Review task
> As a contrast with the previous two review tasks, produce a list of key location factors for modern high-technology locations in Britain. From your own notes add examples of regions and/or companies which illustrate points made above.

Similar trends can be identified elsewhere. In France, for example, the Rhône-Alpes region of France has attracted a great number of high-technology companies, both French and multinational. The factors that are advertised as its advantages include its location and excellent transport facilities, the superb quality of life in its rural and Alpine areas, and the well-developed growth points around Lyons and Grenoble.

Changing global patterns

The **global shift** in industrial location is part of the overall process of **globalisation** which has been described as the shrinking or compression of the world through 'the increasingly dense interconnections between people and places' around the world (*Introducing Human Geographies*, P. Crang and M. Goodwin, Arnold, 1999). Economic activity is becoming more and more organised on a global scale giving a **new international division of labour**, with production, investment patterns and movements and technology transfers all becoming more global. The actual geographical pattern of industrial activity has become more dispersed with a wider range of goods being produced in a greater number of countries. Within this overall statement there are four areas to consider: the newly industrialised countries; the growth of cross-border industrial regions; the development of regional trading groups; the role of transnational corporations.

Newly industrialised countries

A major feature of the changing world pattern has been the rise of the Newly Industrialised Countries (NICs), the best known being those of East Asia (Hong Kong, South Korea, Taiwan and Singapore). Other countries in the region which also developed rapidly are called Near-Newly Industrialised Countries (NNICs) and they are China, Thailand, Malaysia and Indonesia. These were all affected by the economic crisis in East Asia in 1997–8 as a result of which Indonesia, in particular, suffered badly.

The NICs showed a unique level of development and they outstripped every other country with three decades of exceptional growth from the 1960s. The NNICs also maintained above-average growth for three decades but failed to match the performance of the former group, although China's growth has increased since the mid-1980s.

The growth of the East Asia NICs has been attributed to:

- an emphasis on export-oriented industrialisation funded from savings;
- a general growth in world trade in the 1960s;
- initially, at least, low wage costs;
- availability of capital due to high savings levels;
- strong government involvement in and support for industrial development.

The NNICs of South Korea and Malaysia followed a slightly different track.

- Initially their exports were primary products which provided the base income for development.
- They depended greatly on foreign inward investment (FID) for their industrial growth, and this included European and American companies as well as the very important Japanese investments.

> ### Review task
> Examine Figure 13.9 and compare the changes between 1970 and 1995 for the two countries. How does the evidence support the points made above? Look carefully at the different segments in each pie chart and identify the exports which have grown and those which have declined in relative importance.

Growth of cross-border industrial regions

The growth of new industrial regions and the decline of old established ones is generally recognised. What

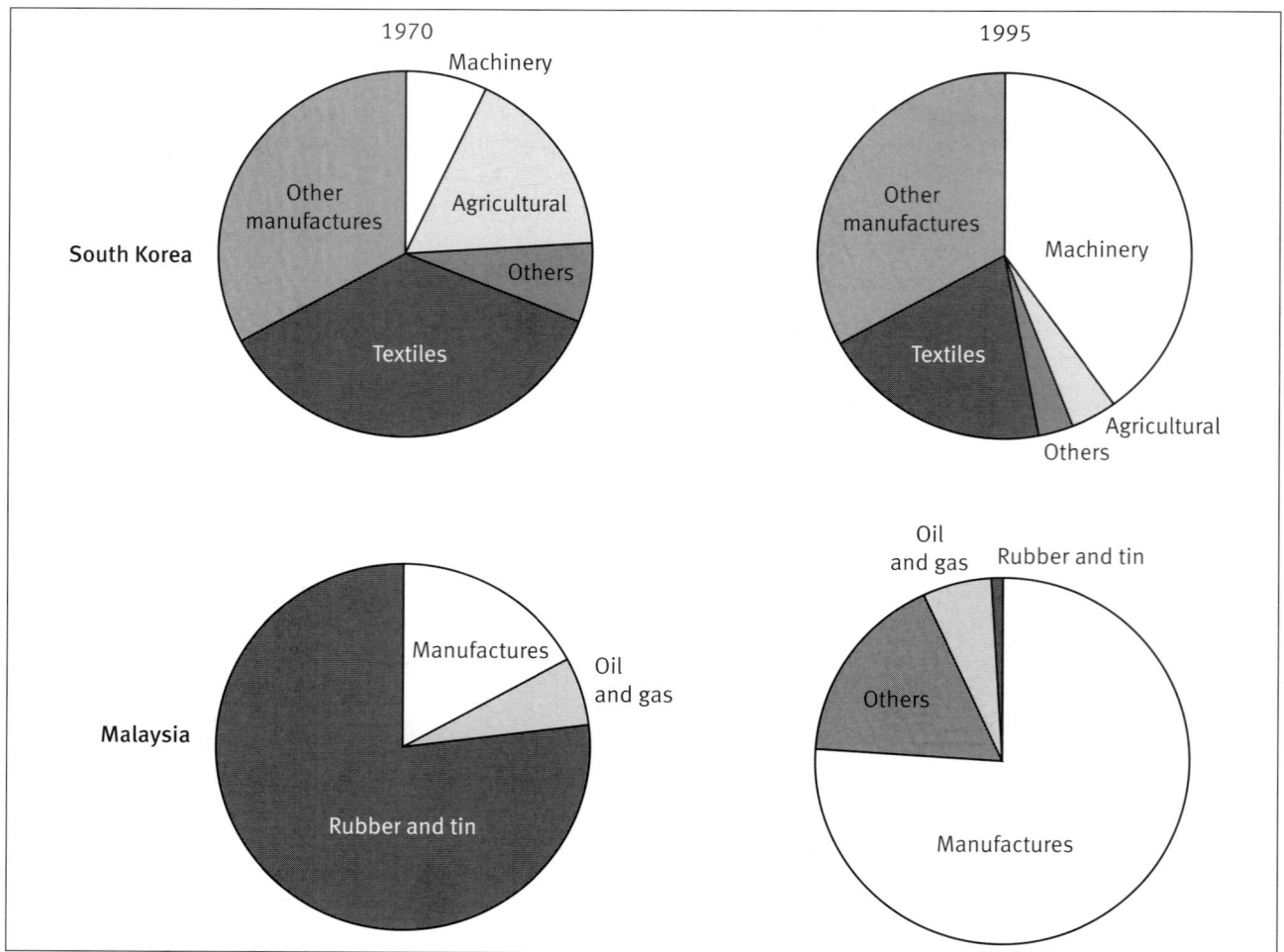

Figure 13.9 Export commodities of South Korea and Malaysia, 1970 and 1995

is not so commonly recognised is the way new industrial regions that cross national boundaries are developing. In many ways this matches how individual companies are organising their production across national boundaries. However, the difference is that these regions form natural economic territories across a national border, with trade carrying on across the borders and within the region. These regions have a core and a periphery. Figure 13.10 shows the cross-border industrial regions or **sub-regional economic zones** which have developed in East Asia. They have developed without any formal government agreements, although political encouragement has been acknowledged. Each of the areas is made up of neighbouring provinces of different countries and they are linked together by trade, investment and movements of people. In some of them local free-trade zones have been established by local government to help spread development from the core to the periphery.

Development of regional trading groups

Besides globalisation and the removal of trade barriers under the World Trade Organisation, regional groupings of a trade, economic and partly political nature have formed. The three major groups are the European Union, NAFTA (North American Free Trade

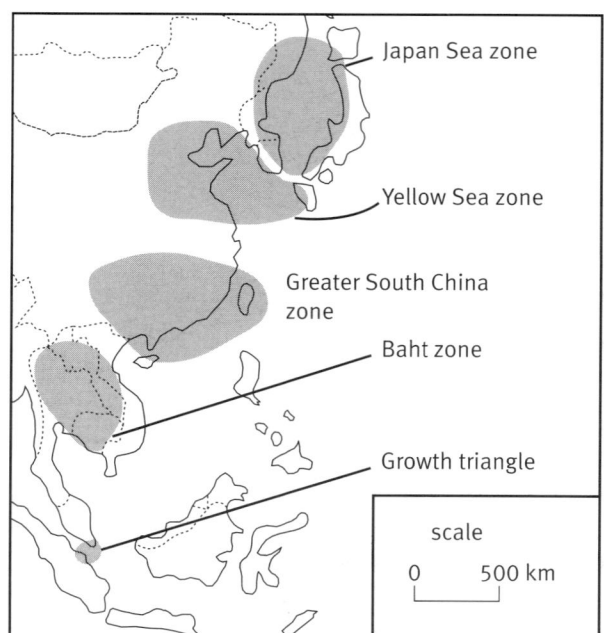

Figure 13.10 Cross-border industrial regions or sub-regional economic zones in East Asia

Agreement) and APEC (Asia-Pacific Economic Co-operation) (Figure 13.11). Membership of the second and third of these organisations is overlapping but there are, in any case, differences between them in that APEC has a much looser and more informal structure, NAFTA is a strictly trade organisation, whereas the EU is both a formal organisation with a wider political as well as economic role. The worry of

Review task

Approximate percentage share of foreign transnationals in manufacturing output (selected East Asia countries, early 1990s)

Country	%
Japan	3
South Korea	13
China (excluding Hong Kong)	14
Indonesia	16
Hong Kong	17
Taiwan	18
Malaysia	41
Singapore	70

Comment on the differences shown and use Figure 13.13 to suggest some of the implications for the countries concerned.

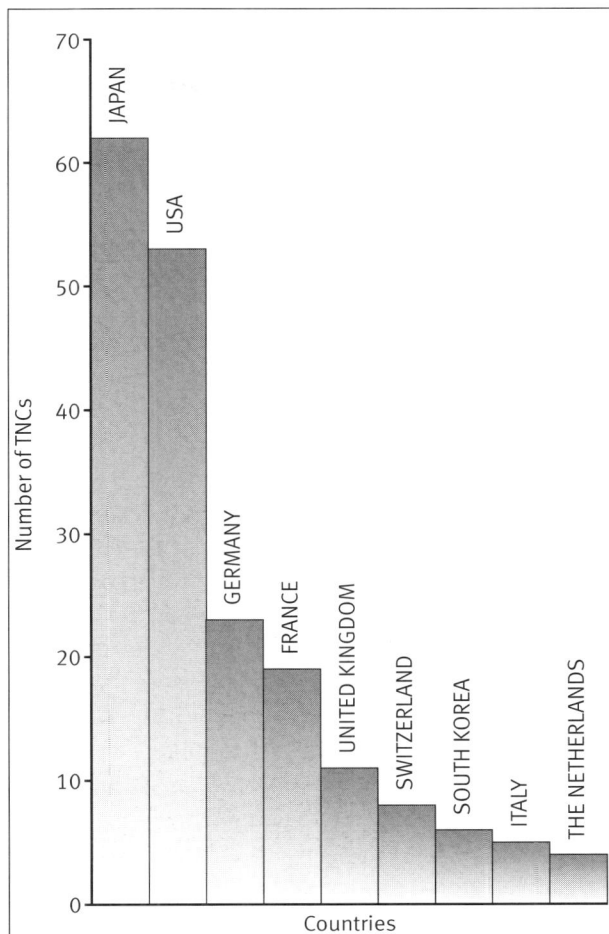

Figure 13.12 Headquarter countries of the 200 largest TNCs

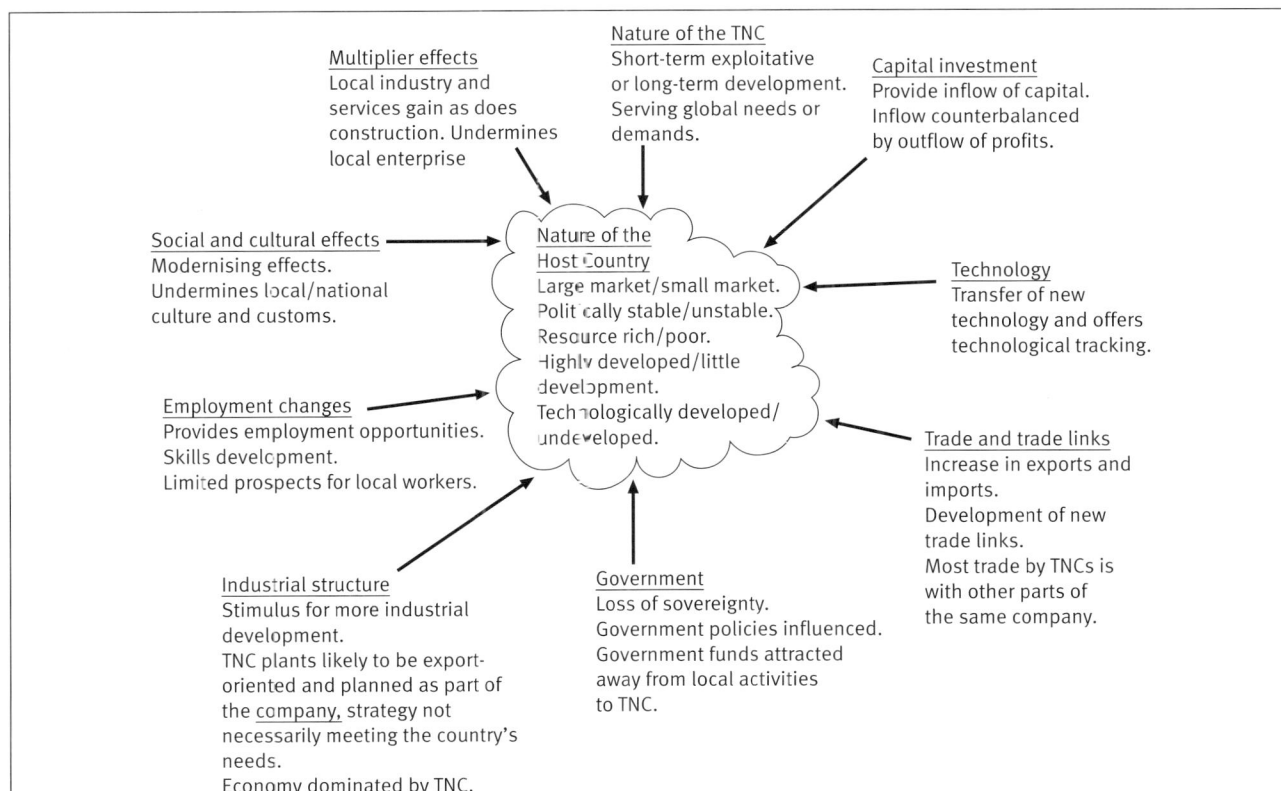

Multiplier effects
Local industry and services gain as does construction. Undermines local enterprise

Nature of the TNC
Short-term exploitative or long-term development. Serving global needs or demands.

Capital investment
Provide inflow of capital. Inflow counterbalanced by outflow of profits.

Social and cultural effects
Modernising effects. Undermines local/national culture and customs.

Nature of the Host Country
Large market/small market. Politically stable/unstable. Resource rich/poor. Highly developed/little development. Technologically developed/undeveloped.

Technology
Transfer of new technology and offers technological tracking.

Employment changes
Provides employment opportunities. Skills development. Limited prospects for local workers.

Trade and trade links
Increase in exports and imports. Development of new trade links. Most trade by TNCs is with other parts of the same company.

Industrial structure
Stimulus for more industrial development. TNC plants likely to be export-oriented and planned as part of the company, strategy not necessarily meeting the country's needs. Economy dominated by TNC.

Government
Loss of sovereignty. Government policies influenced. Government funds attracted away from local activities to TNC.

Figure 13.13 Possible impacts of transnational corporations

the LEDCs is that there will be barriers to their exports to the various blocs which will favour trade with countries from within that bloc. It is also, of course, a concern for those countries in, say, East Asia, that trade with the other areas will become more difficult.

The role of transnational corporations

Transnational corporations or TNCs (sometimes called multinationals or MNCs) are, according to the United Nations, businesses which 'possess and control means of production or services outside the country in which they were established'. There are about 37 000 major transnational companies but the bulk of world trade is covered by around 200 of them. They have been changing rapidly by expanding through mergers and takeovers and have become 'the powerful players in the global village'. In the 1960s transnationals accounted for 17 per cent of the world's Gross Domestic Product but by 1995 that figure had reached 32 per cent. The major transnationals are conglomerates which are involved in all sectors including mining and agriculture, manufacturing, financial services and transport. In terms of location they have headquarters in just nine countries (Figure 13.12).

Restructuring of activities within transnationals has played a large part in the rearrangement of the pattern of world production. Changes are made within the context of the corporations' activities with specialisation of production and transfer of materials and components taking place entirely within the overall TNC organisation.

The impacts of transnational corporations on individual countries can obviously be considerable especially as annual sales of some TNCs exceed the imports and exports of many LEDCs (Figure 13.13). This gives considerable power to TNCs in negotiations with governments. It also applies to MEDCs where, as has already been stated above, government assistance is often a deciding factor in locating a new plant.

The **transnational organisation of manufacturing on a global scale** and increased competition has led to a restructuring of existing plants with the closure of uneconomic (or less economic) plants and the expansion of more economic ones, coupled with the establishment of new plants in new locations.

The establishment of **new plants in new locations** reflects two factors:

1 The need for a presence in particular markets – a good example is the Japanese motor industry in the European Union, especially in the UK. This gives the companies a presence in the European Union.
2 The desire to maximise profitability by locating in areas with lower production costs, particularly lower labour costs. This again can be illustrated by the Japanese motor industry's spread into South East Asia.

Both points can be illustrated by examining Toyota. It has established integrated manufacturing systems in each of its three main markets in North America, Europe and Asia. For Asia the pattern of parts and components production and movement is shown in Figure 13.14. In addition to low labour costs, the high value of the yen has increased the cost of production in Japan, making overseas locations more attractive. Also, all the countries grant special deals of various kinds to overseas investors. This last point crops up time and again around the world as governments compete to attract foreign investment. (See page 138.)

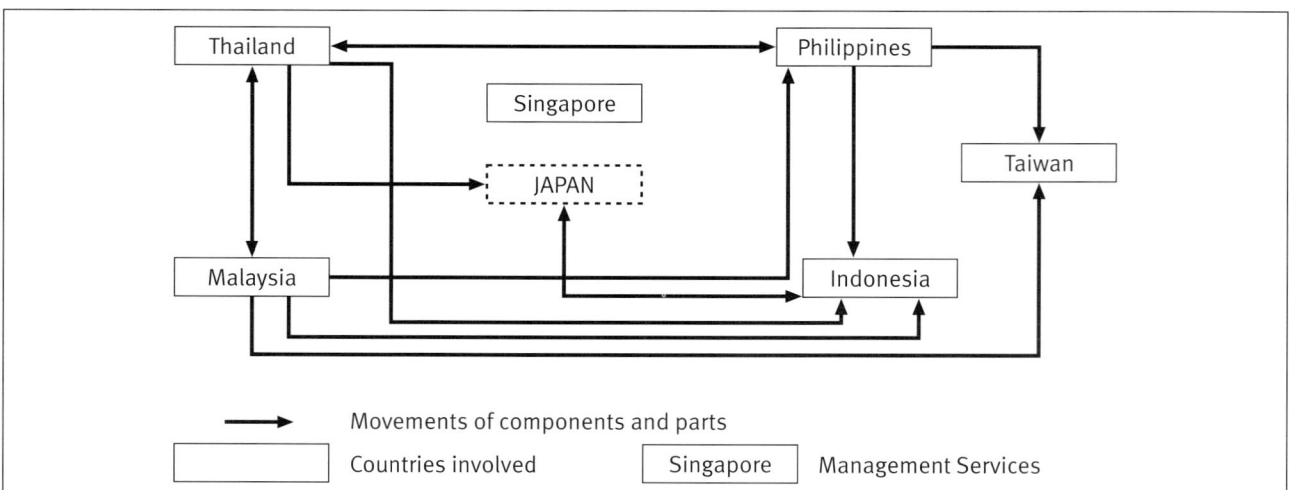

Figure 13.14 Toyota's production in Asia

Consequences of manufacturing industry's growth and decline

Economic and social effects

Industrialisation has a range of positive and negative effects. On the **positive** side it leads, obviously, to increased employment opportunities, wealth creation and through the multiplier effect to an improvement in overall infrastructure and broader economic opportunities as a result of the growth of other manufacturing industry and services. On the **negative** side, there has often been rapid urban growth with poor housing and cramped, unhealthy conditions, health problems related to industrial activity itself and short- and long-term land use problems. The economic and social problems, which are most apparent in regions of economic decline, are effectively the reverse of the multiplier effect.

Environmental effects

The environment also has a dual effect on industrial location. First, it affects industrial location, due partly to the pattern of resources and partly to the attraction of pleasant environments for new company location. Second, industry affects the environment, usually in a damaging way, and causes long-term and short-term damage to the environment itself, creating similar problems for people.

The levels of controls on environmentally damaging activities vary both within and between countries. Controls on emissions, dumping of dangerous substances, limitations on visual impact and the need for landscaping are some examples and there is evidence that variations in the levels of controls do affect location choice especially for industries which are potentially the most damaging environmentally. For example, the environmental damage created by the 'maquiladora' factories on the Mexican side of the US–Mexican border contrasts with the effects of the factories owned by the same companies in the USA,

Britain or Germany, where environmental controls are stronger, though still variable.

The concept of **externalities** is useful in examining environmental effects. These are the effects which are not included in the costs and prices of an economic activity; they are, instead, borne by the community in the case of local-scale impacts, but by whole regions in the case of broader impacts. Examples of the local scale might be dereliction of land polluted by heavy metals, whereas at the regional scale acid rain illustrates the combined effects of a range of activities. Note that while some externalities could be classed as quite deliberate as a result of a policy of only meeting the minimum standards required by regulations, others become apparent only as harmful after a long time – the damage to the ozone layer is one such case.

An important general point about externalities is that they are spatial in effect. Some are globally damaging, others are regional and many have relatively local effects. Local effects tend to decrease away from the source, for example noise, smell or effluent pollution.

Review task

1 From your studies identify an industry which is a source of environmental damage (or externalities). What kinds of damage (externalities) are evident and how far does the area affected extend (externality field)? If possible, use a local example.

2 Review your course notes, and answer the following with reference to specific cases studies:
   a) What kinds of region are likely to show most environmental damage due to industry?
   b) Give an example of each of the following: water pollution, air pollution, landscape degradation.

PREVIEW

What you need to know:

- **The nature of service activities and occupations**
- **The process of tertiarisation; changing employment patterns**
- **The location of retailing within cities; types of shopping centre**
- **The retailing revolution**
- **Characteristics of and changes in CBDs**
- **The location of offices**
- **Service provision in rural areas**

Also, you will need to be able to:

- **apply your understanding of models and theories of urban geography, such as central place theory, to the provision and location of services;**
- **appreciate the link between de-industrialisation and tertiarisation;**
- **apply your knowledge and understanding of service activities to case studies;**
- **analyse maps and other data on service activities.**

Service activities and occupations

Employment in the service or **tertiary** sector can be divided into three groups:

1. **Producer services** which serve manufacturing and other service industries. They include financial and insurance services (although both of these serve consumers as well), legal services, computing, advertising, market research and freight transport.
2. **Consumer services** such as retailing, catering, recreation and tourism.
3. **Public services** which include education, medical services, social services, telecommunications, postal services, passenger transport and administration by local and central government.

Service occupations also occur in other sectors, for example, office workers in the secondary sector. Sometimes an additional **quaternary sector** is identified, referring to activities involving information, knowledge and ideas, for example information and computer technology. However, it is difficult to isolate this sector as such activities and their associated occupations are found in all three other sectors.

Chapter 13 (page 132) looked at the **sector model** which links the classification of economic activity to economic development. The final stage of this model, applying to MEDCs, involves employment in the tertiary sector becoming very dominant (with over 60 per cent of the workforce being employed in this sector) and is referred to as the **post-industrial stage**.

The process of tertiarisation

Page 132

The move from an economy dominated by manufacturing to one dominated by the service sector is the process of **tertiarisation**. It is closely linked with **de-industrialisation** (see Chapter 13). Figure 14.1 shows sectoral change in UK employment and how the process of tertiarisation has accelerated in the last decades of the 20th century. Reasons for this are as follows:

- Decline in manufacturing employment (refer to Chapter 13).
- The increased importance of producer services. This reflects major changes in the business world such as the wider use of electronically based information and communication technologies, and thus a rising demand for many specialist services.

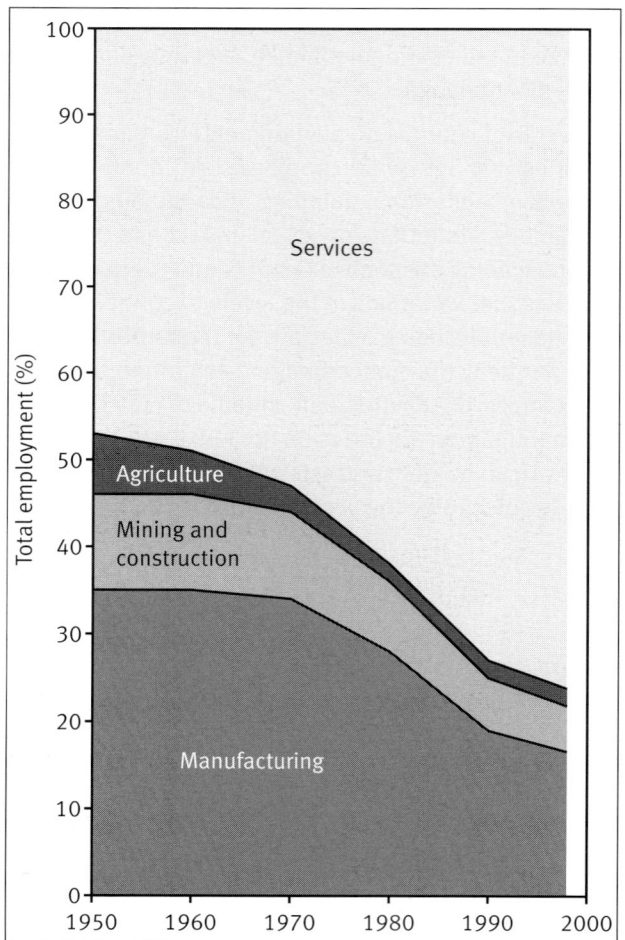

Figure 14.1 Sectoral change in employment in the UK

- The increased international operation of companies. This increasingly applies to service industries such as banking and advertising. Thus services are in effect exported.

- Movement of service occupations from the secondary to the tertiary sector. Manufacturing companies increasingly contract out services, ranging from advertising to cleaning, so contributing to the growth of producer services.

- Increased employment in certain consumer services, in particular catering, recreation and tourism, together with the consumer side of financial services; all largely a result of growing incomes.

- Increased employment in some public services such as education and medical services, especially in the 1960s and 1970s.

Not all services have shown a large expansion in employment. Employment in public transport and merchant shipping has declined. Employment in retailing in the UK has grown only slightly because of the rise of larger shops with their huge economies of scale and low staffing in relation to sales.

Changing employment patterns

The increased importance of the service sector in MEDCs as been accompanied by significant changes in the labour force, often described as a move towards a flexible labour market. Although most of the changes listed below have occurred in manufacturing, they have been most marked in the service sector:

- Part-time employment has become more important. Employers can therefore meet short-term changes in demand from customers (for example, in retailing and the fast-food industry).

- The increased use of temporary workers a so enables companies to have a more flexible labour input.

- An increase in self-employment has been particularly great in the service sector, partly a result of government encouragement.

- The female labour force grew in the late 20th century and is particularly important in the service sector (1998: 28 per cent of manufacturing employment, 57 per cent of service employment in the UK).

- Low-paid jobs are particularly important in the service sector. Low pay affects females more than males and part-timers more than full-timers.

Regional differences in service employment

Employment patterns vary considerably within the UK as Figure 14.2 shows. The service sector is particularly large in London and the South East. The concentration and growth of producer services and consumer financial services in these regions reflects demand from growing high-technology industry and existing services, higher levels of prosperity and the importance of London as an international financial and commercial centre. Many companies which export services are London-based and a very large proportion of companies have their headquarters in London and

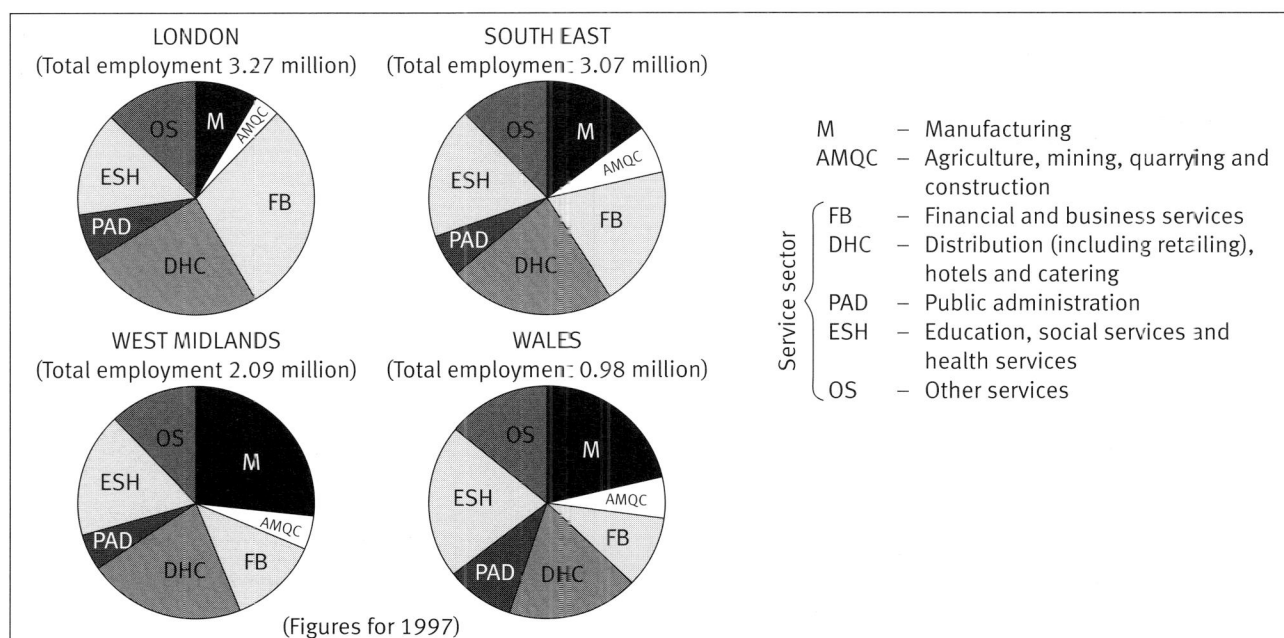

LONDON (Total employment 3.27 million)

SOUTH EAST (Total employment 3.07 million)

WEST MIDLANDS (Total employment 2.09 million)

WALES (Total employment 0.98 million)

(Figures for 1997)

M	– Manufacturing
AMQC	– Agriculture, mining, quarrying and construction
FB	– Financial and business services
DHC	– Distribution (including retailing), hotels and catering
PAD	– Public administration
ESH	– Education, social services and health services
OS	– Other services

Service sector

Figure 14.2 Contrasting employment patterns for four UK regions

the South East. In recent years the growth of producer and financial services has been less in London than the rest of the South East because London has always had a very high proportion of service occupations and many services have decentralised from London into the surrounding region. Public services are more evenly spread through the UK; they have contributed little to the growth of the service sector in London and south-east England.

Review task

1   Check your understanding of the various types of services. To fully understand the process of tertiarisation, you should be familiar with de-industrialisation – refer back to Chapter 13.

2   Refer to Figure 14.2. Describe and explain the contrasts in employment patterns between the four regions.

Retailing

The traditional shopping hierarchy in cities

In the same way that a hierarchy of service centres exists within a region (see Chapter 11), a hierarchy of shopping centres exist within a city. As the same key terms and underlying ideas are relevant here, check your understanding of the following before reading further: **range, threshold, low- and high-order goods and services, convenience goods, comparison goods, sphere of influence**. Also, check your understanding of the ideas of Christaller and Reilly's law of retail gravitation.

Pages 115–17

The locations of shopping centres within a city are shown in Figure 14.3. The traditional hierarchy consists of the CBD, **district centres** (major suburban centres), **neighbourhood centres** (parades of 15 to 30 shops) and, at the bottom, small **community centres** and individual 'corner' shops.

District centres are typically located where ring roads cross radial routes, sometimes former village centres. They are characterised by the presence of shops selling comparison goods (many owned by multiples) besides covenience goods shops, a supermarket belonging to a major chain and services such as banks, building societies and estate agents. Older inner city district centres often have the form of **shopping ribbons** along major radial roads. They may show signs of stagnation or decline, with vacant shops and few multiples (reflecting inner city poverty

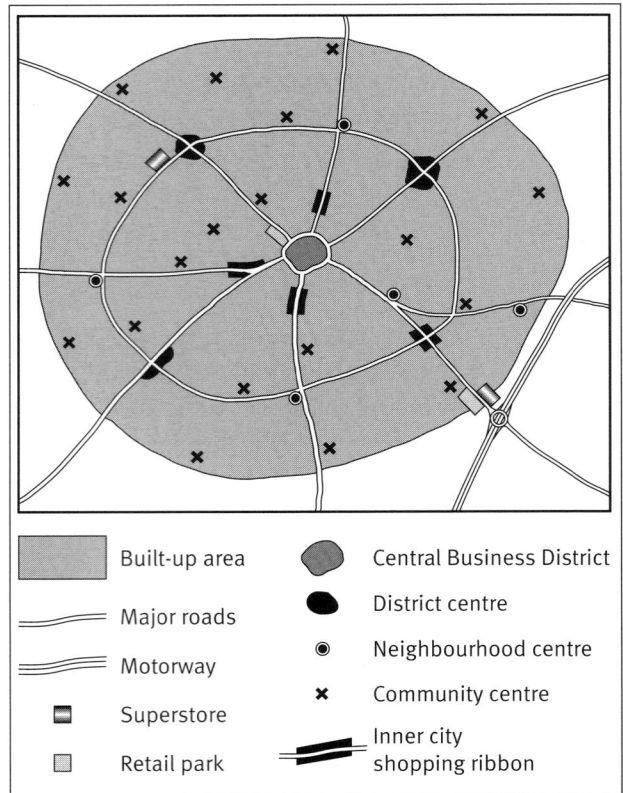

Built-up area	Central Business District
Major roads	District centre
Motorway	Neighbourhood centre
Superstore	Community centre
Retail park	Inner city shopping ribbon

Figure 14.3 The pattern of shopping centres in a city

and population decline). They also have shops and services providing for local ethnic groups and businesses deriving custom from passing traffic.

EXAM TIP

The hierarchy of service centres in a region is sometimes described as *inter*-urban; within a city, the hierarchy is described as *intra*-urban. Make sure that you do not confuse these terms.

The retailing revolution

As Figure 14.3 shows, retailing locations other than the centres of the traditional hierarchy are found in cities. They result from changes in retailing over the last 30 years, changes so great that the term 'retailing revolution' is often used. The 1970s saw supermarket chains beginning to develop superstores, the 1980s saw a boom in retail parks selling non-food items and the late 1980s and 1990s the growth of large regional shopping centres. Various social, economic and technological changes underlie the retail revolution:

* Changing ownership of shops. Large companies with national chains of shops – the multiples – dominate retailing (over 75 per cent of food sales). Independent traders have declined. The multiples

have the advantages of economies of scale through purchasing, national distribution networks and operating in large stores.

- Suburbanisation and counter-urbanisation of more affluent people encourages the decentralisation of retailing.

- Increased standards of living give people more purchasing power and encourage greater car ownership. The increase in home ownership encouraged the rise of the DIY chains.

- Changing patterns of accessibility. Construction of ring roads and motorways at the edge of cities, combined with greater car ownership has made suburban locations more accessible to shoppers, in contrast to more congested city centres.

- Changing shopping habits result from social changes such as more working women and technological changes such as the widespread ownership of freezers. Thus food shopping tends to be carried out less frequently and in bulk, a practice also encouraged by the use of cars.

- The development of shopping as a leisure activity, a trend encouraged especially by developers of regional shopping centres (see below).

- The development of new products has resulted in new types of outlet, for example, computer stores. Increased affluence has fuelled demand for a wider variety of products.

- Changing planning strategies have had an impact on the location of new retail developments, for example a relaxation of planning controls in the 1980s and early 1990s encouraged 'out-of-town' developments. In the late 1990s controls were tightened again.

- Land availability. The large sites required by superstores and retail parks are most likely to the found at the urban fringe.

Superstores

Superstores (outlets with over 2500 m^2 of selling area) have increased greatly in numbers over the past 30 years. They are owned by the major supermarket groups who have over three-quarters of UK food sales. Many superstores also sell household items and clothes. Most are found in edge-of-city locations, but some have been built at the edge of existing district centres or smaller CBDs. Tighter planning controls are restricting further development of superstores. Major chains have responded by developing smaller stores in existing shopping centres – Tesco Metro and

Sainsbury's Local – a further threat to independent food shops.

Retail parks

Retail parks are complexes of warehouse-type buildings selling non-food items such as DIY goods, furniture and electrical goods. Their development was especially rapid in the late 1980s and early 1990s. Most are in suburban locations, but a significant number are found on the edge of CBDs, often on former industrial land. Some larger, more recent developments involve a combination of retail warehouses, a superstore and leisure facilities such as multiplex cinemas and fast-food outlets.

Regional shopping centres (RSCs)

Large regional shopping centres, also known as out-of-town centres, dominate retailing in the USA. In the UK the first such centre opened at Brent Cross in north London in 1976. The Metrocentre (Gateshead), opened in 1986, was the first super-RSC (with a size of over 100 000 m^2). There are now five other super-RSCs: Merry Hill (Dudley), Meadowhall (Sheffield), Trafford (Manchester), Lakeside (Thurrock, Essex) and Bluewater (Dartford, Kent) and several smaller RSCs. Key features of the super RSCs are:

- They have at least 8000 car parking spaces.

- Shops are arranged along air-conditioned malls on two or three levels. There are leisure facilities such as restaurants and cinemas.

- All except Merry Hill have direct access from motorways or major trunk roads.

- They have had a major impact on existing shopping centres (in the case of Merry Hill, the nearby centre of Dudley has lost most of its major stores) and attract shoppers from considerable distances.

- All except Trafford were built on derelict land. Metrocentre and Merry Hill were built in Pages 124–5 Enterprise Zones, Trafford and Meadowhall on UDCs. There were therefore tax concessions or other incentives for development.

The relaxation of planning controls in the 1980s and early 1990s enabled RSCs to be developed. Few, if any are likely to be developed in the near future, both for environmental reasons and their impact on existing shopping centres. A specialist type of RSC is the **retail factory outlet**, a complex of shops selling 'designer' clothes and other products at a discount. Examples of such centres include Bicester Village, Oxfordshire, and Cheshire Oaks near Chester.

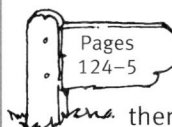

It is important that you can relate patterns of retailing to case studies. In the case of a small or medium-sized urban area this can be done for the whole area. In the case of a larger city, it may be easier to concentrate on one sector stretching out from the CBD.

1. Draw a sketch map to show the location of the CBD, district centres, inner city shopping ribbons, superstores, retail parks and other major retailing locations. Note the reasons for their locations and the changes that have taken place over recent years. Consider the impact of new developments. (Do not consider the CBD at this stage.) This information can be summarised by means of detailed labels around your map.

2. Check your knowledge of other relevant examples that might not be covered by your main case study. (For example, many urban areas will not include a regional shopping centre.)

3. The social and economic causes of changes in retailing have been noted above. It is also important to consider the social, economic and environmental effects. Draw up lists of such effects for each of these three categories. Wherever possible note actual examples from your case studies.

The Central Business District (CBD)

Characteristics

The CBD acts as the heart of the city and has distinctive characteristics:

- It is a major concentration of retailing, with department stores and specialist shops which have high threshold populations and wide ranges.
- It has a major concentration of offices.
- Land values and rents for buildings are the highest in the city.
- The CBD as the focus of the transport system is a particularly accessible location.
- The many high buildings reflect the wish to maximise the use of expensive land.
- There is an absence of manufacturing and there are few residents.
- Pedestrian and traffic flows are very high.
- Land use within CBDs is zoned, partly reflecting varying land values and rents. Office zones tend to

be away from the central location demanded by the major shops. A CBD **core** (with the main concentration of shops) can be distinguished from the surrounding **frame** which would include land uses such as railway stations, car parks, car showrooms and warehouses.

- CBDs usually show **zones of discard** (or retreat) which are in decline and **zones of assimilation** (or advance) where the CBD is expanding, replacing non-CBD land uses.

Changes in CBDs

CBDs have undergone considerable change since the late 1960s. As retailing expanded in the suburbs, it stagnated or declined in city centres. Some smaller CBDs have had great difficulty in halting the decline (such as Dudley near the Merry Hill RSC). Many CBDs have experienced major planning initiatives aimed at regeneration: for example, new indoor shopping malls, new office developments, improvements in public transport, pedestrianisation of main shopping streets, security improvements such as CCTV and widespread environmental improvements. Multipurpose developments on the fringe of some CBDs have aimed both to regenerate central areas and build up the reputation of the city as a whole (Figure 14.4). The development of urban tourism is an important element in such developments.

It is important to be familiar with a case study of a CBD.

1. Work through the list of characteristics above and apply them to your case study, noting examples of buildings and locations within the CBD.

2. Build up a simple map of zones within your case study CBD. Identify a core and frame and zones of discard and assimilation. Label influences on the pattern of zoning.

3. List the recent changes in your case study, considering whether they fit into a policy of regeneration. Does your CBD face particular competitive pressures from new retailing developments elsewhere or particular planning problems?

Offices

Over one-third of all jobs in the UK take place in offices. They are needed to provide a very wide range of services and to plan and manage organisations in

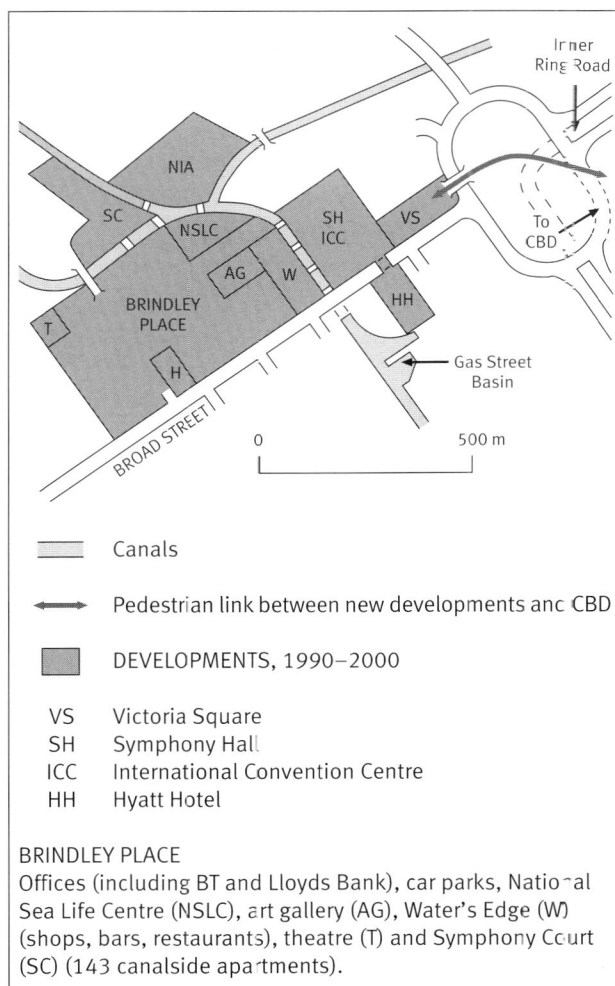

Figure 14.4 Recent developments on the western edge of Birmingham's Central Business District

Canals

Pedestrian link between new developments and CBD

DEVELOPMENTS, 1990–2000

VS   Victoria Square
SH   Symphony Hall
ICC International Convention Centre
HH   Hyatt Hotel

BRINDLEY PLACE
Offices (including BT and Lloyds Bank), car parks, National Sea Life Centre (NSLC), art gallery (AG), Water's Edge (W) (shops, bars, restaurants), theatre (T) and Symphony Court (SC) (143 canalside apartments).

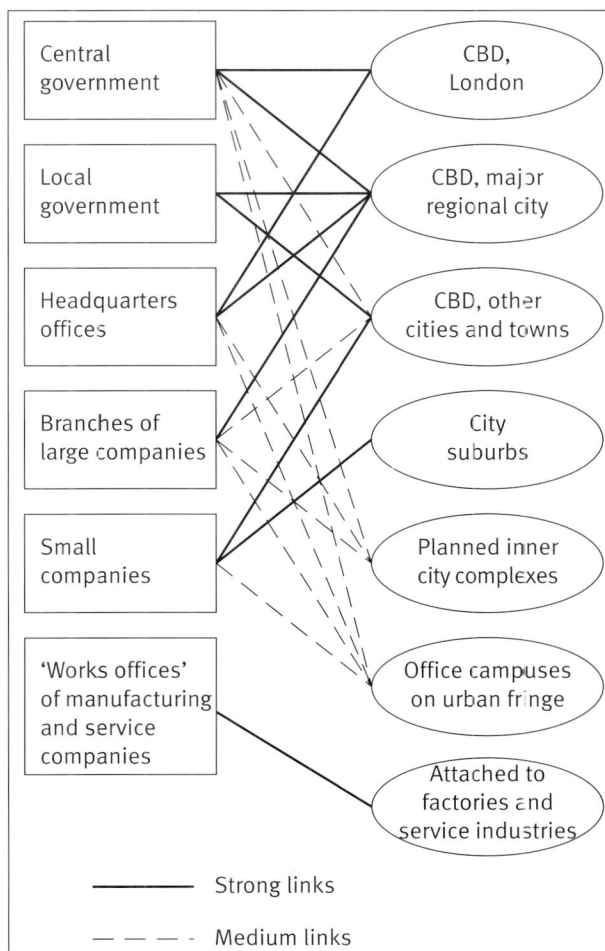

Figure 14.5 Types of office and their location

all sectors of the economy. Figure 14.5 shows the relationship between types of office and their location. Since the 1970s considerable office development has taken place in capital and other major world cities, including London. This reflects factors such as the increased internationalisation of business, the increasing size of many companies and the growth of service industries, especially financial services. Also, the introduction of computer technology has led to the need for new designs of offices.

Offices in the Central Business District

The CBD has traditionally been the most important location for offices. Reasons for this are: the **accessibility** of the CBD for staff and clients, **prestige** (for example, the City of London) and for **functional links** (regular and easy personal face-to-face contact is important for many businesses). Within large CBDs offices usually occupy a distinctive zone. Other offices are to be found above shops. Rents for office space vary considerably depending on the type of building

and location. Newer blocks with air conditioning, space for computer cabling and underground parking would have high rents. On the other hand, office users vary in their ability to pay rent. Thus headquarters offices and regional offices of large companies including banks and insurance companies will tend to locate on central, prestigious sites, while smaller organisations will locate further from the peak land value.

Office decentralisation

From the late 1970s onwards the attraction of CBDs for office activities has become less. Reasons for this include: rising CBD rents, increased traffic congestion in city centres, insufficient car parking spaces and the unsuitability of many old office buildings for modern business. In central London there were problems of staff recruitment and the need to pay higher salaries. The rise of modern electronic communications has enabled companies to have a greater choice when locating offices.

Decentralisation has taken place on two scales:

1 From central London to outer suburbs (especially along the M4 and to Croydon) and other towns in the South East where easy access to London remains (Figure 14.6). Many government offices have moved much further, to the North and to South Wales.

2 From CBDs to suburban and edge-of-city locations, including suburban district shopping centres, greenfield sides and business parks (Figure 14.7). Such locations have advantages of lower rents, space for car parking and future expansion, good road links, greater security and landscaped environments.

Major office developments have also taken place in inner city locations, usually as part of a major regeneration scheme. In east London, the London Docklands Development Corporation attracted developers to build a major concentration of offices centred on Canary Wharf. Other examples include Salford Quays (formerly part of Manchester docks) and the Waterlinks area of Birmingham Heartlands. Some inner city developments are, in effect, an expansion of the CBD into the surrounding 'zone in transition', for example, Brindley Place, Birmingham (Figure 14.4). Many of the inner city developments like Brindley Place and Salford Quays are part of multipurpose developments which include leisure facilities ranging from museums to restaurants, hotels and apartment housing.

Review task

You should also be familiar with case studies of changes in office location. Consider, in turn, examples of developments within the CBD, in the inner city (your knowledge of planning developments in an Urban Development Corporation may apply here), in suburbs and at the city edge.

Decentralisation of services

Besides retailing and offices, other services such as those in the leisure industries have decentralised. These include multiplex cinemas, fast food restaurants, sports and fitness centres and hotels. Many of these establishments are owned by national chains or multiples. They require large sites for buildings and car parking. Suburban sites close to affluent residential areas and with good road access are particularly favoured. Figure 14.7 shows the A34 axis in Solihull near Birmingham. Here a major radial road which joins the M42 and is close to large recent housing developments has attracted decentralised services (retailing, offices and leisure facilities) together with warehousing and Page 126 industry. Considerable pressure has therefore been placed on the urban–rural fringe (see Chapter 12).

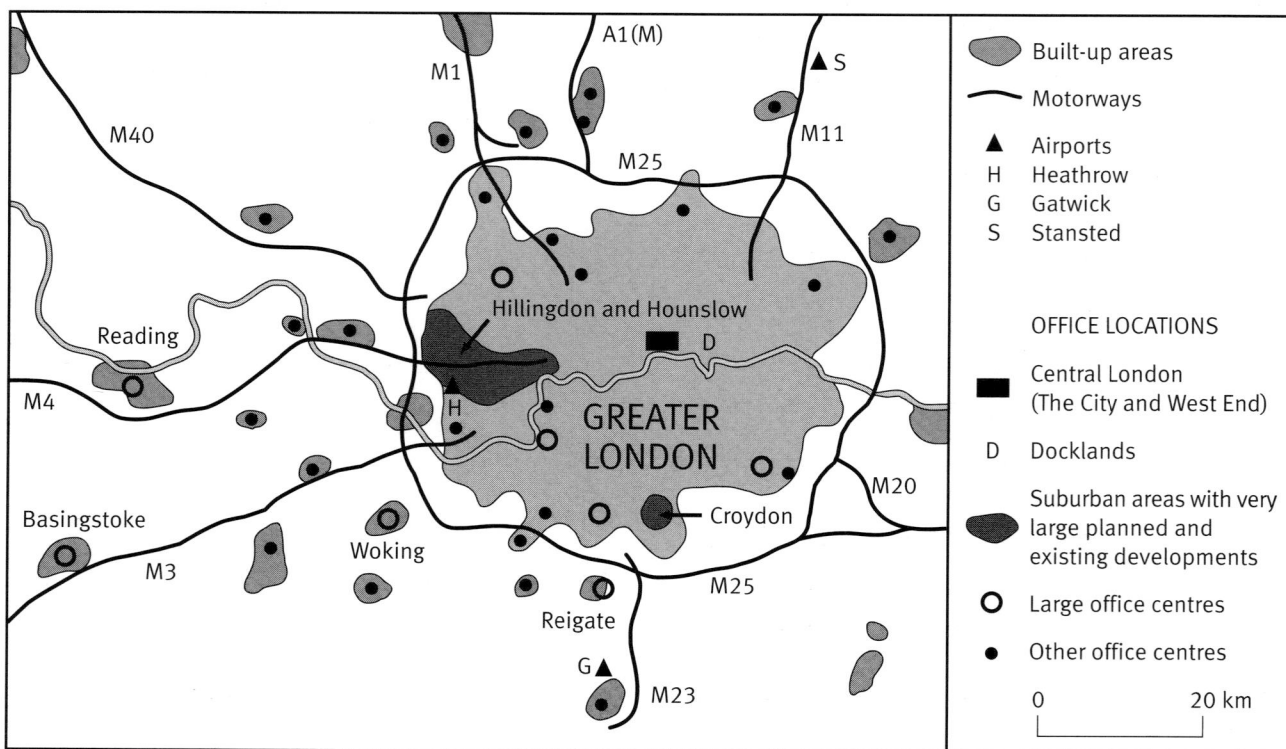

Figure 14.6 Office development in the London region

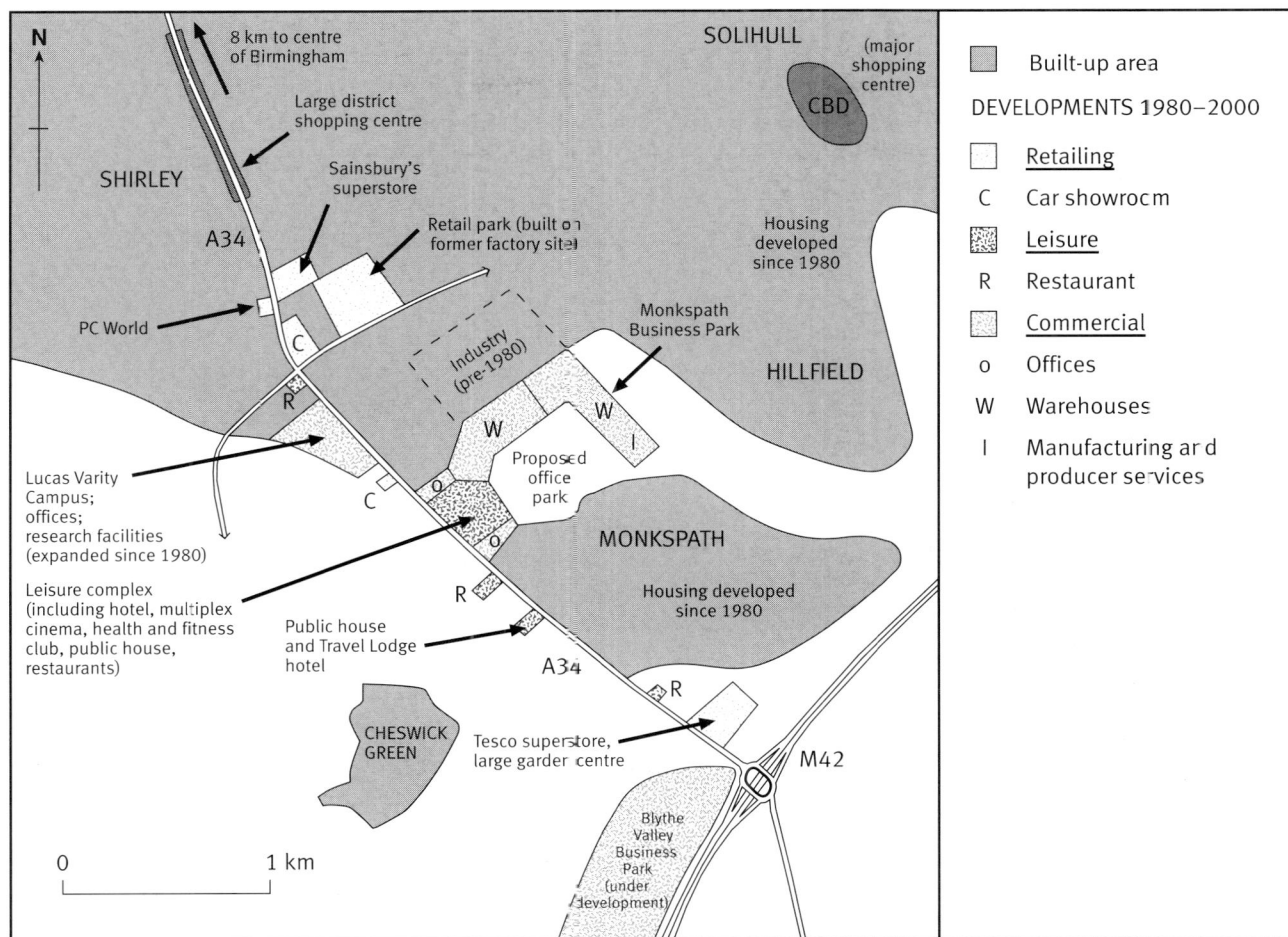

Figure 14.7 Decentralisation of services, Solihull

Commuter villages close to urban areas

- Increasing population enables shops to survive, but they face increasing competition from superstores and other edge-of-city retail developments. Commuters do most shopping outside the villages. Some specialist shops and services may open in large, prosperous commuter villages. Some may attract people from nearby urban areas, for example public houses and restaurants.

- Some service provision expands: doctor's surgery, primary school.

- Public transport may remain limited because of commuters' dependence on cars.

Villages in remoter areas

- Small, declining populations cannot support services such as post office/general store, doctor, primary school.

- People with cars, especially newcomers, rely on long-distance shopping trips to 'stock up'.

- Public transport provision declines or even disappears.

Villages in tourist areas

- Permanent population may not be large enough to support services such as doctor or primary school. Slight expansion of population and the tourist trade enables some services and shops to survive and even expand.

- Tourist shops replace shops providing for local needs.

- Prosperous, car-owning retired incomers and second home owners rely on urban areas for most of their food shopping and other services.

Figure 14.8 Changing service provision in rural areas

Access to services in urban areas

Within urban areas the decentralisation of retailing and consumer services can cause problems for peoples' access to services. Decentralisation of services has the following implications:

- Many inner city areas lose services and are seen as unattractive locations for many new retail developments – as their populations are declining and have low purchasing power. Retailing that remains may offer less choice and higher prices.

- People without access to cars are at a disadvantage as new suburban and edge-of-city retail developments are geared to car owners. (Consider which groups of people will be most affected.) Access to services may be poor for people living on some edge-of-city council estates. Local services may be limited, again reflecting low local purchasing power. New edge-of-city developments are likely to be in more prosperous districts, while public transport routes tend to radiate from city centres rather than link together peripheral locations.

Access to services in rural areas

To state that there is a general decline in service provision in rural areas in the UK is an oversimplification. Service provision has changed in various ways and the provision of certain key services has declined in many villages. People without cars are at even more disadvantage than their counterparts in cities. Figure 14.8 shows how services have changed since the 1970s in three types of village: commuter villages close to urban areas, small villages in remoter areas and villages in tourist areas. It is clear that the smallest villages have suffered most from service decline, losing key services such as post office/general store and primary school.

However, large villages may also lose key services. For example, many banks in rural areas have closed recently, partly because more banking operations are taking place by telephone or on the internet; small branches therefore become less commercially viable. Their closure can leave people without easy access to cash. Local businesses are affected both directly

(being unable to bank cash takings) and indirectly, because when people are forced to travel to the nearest town to visit a bank, they may well do their shopping there at the same time.

The most common planning response to the problem is the identification of **key villages**, usually larger villages, where services will be concentrated and neighbouring villages served. Such a policy can be effective with public services such as education and health but may only have an indirect impact on retailing (thus concentration of new housing development in a key village will help to support existing shops).

The services available to people living in small towns in rural areas may sometimes stagnate, but may also change significantly. Many small towns do not have the threshold population necessary to support the shops belonging to multiple retailers (for example, clothing stores). Although independent traders may survive, people do not have access to the wider variety of goods and lower prices available in larger urban areas. Therefore major shopping trips may be made to larger, but distant towns. As in larger villages, key services such as banks may close. When a major supermarket chain does locate in a small town the impact may be considerable: for example, the location of a Tesco store in the Cotswold town of Stow-on-the-Wold resulted in the closure of local grocers and butchers, so reducing the choice of shops available. The vacated shops were taken over by shops geared to the Cotswold tourist industry and antiques trade.

Review task

1   **Consider the decentralisation of services in your case study urban area. List the main examples of this process. Note the ways in which access to services has changed for people in the inner city and the suburbs.**

2   **Apply the changes to services in rural areas, listed in Figure 14.8, to examples that you have studied.**

PREVIEW

What you need to know:

- **Ways of categorising tourism and recreation**
- **The growth of tourism at national and global scales**
- **Resources needed for tourism**
- **Changing social, economic and environmental circumstances of tourism**
- **Pressures and changes affecting domestic tourism**
- **The growth and impact of international tourism**
- **The role of tourism in LEDCs**

Also, you will need to be able to:

- **assess the effects of political, social, economic and environmental factors on the growth, development and change of tourism;**
- **understand the impact of tourism at a range of scales from local to global.**

EXAM TIP

Tourism, recreation and leisure activities generally are important economically, environmentally, socially and culturally (Figure 15.1). Therefore, even if your course does not have a tourism module certain aspects of the topic are likely to occur elsewhere.

The topic covers activities which are widespread spatially from local to international scale, and in time from a few hours to many weeks, so the breadth of the topic means that you have to be prepared to apply your knowledge and understanding of its basic ideas and content to a great variety of situations.

Definitions

Recreation and tourism have different definitions but they overlap and there is no consistent separation of the two. Day trips, in fact, make up a significant element in the economic value of tourism.

- **Recreation** covers leisure-time activities people take part in outside the home.
- **Tourism** requires travel from the home and is generally regarded as needing at least one overnight stop.

- Travel and temporary stay away from home, experiencing new places and situations without the need to work or follow normal routines.
- Tourists buy products or services from tourism businesses directly or as a package through travel agents and tour operators.
- Tourism services provide information and training, guiding, brochures.
- The tourist industry is made up of a small number of very large companies and a large and varied number of medium and small businesses.
- Tourism is continually changing as new destinations and activities become popular.

Figure 15.1 The nature of tourism

An alternative distinction is one made between **mass tourism**, in which there is little differentiation between holidays of the same type, and **alternative tourism** which is more flexible and more individually designed.

Within and between both sets of categories the boundaries are rather blurred with considerable overlap. Figure 15.2 summarises types of recreation and tourism.

Growth of tourism and recreation

The tourism industry worldwide has grown enormously in the second half of the 20th century with 25 million international holidays in 1950 rising to about **500 million in 1995**. On top of these numbers are the domestic tourists who, in the UK alone, accounted for £12 430 million spending by residents who stayed overnight, with a further £9500 million spent on day trips.

The basic distinction is between domestic (or intra-national) and international.

Domestic
- urban-based leisure, recreation and tourism
- seaside resort-based
- countryside tourism and leisure
- historical and cultural interest

International
- resort-based sea and sun
- weekend city trips
- activity and exploration
- long haul
- niche tourism

Figure 15.2 Types of recreation and tourism

Climate
Coasts and sea
Landscapes
History
Culture
Purpose-built attractions

Figure 15.3 Resources for tourism

In numbers and in money terms the industry is dominated by the MEDCs, but the whole world is involved. There is hardly a country that is not a tourist destination and those that are sources of tourists are also destinations for others.

Journeys related to tourism and recreation are a response to the tourism resources available (Figure 15.3), but the **growth of tourism and recreation** has a wide range of reasons:

- increased disposable income

- increase in available leisure time

- cheaper, quicker travel by air

- inclusive package holidays

- growth of large-scale tour companies.

Mass tourism in Europe started in the 1950s with the Mediterranean being the focus. The USA is so large that there the same process was a domestic one with states like Florida being the draw to tourists. As destinations have spread so tourism has become a global industry, although many newer destinations are still 'mass produced' standardised destinations of 'sun, sea and entertainment'.

International tourists still come mainly from western Europe, USA and Japan and the continued demand for new and different locations has fuelled the spread of the industry. This has, at the same time, increased the competition between destinations to attract unpredictable numbers. Two sets of events in recent years illustrate the vulnerability of tourism-dependent countries. The first was the effect of attacks on tourists in Egypt which led to a collapse in numbers visiting the country, and the second was the Kosovo crisis in the Balkans which led to mass cancellations of holidays throughout the region.

The product cycle and tourism

Change does not always mean growth and, although the tourist industry as a whole continues to grow, individual resorts, districts, countries and even regions undergo decline. The first tourist areas that developed to meet the demands of mass tourism, like the resort regions of the Mediterranean, suffered as a result of being first. The industry is partly a 'fashion' industry and image is important. Tourists want new experiences and new 'exotic' destinations, even if the actual holiday is of the same type. Other factors, such as the relative decline in the cost of long-haul flights have made resorts in South-East Asia or Florida competitive for European tourists. The result of this decline in popularity, whether of British seaside resorts or Mediterranean resorts, led to action to refurbish and modernise and to build new attractions and conference centres to attract a different clientele. Most importantly, given the nature of the business, emphasis has been given to the creation of a new marketing image. Figure 15.4 summarises this process.

Review task

List some reasons why dependence on tourism can be a risky business. Use examples you have studied.

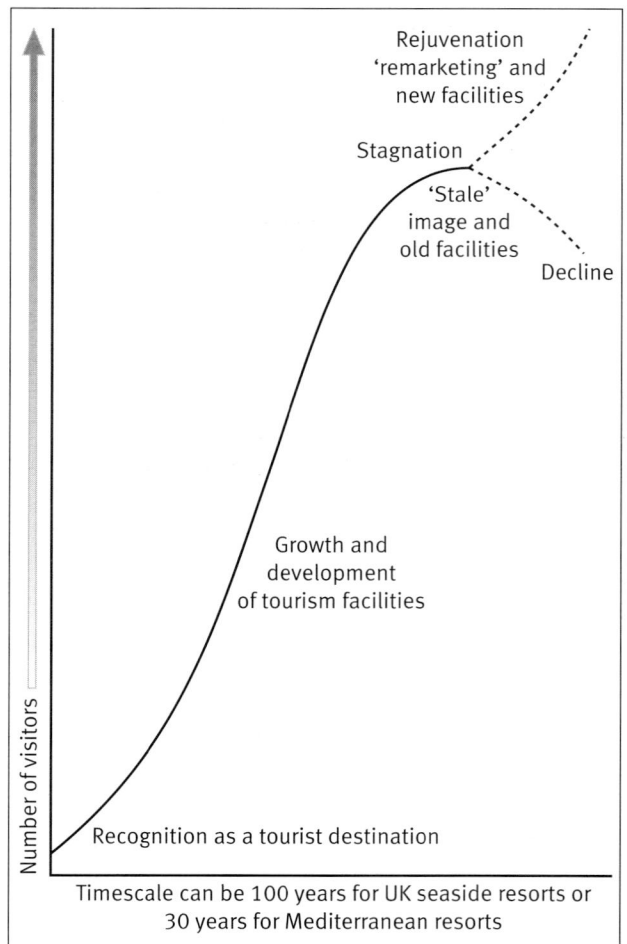

Figure 15.4 Tourist industry product cycle

Scale, nature and pattern of tourism

The **scale** of tourism and recreation is growing as we have already seen in the last section. At the global level there are great differences in tourist activity between MEDCs and LEDCs in that the former have well-developed domestic and international tourist industries whereas the latter have, in general, very little in the way of domestic tourism even where international tourism is a vital part of the economy. This emphasises the relationship between tourism and affluence.

Domestic tourism

Domestic tourism is much more significant than is generally realised. Although an enormous amount of attention is given to international tourism, domestic tourists in MEDCs account for greater numbers and greater spending than international tourists. Spending is greater than would be expected because day visitors are included.

Domestic tourism in the UK went through a number of phases (Figure 15.5), and in many ways they can be matched against the experiences of other developed countries. In addition, the phases help in understanding the lack of domestic tourism in many LEDCs.

The **nature** of tourism is continually changing. In the 1960s the standard holiday in the UK was one or two weeks away from home probably at a seaside resort, often one that was easily accessible by rail. This main holiday might have been augmented by day trips at bank holidays or weekends. The decline of the seaside resort as a primary destination is linked to changes in the number and length of holidays as much as the growth of new holiday destinations, particularly overseas.

Review task

Summarise the pattern of holidays that has replaced the annual one or two weeks at a seaside resort.

Increased mobility has meant that day trips can take in activities and places previously only possible as longer stays. This has meant that:

- overnight stays are not necessary for many purposes;
- day trips add significantly to the economic value of domestic tourism;
- short breaks to relatively distant places are more feasible;
- day trips and short breaks and even extended holidays overlap.

Classifying domestic tourism by length of stay is only one approach. Another is by type of destination:

- attractions, such as, theme parks, stately homes
- seaside
- countryside
- cities.

Before 1750	Many local leisure activities for the bulk of the agricultural population but only wealthy people travelled. Spa towns developed all this time.
1750–1830	During the period of the Industrial Revolution the bulk of the working population had less leisure time as working hours were long and there were few days off. Better-off people did begin taking 'trips' to the seaside.
1830–1900	Domestic tourism began with the introduction of shorter working weeks and annual works holidays and, as incomes rose, travel became possible with the spread of the railways. Day trips started the growth and then holiday weeks became popular as seaside resorts grew. Better-off people started travelling to the continent including Switzerland.
1900–1950	Two world wars and an economic depression interrupted the steady growth of domestic tourism. By 1945, 80 per cent of workers had paid holidays and many headed for seaside resorts by rail, and seaside resorts reached their peak. International tourism also began to spread.
1950 to present	Increased affluence, greater car ownership, more leisure time saw change and growth. Domestically, seaside resort holidays declined, camping and caravanning grew, shorter off-season holidays became popular and more varied types of holidays developed. Internationally, the general population were able to travel first to the Continent, particularly Spain and later to long-haul destinations. These became the main holidays and domestic holidays were second or third holidays and short breaks. Winter sports holidays became a major part of mass tourism.

Figure 15.5 Phases of domestic tourism in the UK

Review task

If possible, do this as a group task with some friends; more ideas are likely to come that way. Extend each category above into different types of area or different activities possible. What do you notice about the problems with any such classification?

Changes in the scale and nature of domestic tourism have led to changes in the **pattern**. Domestic tourism is now more complex with the effect that:

- Former peaks of business have declined especially in seaside resorts.

- The tourist season is now longer, being made up of more short breaks taken throughout the year.

- The business is spread across different areas and activities as tastes and interests change.

- There is increased competition between areas to attract tourists.

- Business tourism involving meetings and extended conferences has developed strongly with many seaside hotels and major city hotels concentrating heavily on this aspect of tourism.

For the future, much depends on changes in the population structure as well as changes in the approach people take to holidays. Numbers of elderly and early retired people have provided a market for tourism available all year where people are reasonably well off. This age group does not necessarily want sedentary holidays and are as likely to demand varied holidays as any other age group.

The points made about domestic tourism in the UK apply in a general sense to all developed countries. The main difference is probably the point made elsewhere that in very large countries domestic tourism actually involves distances that would demand international travel in much of Europe. The USA is a prime example of this.

International tourism

Tourism has become the world's largest industry, providing 7 per cent of the world's employment and accounting for 12 per cent of consumer spending. For many individual countries it makes a significant contribution to the overall economy and is the largest single employer. This latter fact is often hidden by the fragmented and diverse nature of the industry itself.

International tourism, or global tourism as it is often called, began with the growth of cheap air transport. Remember, when looking at statistics for international travel, that the USA is under-represented because of its size and that Europe is over-represented, given that individual countries may well be smaller than many states in the USA.

The demand for tourism comes from the MEDCs; these are the countries with people who have the disposable income and the available time to travel to other countries for their holidays. Other factors are population size, and for large countries particularly, tourism variety on the domestic scene.

The major destinations for tourists are near the major sources of tourist flows (Figure 15.6). For example, the majority of visitors to the Mediterranean tourist areas are from northern Europe.

Other factors besides distance are:

- attractions, whether natural or constructed

- communications, both to and within the destination region

- relative costs and standards of living

- cultural links

- political factors.

Consequences and impacts of tourism

The consequences and impacts of tourism are economic, social, cultural and environmental. In the case of each there are gains and losses, so there are gainers and losers. Figure 15.7 summarises these gains and losses. Note that some apply primarily to MEDCs, some primarily to LEDCs and some to both groups. Notice also that the gains and losses apply at different scales, in that many points can be valid at village level or at national level.

Problems, conflicts and management

The issues involved in this section are:

1. Problems, which arise as a result of tourism's success, as well as from failure, or decline.

2. Conflicts, which occur as result of clashes between different activities, such as agriculture and tourism, or between groups with different values or interests, such as conservationists and tourists.

3. Management, which is concerned with the prevention and resolution of problems and conflicts.

Figure 15.8 summarises some of the problems and conflicts typical of tourism in the different types of destination in MEDCs.

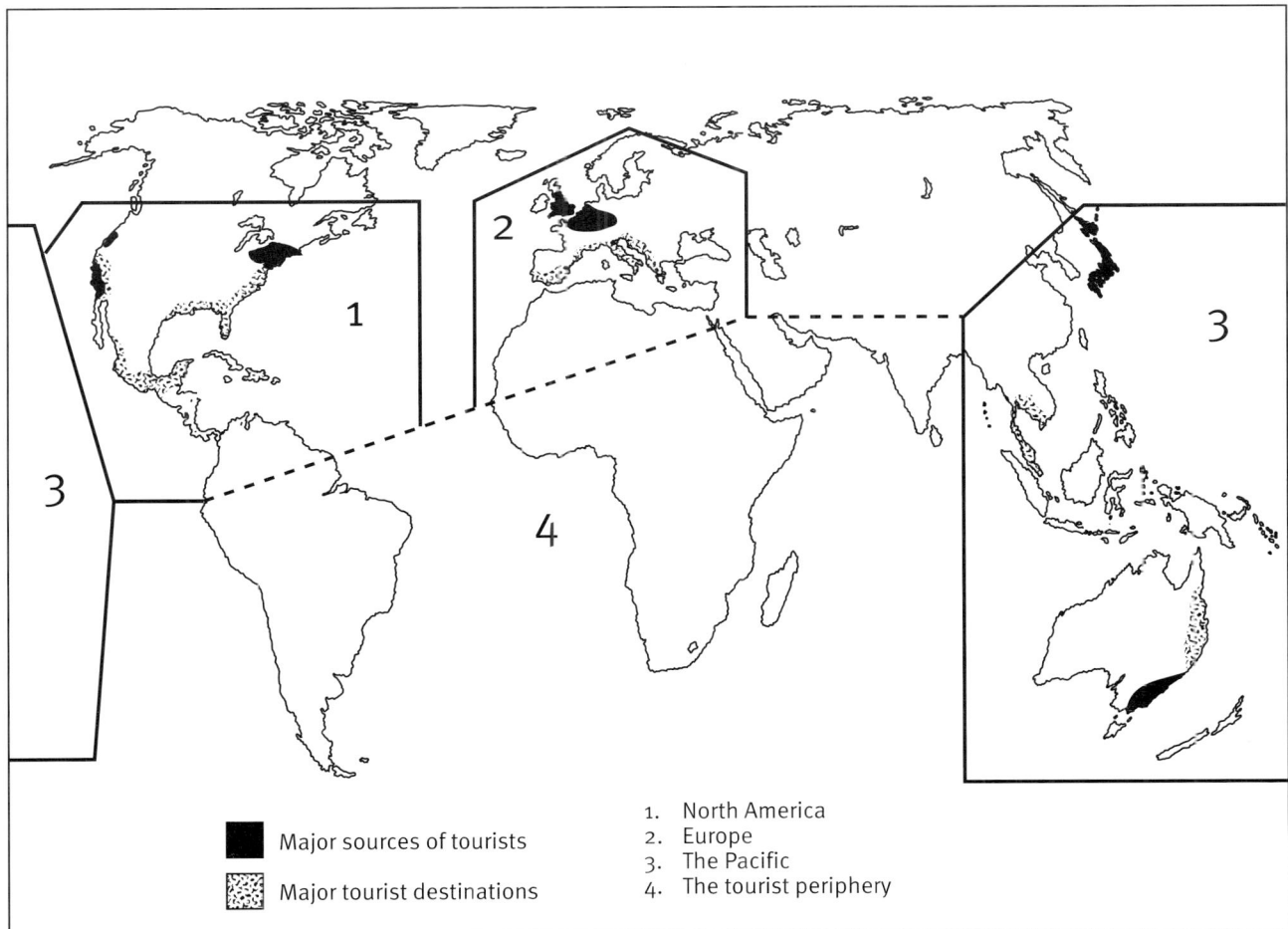

Major sources of tourists

Major tourist destinations

1. North America
2. Europe
3. The Pacific
4. The tourist periphery

Figure 15.6 World tourist regions

Management of tourism-related issues is closely linked to land use matters. Major points, and these are particularly relevant to the UK, involve:

Towns and cities

- The role of tourism in the regeneration of declining urban areas, by finding new uses for old industrial and commercial buildings, maintaining the character of converted buildings.

- Provision of tourist facilities in 'international' cities.

- In tourist cities there is generally a core tourist area, usually the historical core, which may suffer from the development of unsympathetic tourist attractions. Benefits for residents may not be particularly apparent.

- Local job opportunities associated with tourist developments have a multiplier effect in the local economy, but ownership of the developments, if not local, may take money out of the local area.

Seaside resorts

- Regeneration of old seaside resorts has involved broad economic as well as land use planning. Diversification, with light industry estates, office development and conference centres, has been possible for some, especially those in the south of the country.

- Refurbishment using local financial resources as well as external aid from government or the EU where available, together with 're-imaging' themselves as a touring centre, is an approach taken by many resorts.

Countryside

- In the countryside tourism has a very uneven pattern creating great environmental pressures in 'honeypots'.

- The development of tourism in the countryside is seen as an economic benefit and, in many cases, as an environmental benefit with tree planting, nature reserves and country parks counter-acting urban sprawl.

Economic	
Gains	**Losses**
Increased income, nationally and locally	Government expenditure diverted from other uses
Increased foreign earnings	Foreign earnings reduced by outflow
Increased employment opportunities	Pressures on inadequate infrastructure
Generates other economic activities	Labour attracted from other important sectors and regions
Infrastructure developments support other activities	limited skilled work for locals

Social	
Gains	**Losses**
Develops international understanding and improves attitudes towards different nations and ethnic groups	Creates antagonism between poor locals and rich visitors
Stimulates education and training	Increases emphasis on material possessions
Improves the role of women	Increase in prostitution and crime
	Cultural differences puts strains on family structures

Cultural	
Gains	**Losses**
Fosters local culture	Turns local culture into a commodity
Introduces new ideas and understanding	Local traditional values put under pressure
	Cultural differences puts strains on family structures

Environmental	
Gains	**Losses**
Enhances value and therefore protection of built and natural environment	Lack of integration of new building with the area
Possibility of infrastructure improvements enhancing environmental health for locals	Damage to local environment and habitats
	Range of environmental problems caused by pressure of tourist numbers

Figure 15.7 Gains and losses with tourism

	Problems	Conflicts
Towns and cities	Traffic congestion Day trippers not overnighters Season Tourist services	Tourist pressures cause antagonism from locals Conversion of buildings to tourist uses
Seaside resorts	Decline in popularity and seasonality Competition from the Mediterranean Wider horizons Alternative attractions Changing holiday patterns Lack of development resources	Use of resources for locals or visitors
Attractions	Highly localised Traffic quantities	Local residents, especially in rural areas
Countryside	Damage and disturbance Development pressures Competition for resources Congestion	Residents and tourists Conservation and tourist activities Local customs/traditions and tourist impact

Figure 15.8 Problems and conflicts

- Management of these issues occurs at different levels, from national policies down to local management at the scale of nature reserves or country parks. At all scales it has several aims: to increase opportunities for leisure and recreation; to maintain and enhance the quality of the environment; to improve the level of rural economies in a sustainable way.

National Parks

Tourism is major source of pressure on human and natural environments. In cities, tourist numbers are as great a problem as in rural areas, although it is in the latter that the impact often appears to be greatest. This is especially so in more densely populated countries where there is great deal of competition for land. National Parks illustrate the issue.

In England and Wales they were established to conserve landscapes of great beauty and to improve access to them for the largely urban population. Immediately the issue arises that the National Parks are not nationally owned. Most of the land is privately owned and is used for a variety of activities. Tourism and recreation adds further competition. The key element of planning in national parks is conflict: between different land uses and occupations, between conservation and development, and between residents and visitors.

England and Wales does not meet the international definition for National Parks which requires conservation of landscape but, besides limited and controlled recreational use, does not allow for any other activities. This is feasible in countries with large amounts of space, but, even in such places, National Parks established adjacent to large concentrations of populations face land use pressures from a variety of sources: development for housing or industry, degradation of the landscape by increased visitor pressure, pollution of air and water from nearby industry, and pressure on land for agricultural use or exploitation of natural resources.

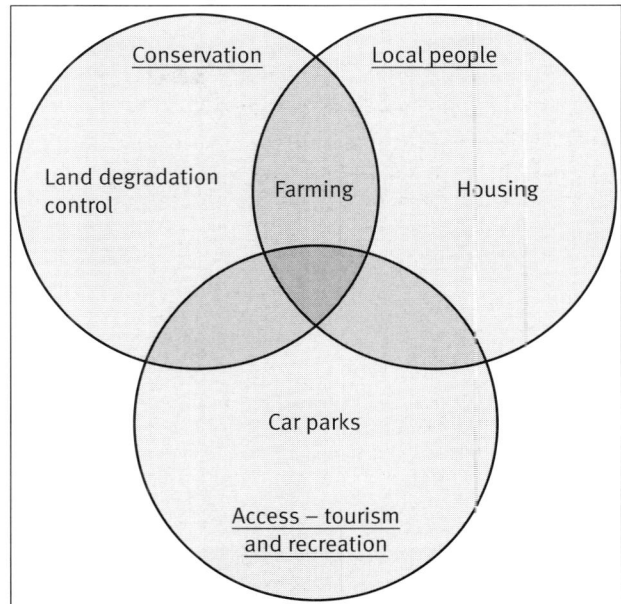

Figure 15.9 National Parks

Tourism in LEDCs

This is usually studied separately for four reasons:

1 Tourists visiting LEDCs are mostly from the MEDCs, and internal or domestic tourism is relatively limited, although this is changing.

2 Tourism tends to be much more localised and in some places tourist areas are almost prohibited to locals except for workers.

3 Tourism is a relatively late development in most countries.

4 There is often little local control over the industry with much control being held by TNCs.

> **Review task**
>
> Make a large copy of Figure 15.9 and refer to examples you have studied in your course to add points to each of the circles. Where points overlap between the major themes place them in the overlap areas. This is an ideal way to review a case study of one National Park or similar type of area.

> **Review task**
>
> 1 Since this aspect of tourism covers an enormous amount of the world the general points that follow need to be understood in the context of case studies. Pencil in references to the case studies you have studied in your course. In the same way, on your own notes, pencil in the general points from this section which apply.
>
> 2 Look back at earlier sections of this topic and identify the factors which have stimulated demand for holidays in LEDCs. Remember to give examples!

The attraction of LEDCs

A simple list of the attractions of LEDCs could be made from travel advertisements. They could include:

- 'tropical paradises of golden beaches, palm trees, sparkling blue waters and sunshine'
- exotic environments and cultures
- ancient civilisations
- 'unusual' or 'exotic' wildlife

As many countries are promoting similar experiences in direct competition to each other, they are vulnerable to variations in numbers of visitors for any of a variety of reasons beyond their control. This becomes less as distinctive images of the country are created in the tourists' minds.

> **Review task**
>
> **Think of as many LEDCs as you can which fit any or all of the features listed above. Which of the features are less and which are more susceptible to the effects of competition from other areas, or changes in fashion? List some of the 'reasons beyond their control'.**

Benefits from tourism

Tourism has come to be regarded as an important route to development with the potential to provide:

- A source of income and foreign exchange
- Improvement to the infrastructure
- Overseas investment
- Job creation
- Demand for local goods and services
- Encouragement of entrepreneurial activity
- Regional development, if tourism can be directed to less-developed areas

Tourism is different from other economic activities in that the customer comes *to* the country rather than a product being exported *from* the country. The list of potential benefits is wide ranging as so many of the multiplier effects could benefit the host country.

> **Review task**
>
> **Consider each item on Figure 15.10. Each one could have a positive or a negative effect on the economic impact of tourism. Jot down possible effects. It would be useful to do this on a large copy of the diagram; you could then add other points covering social, cultural and environmental issues as you go through the next few sections. Don't forget your case studies!**

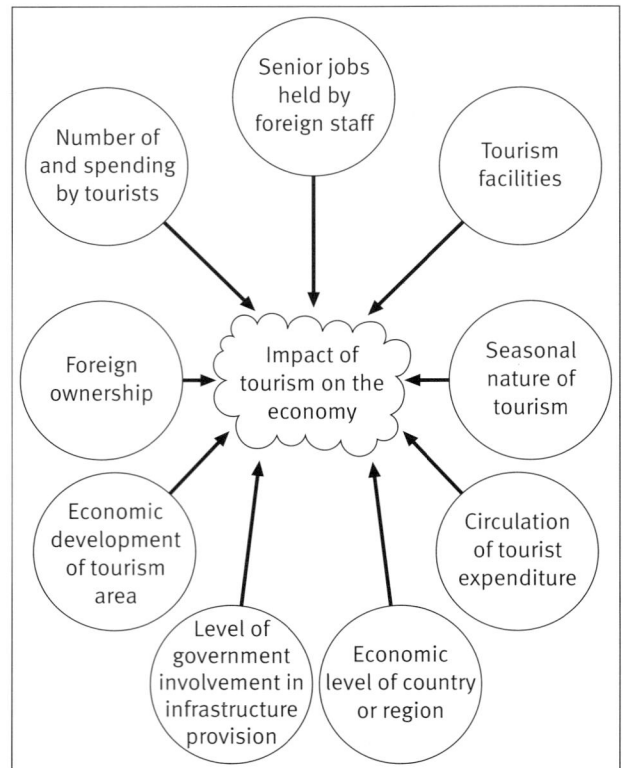

Figure 15.10 The economic impact of tourism

Costs of tourism

On the negative side, tourism has economic, social and cultural and environmental costs, some being more obvious than others. These result in a loss to the host country.

Economic costs

Direct costs include:

- the promotion of tourism;
- the provision of infrastructure including airports, hotels, transport facilities and basic services like water supply and electricity.

Indirect costs include:

- long-haul air travel takes up much of the cost to the tourist and the airline is almost always based in an MEDC;
- most hotels are owned by transnational companies and profits are returned to their home countries;
- tourists demand foods and drinks familiar to them in preference to local food and drinks and these have to be imported and form a drain on foreign currency;
- problems of dependence on a single 'product';
- difficulties associated with a 'seasonal' activity.

Social and cultural costs

Local cultures and traditions are affected favourably and unfavourably by:

- generating work for local musicians and craftsmen;
- turning their work into a standardised product to suit the international market.

The 'demonstration effect' of affluence displayed by tourists causes social and cultural conflicts, involving dress standards and behaviour.

Environmental costs

Wide-ranging environmental effects have been noted where:

- resort development has been located in and damaged sensitive environments;
- productive agricultural use has been replaced by tourism development.

Long-term sustainability can be achieved only if the natural landscape is treated as a resource and managed effectively. Planning for 'multiple use' is seen as an important way forwards for places with limited resources.

Conservation generally receives a low priority unless there is a direct link to tourist demands. Some countries have used the 'carrying capacity' concept to set limits to tourist numbers.

Review task

Produce a diagram similar to Figure 15.10 for social and cultural effects and for environmental effects.

Tourist patterns

Tourist movement can be measured in numbers of arrivals but it is much more significant to measure the income generated.

On the basis of income, Asia and Mexico and the Caribbean are the major beneficiaries. Africa, and the rest of Latin America and the Pacific islands are far less significant.

Based on numbers of arrivals, the leading destinations include Mexico, China, Malaysia, Hong Kong, Thailand and Singapore. However, many places with small numbers of tourists depend greatly on tourists and this applies especially to island economies like Barbados or the Seychelles. Such places are particularly vulnerable to changes in demand and competition from other destinations, as well as natural disasters and political unrest.

Review task

1 Make an outline summary of the chapter with no more than 12 points. Select case studies from your course notes to illustrate each point – you can use the same case study as many times as you like. Present the summary as a table, as summary notes, or best of all, as a Mind Map.

2 Refer back to 'What you need to know' at the start of the chapter. For each part jot down the main points you would expect to write about in an answer to an examination question.

What you need to know:

- **The meanings of development and the ways of measuring it**
- **Global patterns of development and the terminology used to describe them**
- **Spatial inequalities and differential rates of development at global, regional and national levels**
- **Models of development**
- **Obstacles to development**
- **The significance of interdependence and sustainability**

Also, you will need to be able to:

- **use a variety of types of data;**
- **apply your understanding to case studies.**

Defining and measuring development

At one time, development was taken to mean only economic development so that modernisation would be achieved by rapid economic growth. In the late 1960s and early 1970s, this definition was extended to include not just economic growth but also adequate food and jobs, reduced income inequality and self-reliance. Another definition identified three key values of life sustenance, self-esteem and freedom as being at the core of development. In real terms this meant a greater emphasis on basic needs, ensuring greater equality of access for all people.

Changing definitions of development have meant changing **indicators of development**. It was, and still is to a certain extent, common to use **gross national product per person (GNP per person)** and **gross domestic product per person (GDP per person)**, to compare countries and measure the level of development. These total measures of wealth give little indication of the actual standards of living of the majority of people. The World Bank classifies countries by GNP but recognises that 'classification by income does not necessarily reflect development status'.

A range of individual indicators can be used, and in combination they enable comparisons to be made which highlight differences even where GNP per person is the same (Figure 16.1). However, 'it is impossible to come up with a comprehensive measure – or even a comprehensive set of indicators – because many vital dimensions of human development are non-quantifiable' (UN). Nevertheless, a range of composite measures have been introduced (Figure 16.2). These include the Physical Quality of Life Index, the Index of Sustainable Economic Welfare, the Human Poverty Index and the **Human Development Index (HDI)**.

The HDI was developed by the United Nations and extends the definition of development from a narrow focus on income to one which measured the extent to which people can live 'long, healthy and creative lives'. The Index, therefore, was based on life expectancy, adult literacy and income. Since its introduction in 1990 it has been modified to take account of criticisms but is now recognised as a valuable tool, in that it draws attention to important issues quite effectively. The HDI measures overall progress in a country in achieving human development, based on national-scale data so it does not show the differences between rural and urban areas, between different regions or different genders. However, a number of individual countries have produced HDI reports as part of their own development planning.

As you would expect there is a strong relationship between GNP and HDI rankings, but it is far from perfect (see Figure 16.3).

Changing terminology

The terms used to describe different parts of the world in development terms has changed over the years, reflecting changes in attitude to, and understanding of the nature of development. 'Poor' and 'rich' were

Country	GNP per capita ($)	Life expectancy	Energy per capita (kg oil equivalent)	Adult literacy (%)	Infant mortality (‰)
UK	21 400	76.8	3574	99	7
Egypt	1290	64.8	638	51	51
Mali	250	47	–	31	118

Figure 16.1 Selected measures of development

Gross National Product per person (GNP per person)	Total value of goods and services produced within a country with income received from other countries and less similar payments made to other countries, divided by the number of people.
Gross Domestic Product per person (GDP per person)	Total value of goods and services produced within a year, divided by the number of people.
Physical Quality of Life Index (PQLI)	Comparative measurement of a nation's general well-being, based on three social indicators: life expectancy at age 1, infant mortality and adult literacy rates.
Index of Sustainable Economic Welfare (ISEW)	Takes account of income distribution, environmental destruction and natural resource degradation. Use limited by lack of data.
Human Poverty Index (HPI)	It is an index based on three main areas of deprivation: survival, knowledge and standard of living. The higher the index the greater the poverty in that country.
Human Development Index (HDI)	The index is an aggregate of adult literacy and years of schooling, life expectancy and GDP per person.

Figure 16.2 Alternative indicators of development

Review task

Look at Figure 16.3. List some points about the differences between the GNP and HDI figures for different countries. Give reasons why GNP rankings and HDI rankings differ. Consider unequal distribution of wealth as a starter.

	HDI		GNP per capita	
Countries	**Rank**	**HDI**	**Rank**	**$**
Canada	1	0.96	26	20 020
United Kingdom	14	0.932	22	21 400
Egypt	112	0.612	127	1 290
Zimbabwe	146	0.507	180	610
India	139	0.451	165	430
Uganda	160	0.34	185	320
Mali	171	0.236	194	250

Figure 16.3 GNP and HDI for selected countries (1998)

everyday terms that summed up the approach when development was narrowly defined in economic terms. Nowadays it is unusual for the word 'economic' to be included, although we use the terms More Economically Developed Countries (MEDCs) and Less Economically Developed Countries (LEDCs) to match those used in the course specifications. Figure 16.5 summarises terminology.

Patterns of development

Despite what has been said above, most maps showing the pattern of development use GNP per capita and divide the world into two zones as on Figure 16.4. Quite clearly, even on income grounds some countries are in the wrong place, notably the Newly Industrialised Countries (NICs).

Review task

Brainstorm a list of reasons suggesting why GNP alone is inadequate for the purpose of describing levels of development and, in particular, quality of life.

Attempting the last review task brought you up against one point that is often overlooked: no matter how they are measured, there is no sudden gulf between countries that are developed and ones that are not. If you rank all countries you see a **development continuum**, which is why it was difficult to decide where to draw the line. However, it is still common to refer to the **development gap** as a way of summing up the wide difference between the most developed and the least developed countries, however they are measured. Remember, though, that there is a large number of countries in the middle, and it was from this group that the NICs have come.

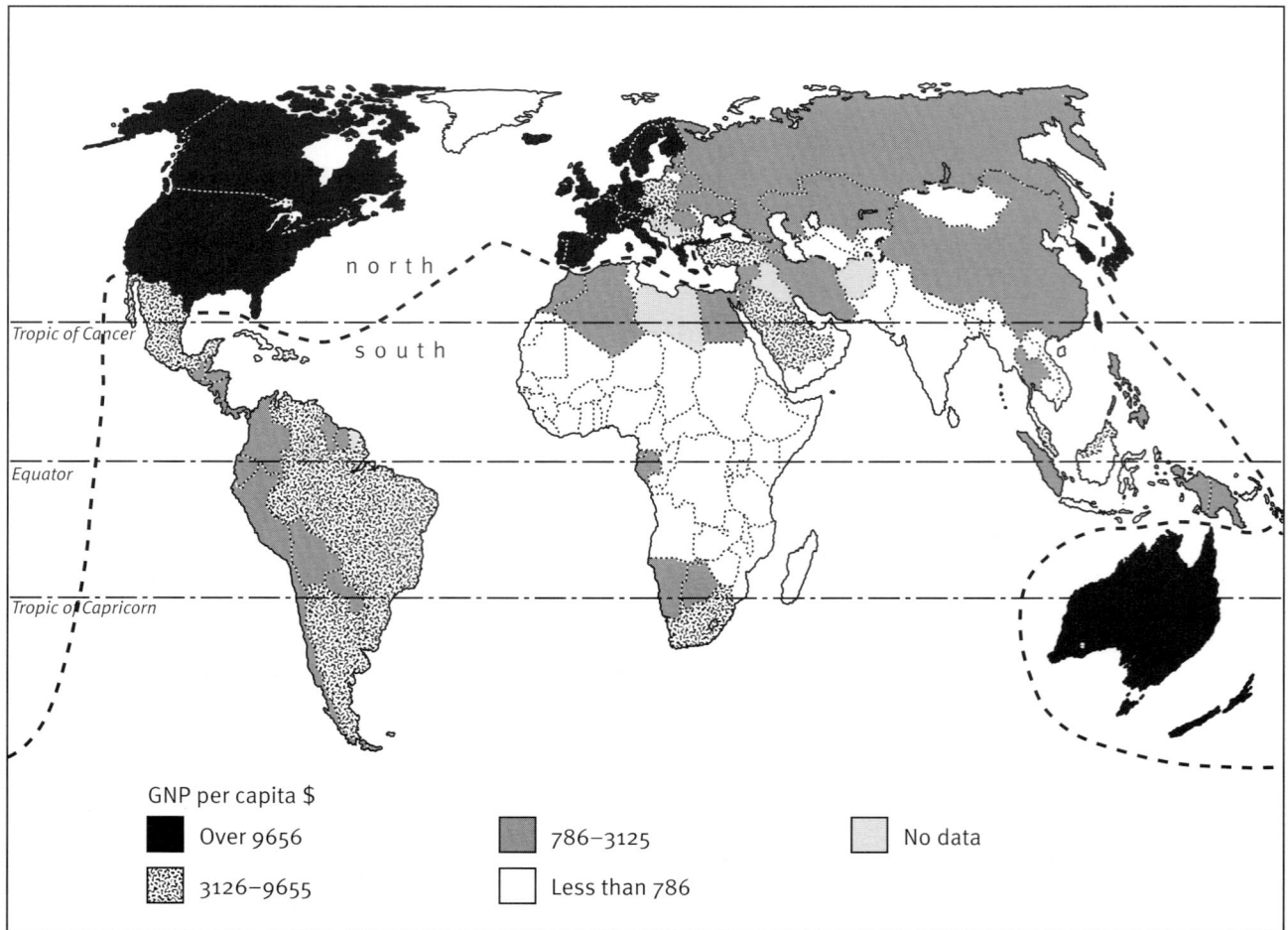

Figure 16.4 The GNP world (1997)

GNP per capita $
- ■ Over 9656
- ▨ 3126–9655
- ▨ 786–3125
- □ Less than 786
- ▨ No data

Review task

As a rough generalisation it has been usual to refer to Africa, Asia and Latin America as the South or the LEDCs, with Japan as the exception, as shown by the boundary line on Figure 16.4. From what you know about NICs, which other countries should be placed in the North with the MEDCs?

Developed	Undeveloped Underdeveloped Developing Dependent	
First and Second World	Third World	
More developed	Less developed	Least developed
North	South	
High income	Middle income (upper and lower)	Low income

Figure 16.5 World development's changing terminology

Models of development

Approaches to development have changed over several decades, and a variety of different models has been developed either to explain differences in levels of development or to show how changes take place.

Modernisation models

These four models followed the assumption that the Western model of development is the ideal and is possible everywhere. They represent the view that national economic planning, together with foreign aid and direct foreign investment would transform traditional societies and that **industrialisation** was a central feature of development.

Rostow's stages of economic development

This was the dominant theory in the 1960s and was based on the assumption that all countries passed through the same linear stages of economic growth. The vital phase was the take-off, when a country began to experience self-sustaining growth (Figure 16.6). The claims for the model were:

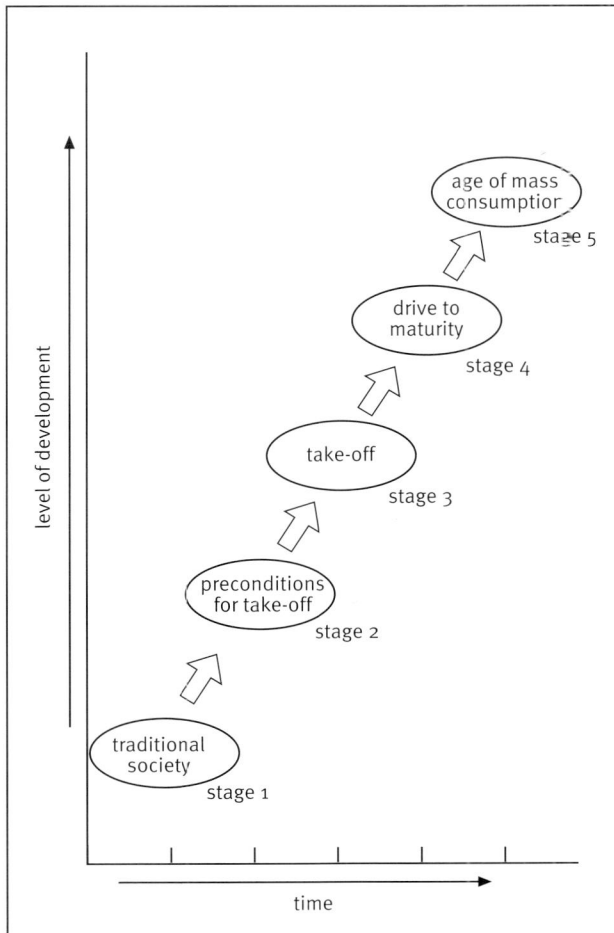

Figure 16.6 Rostow's stages of economic development

- it explained the advantages of developed countries
- it gave all countries an equal chance
- it showed a straightforward line of progress
- it identified economic growth with the Western capitalist system.

The model was criticised for:

- its evolutionary approach with fixed stages for countries to pass through
- having only one route with no alternatives
- ignoring all external influences on a country
- assuming that LEDCs must follow the same path as MEDCs.

The vicious circle of poverty

This was a general and simplified explanation of the interaction of negative effects preventing countries from achieving industrial take-off and ensuring the continuance of low levels of development (Figure 16.7).

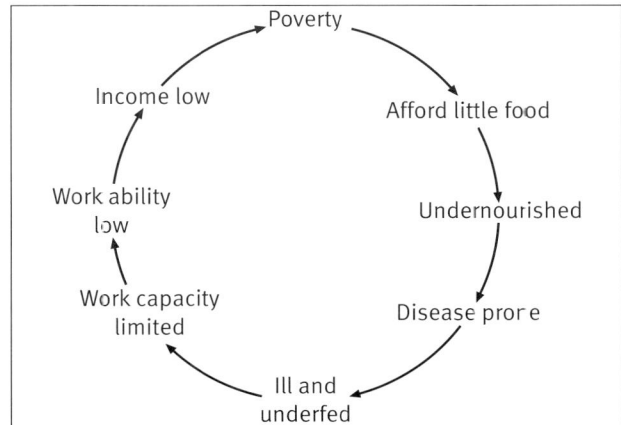

Figure 16.7 Vicious cycle of poverty

Circular causation of cumulative development

This was developed by Myrdal (see Chapter 11) and envisaged the interruption of the circle of poverty by some impetus, such as the building of a factory, the provision of an irrigation dam or new health facilities or, on a larger scale, the development of national transport services. At whatever scale, the **diffusion** of these benefits would produce **spread effects** or **trickle-down effects**. These ideas were adopted in MEDCs as well as in LEDCs and there the focus was on urban **growth poles**.

Page 135

Core-periphery model

This model, which was developed by J Friedmann, envisaged a core region of economic development from which development spread to the rest of the country. Figure 16.8 summarises the Friedmann model of development regions. Note that it includes

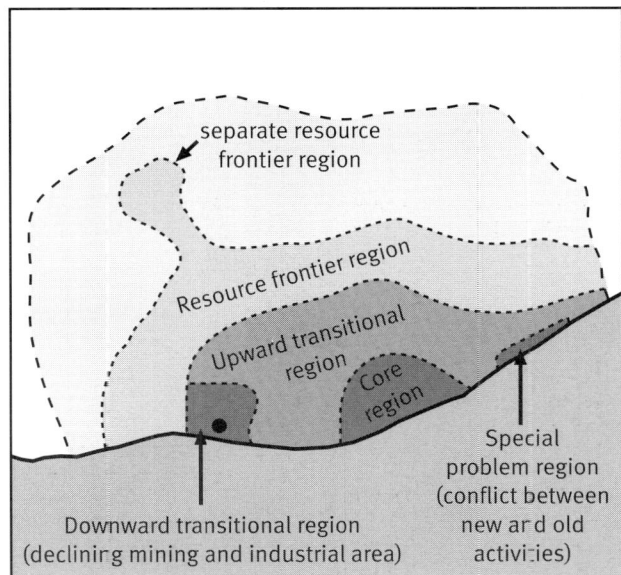

Figure 16.8 The Friedmann model of development regions

declining regions as well as growing regions, and that it can be simplified into the 'core' (the core and upward transitional area) and the 'periphery' (the rest of the country). The model was adopted as the basis for regional development planning in many countries.

A major **problem** of all these models was the way that the trickle-down process was not strong enough to overcome the trend towards polarisation of development in the core regions. In fact, Myrdal had recognised this and thought state intervention would be necessary to make the trickle-down process work.

Dependency models

Other models reflected an approach to development which was based on the idea of **dependency**. Their key elements concern the way LEDCs have no control over external or international economic events and the weakening of their controls over their internal or national economies. This shows itself in four ways:

Trade
Most LEDCs depend on primary products for their exports and import most manufactured goods. They depend on MEDCs buying their primary products more than the MEDCs depend on the LEDCs. With wide price fluctuations from year to year and their weak position in any trade dispute, LEDCs are disadvantaged.

Technology
New technologies mostly originate in MEDCs and have to be imported at a cost which also increases their dependence. Imported technology may be labour saving which might be inappropriate in areas of labour surplus. It would almost certainly be used in the core region of the country, increasing disparities with other regions.

Education and culture
Those trained in MEDCs often return with new attitudes and lifestyles and reject local traditions and products, leading to more imports.

Capital
It is difficult to raise savings to provide the capital for development in countries with high population growth, subsistence level economy and balance of payments difficulties. As a result there is dependence on foreign capital from institutions like the World Bank, or from commercial banks.

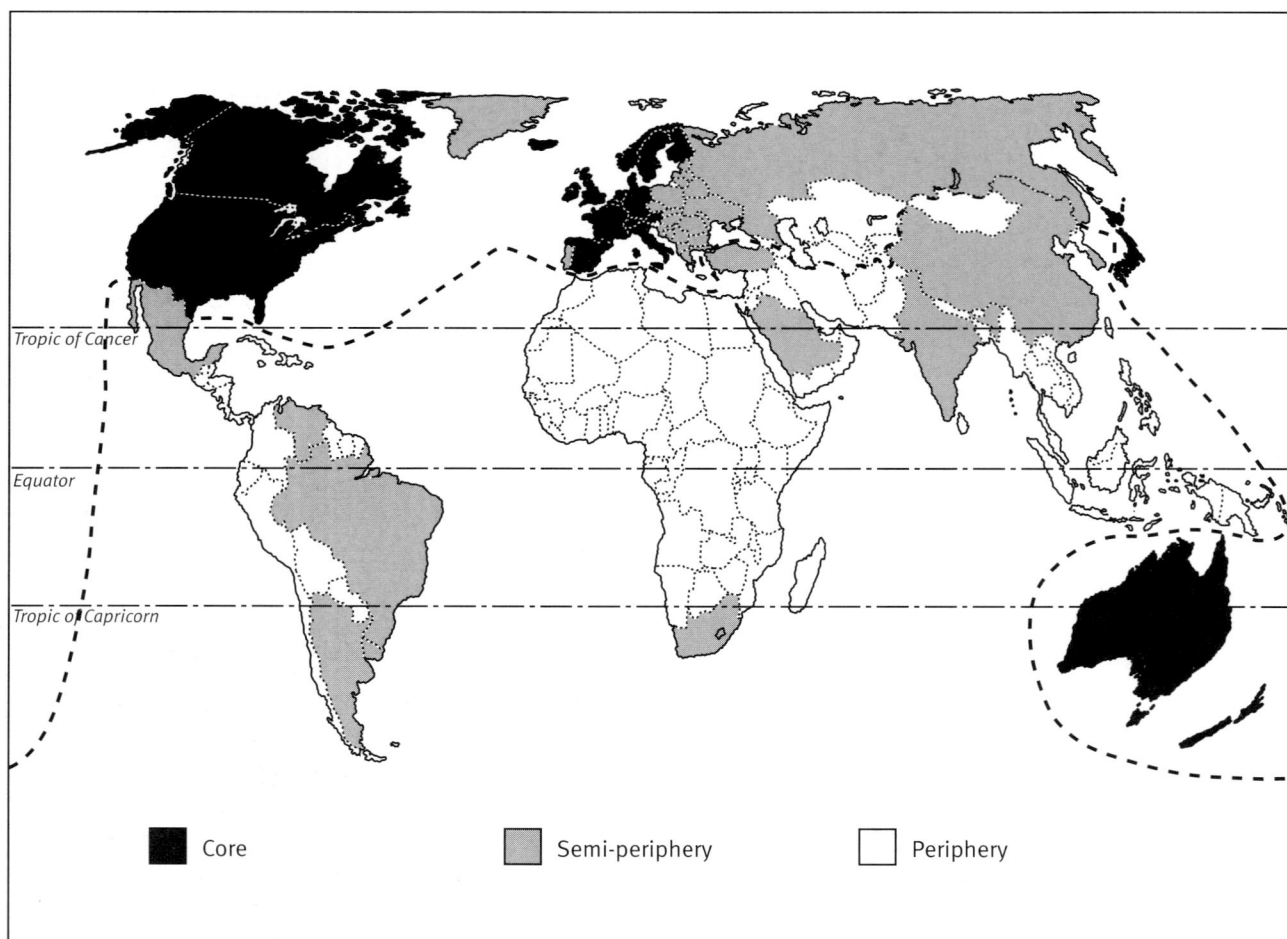

Figure 16.9 The world-system model

Core   Semi-periphery   Periphery

The world-system model

This model extends the idea of dependence to the world scale. It was developed by Wallerstein who saw a world system of a **core, a semi-periphery and a periphery** in which all parts had been integrated economically and politically, and where the core dominated the other parts (Figure 16.9). Note how Figure 16.10 sums up these global disparities of wealth.

A number of **criticisms** have been made against the ideas of dependency models. 'More people today live longer, healthier, and more productive lives than at any time in history' (World Bank, 1992). All the statistics show that there have been massive improvements almost everywhere in LEDCs in, for example, life expectancy (Figure 16.11).

In addition, the NICs of Asia and Latin America (see Chapter 13) show that there are great differences between countries in their paths to development. For some countries, for instance, dependence on primary exports has created development opportunities rather than problems. Oil exporting countries provide good examples from the LEDCs, but it is also important to recognise the MEDCs, such as Australia, which also base their development on primary exports.

A new view of the world

Development is increasingly difficult to measure solely on a whole-country basis. For example, recent communications and information technologies have opened up some well-placed areas in LEDCs to the world market, for example, the Delhi–Mumbai corridor and Bangalore in India. Such areas have been called 'First Worlds in the Third World'.

The improvements that have taken place have been uneven and, in many countries, have increased the gap between more and less developed regions and between richer and poorer groups of people. On the world scale, 'more than 1 billion people still live in abject poverty' (World Bank, 1992).

Disparities also exist in MEDCs with individual regions or districts within regions suffering from decline in traditional industries and resulting widespread unemployment, poverty and social exclusion. Such areas have been called 'Third Worlds in the First World'.

Patterns change according to the **scale**. At the global scale a pattern of rich and poor worlds can still be perceived. Step down to major world regions and the pattern is more complicated, emphasising the differences between individual countries – East Asia is a good example. At the national level, core-periphery ideas may explain overall patterns but individual regions may be linked more to the global

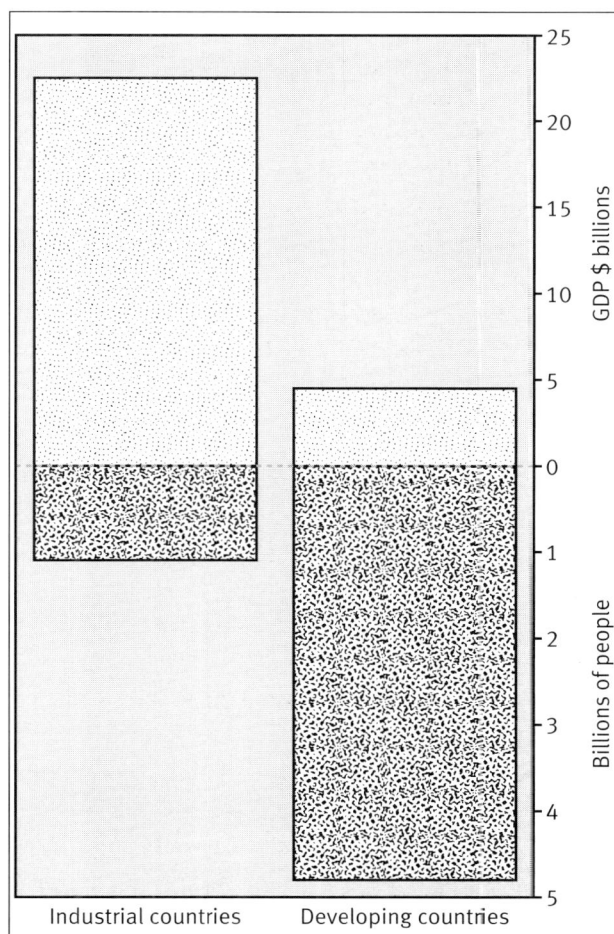

Figure 16.10 **Global shares of population and wealth**

	India		China	
	Males	**Females**	**Males**	**Females**
1965	46	44	53	57
1999	60	61	69	73

Figure 16.11 **Life expectancy changes in India and China**

economic system than the national. Lastly, at the city level, the disparities are often greatest of all with extremely wealthy districts in cities of LEDCs and extremely poor districts in the cities of MEDCs.

A new model of wealth, poverty and power

Figure 16.12 brings together the ideas in the previous four paragraphs. It represents new patterns of poverty and wealth and new patterns of power and influence which cannot be shown easily on a world map because so many different scales are involved. The key features are: the way that the core is represented not only by rich regions but also by rich institutions and TNCs, and the way poorer regions are not only in LEDCs.

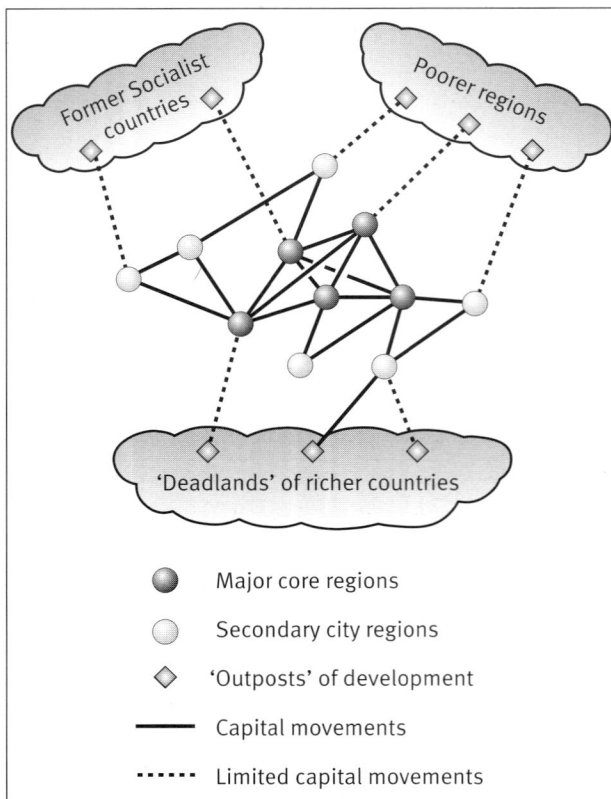

Figure 16.12 **New patterns of wealth, poverty and power**

Review task

1 **Why are the older models of development becoming less and less useful for describing patterns of development? List five points.**

2 **For an LEDC and an MEDC you have studied complete the summary chart of key features using the headings provided (Figure 16.13).**

Structural adjustment or local institutions?

In recent times there has been a number of changes in approach, some in different directions:

1 Large-scale intervention planning has been recognised as having limitations.

2 The most important has been the effect of the changed approach to **global interdependence** through trade and the International Monetary Fund. It is based on the premise that the development of countries will be fostered by a greater involvement in the world trading system.

3 **Structural adjustment** programmes have been instituted to encourage a greater role for market forces and to link local trading systems more firmly into the international trade system. These programmes have involved governments in meeting loan conditions laid down by the IMF/World Bank which reduced government expenditure and intervention in the economy, and had wide social and health implications in many poorer countries.

4 A wide range of interlocking and largely **local-scale developments**, which contrast with the IMF/World Bank's global superiority approach, are evidence of increased interest in the place of indigenous social institutions. One important aspect of this new development is seen in the recognition of the different roles of men and women and, in particular, the important role of women in development.

5 Linked to all of these in different ways and for different reasons is the growth in concern for the environment. One reason was the recognition that development plans which applied economic models to entirely different environments were totally inappropriate and often disastrous for the environment. **Sustainable development** has become an important idea, putting the environment at the heart of development (see page 169).

Interdependence – trade and aid
Trade

The general statement that LEDCs depend on exporting mainly primary products is still valid. However, the pattern varies considerably so some countries are exporters of manufactured goods. The NICs, with deliberate export-directed industrialisation, have been dealt with in Chapter 13. Other countries where manufactured goods provide a large proportion of exports use their advantages in producing low-cost consumer goods, especially textiles, clothing and footwear. Pakistan, Bangladesh, Egypt, India, China and Sri Lanka are examples.

The problem for LEDCs as a whole is that dependence on primary products puts them in a weaker position. Prices of primary products fluctuate greatly and have been moving downwards to the disadvantage of the exporting country. This creates special difficulties since prices of manufactured goods have generally risen. In other words, the **terms of trade** are against most LEDCs.

Other ways in which world trade has worked against LEDCs are the various areas which imposed tariffs on certain imports, such as some agricultural products coming to the European Union. The World Trade Organisation (WTO) and before that the General

	LEDC	MEDC
Name and location		
GNP per capita		
Selected development indicators		
Employment structure		
Main export (category)		
Regional development pattern		

Figure 16.13 Table for review task

Agreement on Tariffs and Trade has taken steps to open up trade and remove barriers of various kinds. In theory, this is to the advantage of LEDCs since it opens up markets throughout the world and should, studies show, lead to a 10 per cent increase in exports. On the other hand, it also opens up LEDCs to greater competition and is a particular problem for small countries which depend almost entirely on one product. A major concern of many is that the greatest gains from more open trade policies will be made by TNCs.

Aid

The use of the term 'aid' gives the incorrect impression of charity. Since the subject is really to do with raising **capital** for development, it avoids misconceptions if this is kept clearly in mind. Development programmes require capital but most LEDCs lack internal sources so have to obtain it from elsewhere. The result is debts which require repayment and the level of indebtedness is a grave problem as interest payments on loans reach over 30 per cent of export earnings for some countries (Figure 16.14).

The sources of capital are shown in Figure 16.15. Note the differences between the sources for low-income and middle-income countries. **Commercial** sources are less likely to lend to the poorest countries which then have to rely on **bilateral** loans from individual governments, and often the money is tied to specific developments which require purchases from the lender country. Loans from **multilateral** sources like the IMF also have conditions which have implications for government spending policies (see Structural adjustment on page 168). These usually involve cutting government spending, damaging programmes that depend on government spending like health and education.

The sources of capital labelled bilateral and multilateral are those often described as **aid**. In addition to those is **voluntary aid** which comes from individuals, national charities or international charities. These play important roles but their significance in terms of capital provision is small, although more and more bilateral government funding is being channelled through non-governmental organisations (NGOs).

Another source of capital is that brought by **TNCs** locating in or expanding activities in a country. This is known as Foreign Direct Investment (FDI) or Overseas Direct Investment (ODI). Although the economic and political power of TNCs is great and they are able to exert great influence on the countries in which they locate, they also provide more work and spread new technologies.

> ### Review task
>
> **Using your course notes, list examples of each type of aid. Remember that examination questions may ask you to refer to a country, so be prepared and choose examples with this is mind.**

Sustainability (see also Chapter 10)

The first real association of environment and development was with the UN Conference on the Human Environment in 1972, when the issues were mainly those of pollution and resource depletion resulting from the industrialisation of Europe, North America and Japan. The 1987 report of the UN World Commission on Environment and Development entitled *Our Common Future*, but usually called the Brundtland Report, had a wider view and was particularly concerned with problems of poverty and human need. It was, therefore, particularly relevant to LEDCs, especially as it showed concern for local-scale issues as well as broader ones.

The Brundtland Report produced one of the most widely quoted definitions of sustainable development: 'development which meets the needs

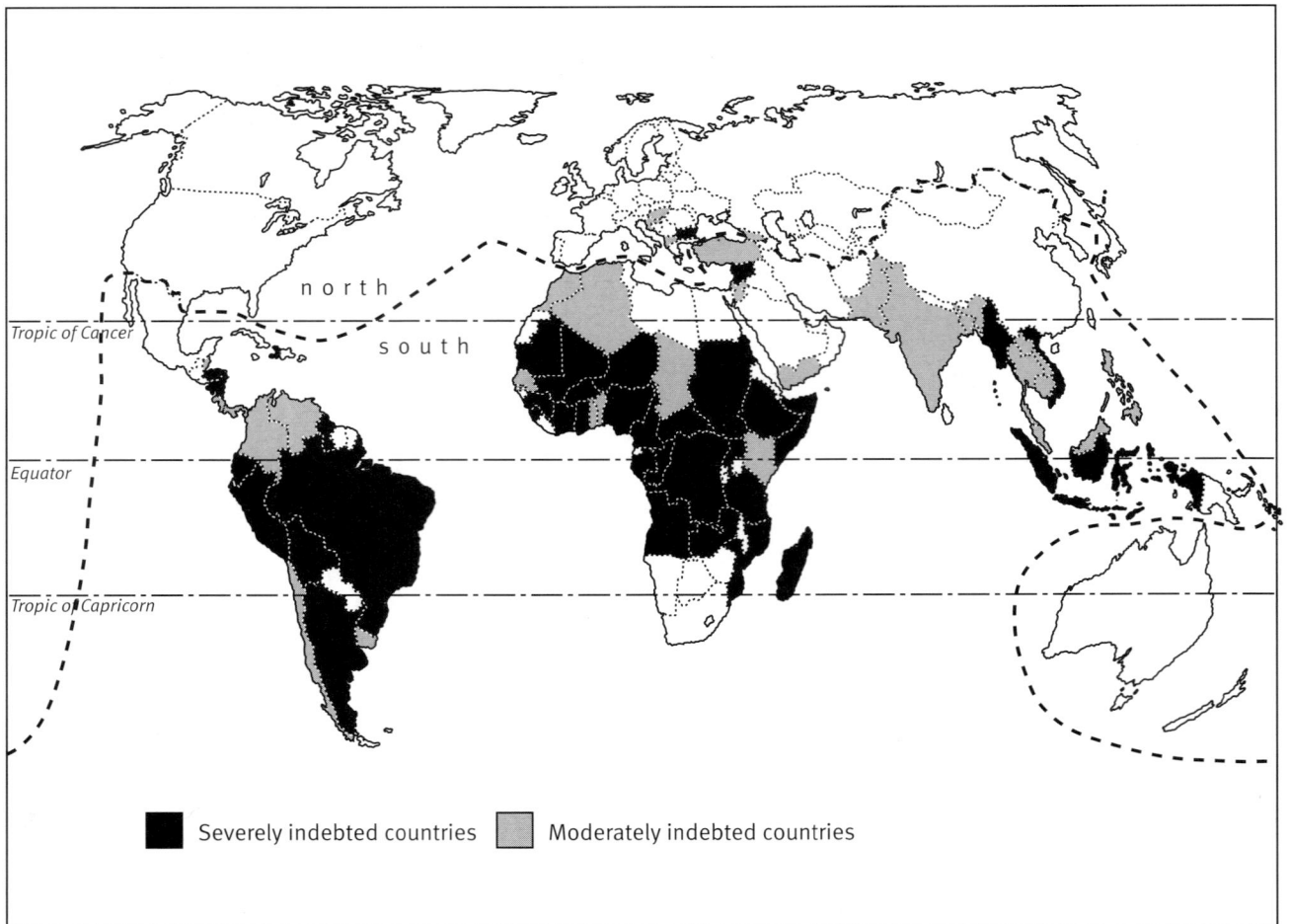

Figure 16.14 Severely and moderately indebted countries (World Bank, 1999)

Severely indebted countries   Moderately indebted countries

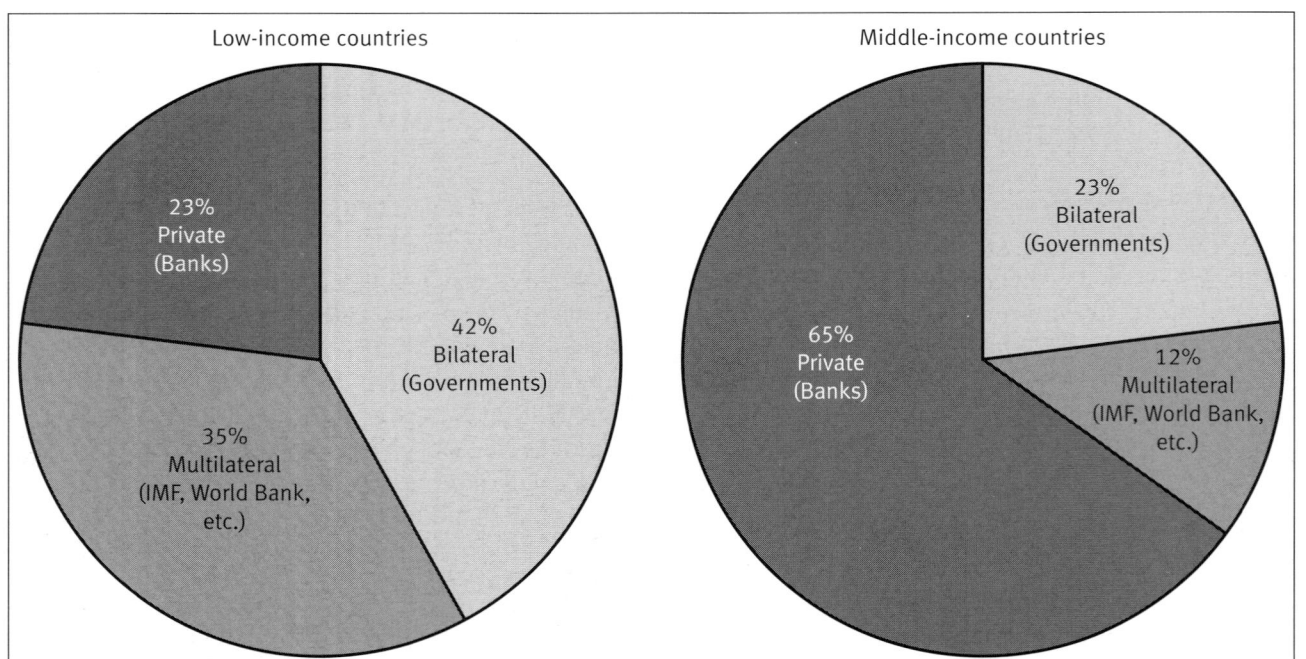

Figure 16.15 Sources of aid

Low-income countries

23% Private (Banks)
42% Bilateral (Governments)
35% Multilateral (IMF, World Bank, etc.)

Middle-income countries

23% Bilateral (Governments)
65% Private (Banks)
12% Multilateral (IMF, World Bank, etc.)

of the present without compromising the ability of future generations to meet their own needs'. Although broad and rather vague, it does bring together issues of environmental degradation and poverty both in the present and the future. At the very least it raised awareness globally of local-scale development issues, but, as Dickenson says, 'it had never ceased to be an issue for a great many people at the local scale'. (J. Dickenson *et al.*, *A Geography of the Third World*, Routledge, 1996)

Review task

Re-read the above quotation by Dickenson. Jot down three or four points which explain what he means. 'Survival' is a clue.

The Rio conference in 1992 had the objective of putting into action the ideas of the Brundtland Report. By this time the global problems of climate change and ozone depletion had come to the fore and highlighted the fact that some problems could not be ignored. Agreement was more difficult to reach with MEDCs stressing overpopulation and deforestation as major causes of global environmental problems, while LEDCs argued that the real causes were emissions by the MEDCs and their overconsumption of resources.

The result of the conference was a series of non-binding principles on environmental and developmental issues, a statement on sustainable forest management and Agenda 21. This last document considered how individual countries could plan for sustainable development at scales from national to local. At national levels the results have been variable, although the term 'sustainable' is used repeatedly in plans. However, many local-scale developments have followed from Agenda 21 throughout the world.

Population, environment and sustainability

The issue of population growth and its impact on development and the environment raises controversy (see the Malthusian view in Chapter 10). However, although much is made of links between environmental degradation, increased poverty and population growth, particularly where desertification is involved (see Chapter 8), agricultural systems can be maintained, and production increased, and they remain sustainable.

The Machakos District in Kenya is one of the best examples of this. In the 1930s the area was thought to be on the verge on an environmental disaster, but over the following 50 years the situation has totally changed: population has risen from 238 000 to

1 393 000 and agricultural production has increased to give food self-sufficiency and surplus cash crops. This has been achieved by extension of terracing, fertilising of the soil largely with manure from stall-fed cattle, the planting and harvesting of trees which also helped reduce the soil erosion which was apparent in the 1930s, more rapid cultivation of land, increasing the proportion of arable land and marketing produce in nearby Nairobi (Figure 16.16). Rather than increased population and development causing an environmental disaster, it has resulted in sustainable development.

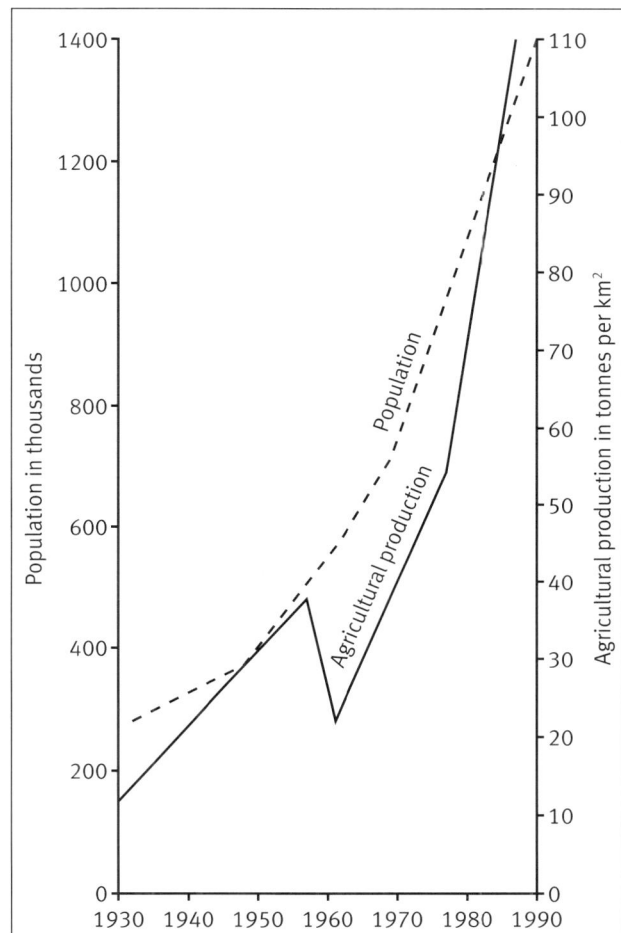

Figure 16.16 Population growth and agricultural production growth in the Machakos District, Kenya

Review task

Increasingly, there is pressure for aid to be linked to sustainable developments rather than those with questionable environmental objectives. Why is this so? Refer to your own course notes and name some examples of both types of development.